Frankenstein's Children

# Frankenstein's Children

ELECTRICITY, EXHIBITION,

AND EXPERIMENT IN

EARLY-NINETEENTH-CENTURY

LONDON

*IWAN RHYS MORUS*

PRINCETON UNIVERSITY PRESS

PRINCETON, NEW JERSEY

Library of Congress Cataloging-in-Publication Data

Morus, Iwan Rhys, 1964–
Frankenstein's children : electricity, exhibition, and experiment in
early-nineteenth-century London / Iwan Rhys Morus
p.   cm.
Includes bibliographical references and index.
ISBN 0-691-05952-7 (cl : alk. paper)
1. Electricity—History—19th century. 2. Electricity—Social aspects—England—
London—History—19th century. 3. Electrification—England—London—
History—19th century. I. Title.
QC527.5.M67 1998
303.48'3—dc21      98–9451

This book has been composed in Caledonia

Princeton University Press books are printed on acid-free paper and meet the guidelines
for permanence and durability of the Committee on Production Guidelines for Book
Longevity of the Council on Library Resources

http://pup.princeton.edu

Printed in the United States of America

1  2  3  4  5  6  7  8  9  10

# CONTENTS

# ILLUSTRATIONS

THIS BOOK had its origins more than a decade ago in one section of one chapter of my Ph.D. thesis. Having started out with the intention of examining the "simultaneous discovery" of the conservation of energy in the first half of the nineteenth century, I soon focused my attention on only one of the putative discoverers: William Robert Grove. Looking at his career brought me to the world of popular electricity in early-nineteenth-century London. That world has provided the focus for much of my research since then, including the present work. I try to achieve two goals in this book. The first is to provide a sensitive cultural history of electricity's place in the first half of the nineteenth century and the role it played in the massive social and economic transformations of that time. Second, but just as important, I have tried to carry out an experiment in the sociology of scientific knowledge. I have tried to apply some of the important insights of recent developments in the social study of science in such a way that the microsociology of science fits into a broader macrohistorical picture.

The backdrop to this book is London in, roughly speaking, the second quarter of the nineteenth century—that is to say, in the years from 1818 or so to 1851. Throughout this period, London maintained a position as the greatest city the planet had seen in terms of population and resources. It is often forgotten that London was also the greatest city of the Industrial Revolution. The metropolis was still England's (and therefore the world's) largest center of manufacturing production. In a very strong sense, therefore, an account of London science during this period is a global history. It is so as a result of the historically contingent position in which the city found itself during those years. London science during this period was "big science" in many ways simply because it was London science. One of my major concerns in this book is to elucidate some of electricity's connections (and those of science in general) to the rise of a global industrial and consumer culture. London at this historical juncture therefore seems a good focus for the study. In this sense, the perspective I aim for in the book is as much synchronic as diachronic, historicist in that it deals with the emergence of global properties from the complexities of a particular local culture.

I take seriously throughout the book the thought that science should be regarded as a species of cultural practice. That is, science is a practical activity, a way of going on, that has connections to and roots in other cultural ways of going on. It is central to the thesis of this book that the production of scientific knowledge is an irredeemably local affair, and that a proper historical understanding of its production therefore requires a detailed understanding of the complexities of the local culture in which it was produced. In particular I have

tried to place early-nineteenth-century electricity and electrical experimentation in the context of a history of labor. In many ways, regarding science as work
is an obvious extension of the constructivist concern with experiment and the
practice of experiment. Focusing on the skills required to produce successful
experimentation also highlights the ways in which those skills are constructed,
demarcated, and evaluated. These are all concerns that the history of science
shares with the history of labor. It is worth noting that *skill* is itself a value-
laden term. To describe an early-nineteenth-century natural philosopher as
skillful was not necessarily a compliment.

I look in this book, therefore, at the kinds of work that electrical experimenters from different classes and backgrounds did in order to bring their experimental productions out of the laboratory, and how those practical activities
connected to other ways of going on in related cultural spheres. I look at the
ways in which status was accorded to different practices and practitioners by
their friends and their enemies. I look at how different practitioners tried to
fashion themselves through their labor and their relationship to the products of
their labor. Finally, I look at the material products of experiment and the ways
in which they entered the world of consumption. Seen from the perspective of
the history of labor, the productions of natural philosophy are commodities for
which their producers must find a marketplace.

The book's title requires some explanation. I have chosen Mary Shelley's
protagonist Frankenstein to symbolize a number of themes that are central to
my concerns throughout this work. For my purposes here, I take Shelley's
narrative to be concerned with a number of issues surrounding the relationships among the natural philosopher, his productions, and society. The title is
intended as a somewhat ironic reflection on the difficulties encountered by the
electricians discussed in the book, like Frankenstein himself (and, I would
argue, scientists in general, then and now) in controlling the fate of their experiments in the world outside their laboratories and lecture theaters. The word
*Children* is intended to convey the suggestion that the book's protagonists were
in many ways the inheritors of the romantic natural philosophical tradition that
Frankenstein typified, as well as the more ironic reflection that their progeny
(their experiments) were the equivalents of Frankenstein's "monster" and
proved just as difficult to control. Read in this way, Shelley's narrative, like this
book, charts the emergence of an industrialized, commodified culture from
what was in many ways a background of romantic sensibility.

The book is divided into two parts. The title of the first part, "The Places
of Experiment," is self-explanatory. I look at the different kinds of places in
early-nineteenth-century London where electrical experiments took place.
The main aim of this first part is to center attention on the heterogeneity of
electrical experimentation and the extent to which that heterogeneity was a
product of social and geographical space and of audience. The practice of experimental philosophy during this period, in electricity as in any other field,
was not a straightforward business. There were no clear-cut boundaries or

protocols that defined the ways in which potential experimenters should carry on. Becoming recognized as an experimenter was in many ways a matter of self-fashioning. One of the features of experiment that I highlight, therefore, is its theatricality. Experiments in early-nineteenth-century London were *performed* in every sense of the word. Electricians needed to define themselves in relation to their experiments and to their putative audiences. To do this, they had to fashion themselves in such a manner as to conform to or even construct their potential constituencies' notions of the kind of person an experimenter should be.

Performing in this way required a range of material resources. Successful experimentation was above all else a matter of successfully acquiring and marshaling resources in a way that audiences found convincing. Material resources were not always easy to acquire and marshal either. Experimenters needed access, through patronage, institutional affiliation, or private wealth, to the raw materials, tools, and instruments essential to their work. The story I tell emphasizes the difficulties of manipulating these resources and the ways in which electrical experimenters needed to place themselves in particular networks in order to gain access to them. While public performances deploying this range of material resources were central to the business of experimentation, they were never in themselves sufficient to gain any audience's assent. Establishing the social status of the experimenter went hand in hand with the process of establishing the validity of his experiments. The relationship between the social identity of early-nineteenth-century electricians and the status of their experimental accomplishments was often fraught.

Very broadly speaking, I discern two distinct trends of experimentation emerging in conflict with each other during the early decades of the nineteenth century. These two trends encapsulated very different visions of science and proper scientific practice, rooted to some degree at least in developing class distinctions. On the one hand, elite, largely middle-class gentlemen centered at institutions such as the Royal Society or the Royal Institution fostered an image of science as the province of highly trained, vocationally minded specialists. Their aim was to articulate a highly abstracted view of nature, divorced from instruments and apparatus. Such gentlemen tended to draw a strict distinction between the relatively private business of laboratory experimentation and the public presentation of their productions to socially or professionally elite audiences. On the other hand, another set of experimenters—popular lecturers, instrument-makers, or mechanics—fostered a different image and a different practice. Their cosmologies were closely connected to their instruments, the machines they used to produce a range of spectacular phenomena. Here the distinction between private laboratory practice and public presentation was considerably less clear-cut. I suggest that these competing visions can be linked to the growing early-nineteenth-century conflict between artisanal and middle-class definitions of work and skill in terms of their ownership, status, and relationship to machinery.

I would emphasize here the implicitly gendered nature of these constructions. The exclusion of women from any large-scale, active role in scientific practice during this period is, of course, well-established. What should be noted as well, however, is the extent to which the resources used to establish notions of the legitimate natural philosophical practitioner were embedded in straightforwardly masculine conceptions. Gentlemen of science were unambiguously gentle*men*. Their image of science as a public vocation drew heavily on constructions of the public man of affairs. Equally, the celebration of practical skill and know-how in experimentation central to other views of proper scientific practice drew on artisanal notions of the property of skill that were crucial to working-class images of manhood. It was precisely through possession of a recognized craft that artisans and mechanics defined themselves as masculine. Essentially, neither of these views of the proper experimenter could allow any significant or active role for women. That fact is taken for granted throughout this book, and masculine pronouns are consequently deliberately employed.

In the second half of the book, "Managing Machine Culture," I move the argument forward, while broadening its scope by looking at the commodification of electrical technologies during the 1840s. While I argue in the first half of the book that electricity was already, through its place in popular exhibition culture, coming to be regarded as a commodity, a more literal commodification took place from the 1840s onward. Natural philosophy was more and more coming to be seen as a source of material advancement. Increased scientific knowledge was seen by many contemporary commentators as a vital resource for the domination and manipulation of nature that they regarded as a prerequisite of economic and social progress. Particularly in the context of debates concerning political economy and the role of machinery, Nature herself (and again the gendered language is deliberate) was coming to be seen as a direct source of economic production. The powers of nature, rather than artisans or laborers, were increasingly regarded as the ultimate producers of a whole range of new commodities.

The process, however, brought into sharp focus a whole range of issues, raised also in the first half of the book, concerning the nature of invention, its relationship to discovery, the relative status of the inventor, and his relationship to the experimental natural philosopher. Many gentlemen of science, regardless of their claims concerning the economic utility of the sciences, drew a sharp distinction between their own activities aimed at the public good and those of the inventor aimed at private profit. Exploiting natural philosophy for private gain was held to be incompatible with the status of a natural philosopher. On this view, the philosophical business of discovery was to be sharply distinguished from the entrepreneurial activity of invention. Others, however, saw nothing incompatible between discovery and invention, asserting that their rights as inventors gave them a claim to philosophical status as well. Establishing status in the world of invention could be just as fraught as in the world of discovery, and a whole range of strategies were deployed accordingly to make or unmake such claims.

By the end of the 1840s, the integration of electrical culture into an industrial, machine culture was well under way. The rise of large electrical industries in the second half of the century would provide opportunities and resources for electricians on a hitherto unprecedented scale. This seemed to me a fitting point to bring the book to a close, particularly since the coincidence of the Great Exhibition in 1851 represented in many ways the culmination of the exhibition culture that provided much of the focus for my work. There was by this time little question—if, indeed, this had ever truly been a vexed issue—as to which image of the proper man of science would prove dominant. The telegraph and subsequently the electric power and light engineers of the later Victorian age were, however, in many ways the direct inheritors of the alternative, craft-based articulation of experimental practice. Neither did they entirely relinquish their claims to be men of science who had a legitimate right through their intimate familiarity with electrical instruments, machines, and systems to make claims about the operations of electricity in the natural world. Both articulations of electrical culture would have a role to play in the burgeoning consumer, industrial society of the second half of the century.

More friends and colleagues than I could possibly name have provided me with invaluable support and assistance throughout the research and writing of this book. Without their help this project would never have achieved fruition. They have all had to put up with a great deal of rubbish about early-nineteenth-century electricity over the past few years, and if nothing else I am deeply grateful for their forbearance and patience. More than anyone else, I must first of all thank my former supervisor and now good friend, Simon Schaffer. His enthusiasm, insight, knowledge, and, above all, willingness to listen, have always been an inspiration, if at times a somewhat daunting one. I found it at times a little depressing, after having spent so long researching early-nineteenth-century electricity, to find out that he still seemed to know substantially more about this topic than I did! I must also thank in particular Will Ashworth, Henry Atmore, Bob Brain, Dan Brown, Nani Clow, Harry Collins, Roger Cooter, Brian Dolan, Gerry Geison, Graeme Gooday, Bob Hatch, Arne Hessenbruch, Jeff Hughes, Bruce Hunt, Rob Iliffe, Myles Jackson, Frank James, Kevin Knox, Jim Moore, Richard Noakes, John Pickstone, Anne Secord, Jim Secord, Steven Shapin, Otto Sibum, Jennifer Tucker, Sonia Uyterhoeven, Debbie Warner, Andy Warwick, Alison Winter, and Norton Wise for help and inspiration along the way. Other colleagues at the Department of History and Philosophy of Science, Cambridge, the National Museum of American History, the Science Studies Program, San Diego, and here at Queen's University, Belfast, have also provided invaluable assistance throughout my time at their respective institutions. I must also thank the archivists and librarians at the Bakken Library and Museum of Electricity and Life, Cambridge University Library, the Institution of Electrical Engineers, the Smithsonian Institution

Libraries, and the Whipple Library of the History of Science, Cambridge, for their unfailing assistance. I would like to thank Mrs. Lenore Symons, Archivist at the Institution of Electrical Engineers, for permission to quote from manuscripts in their possession. Finally, of course, I would like to thank the editors at Princeton University Press, particularly Emily Wilkinson, Sam Elworthy, and Lauren Lepow, for all their hard work in preparing this book for publication.

# PART ONE

# The Places of Experiment

# Electricity, Experiment, and the Experimental Life

A NEW FLUID appeared near the beginning of the nineteenth century, or rather an old fluid started appearing in a variety of startlingly different ways. Electricity had, of course, been around for at least a century before 1800,[1] but following the invention of the voltaic pile and Oersted's unification of electricity and magnetism in 1820, the mysterious fluid appeared in more and more different guises. New machines, new experiments, and new places appeared through which electricity could be seen and admired. The possibilities that electricity represented were enormous. Throughout the century, commentators marveled at the power that could be harnessed by the subtle fluid. As William Robert Grove enthused to his audience at the London Institution in the early 1840s,

> [h]ad it been prophesied at the close of the last century that, by the aid of an invisible, intangible, imponderable, agent, man would, in the space of forty years, be able to resolve into their elements the most refractory compounds, to fuse the most intractible metals, to propel the vessel or the carriage, to imitate without manual labour the most costly fabrics, almost to annihilate time and space;—the prophet, Cassandra-like, would have been laughed to scorn.[2]

Electricity was becoming a symbol of Victorian progress. It provided a new way of mastering and reordering nature and society. It is, of course, somewhat disingenuous to say that the electric fluid simply appeared. Making the fluid visible was a difficult and labor-intensive process riddled with contingency. Electricity appeared through and was constructed out of a range of new machines and technologies. These machines and technologies were the products of human labor and ingenuity.

This book, this section in particular, aims to look at the places of electricity in London during the first half of the nineteenth century. It is by now commonplace in the historiography of British science to locate what some have called the second scientific revolution during this period. The polymathic William Whewell's coining of the word *scientist* at the Cambridge meeting of the British Association for the Advancement of Science in 1833 is often taken as a decisive moment.[3] As a result much new and interesting historical work has appeared recently, focusing on these first decades of the nineteenth century.[4] These works have shown clearly how diverse were the social ramifications and connections of science during this period. The first half of the century was a time of massive cultural and economic change in Britain. The changes in the contents and contexts of natural philosophy were no less momentous.

The outlines of the account that is now emerging are quite straightforward and well-established. Following the political and social turmoil surrounding the French Revolution and its bloody aftermath, new groups emerged who wished to harness natural philosophy as an instrument of social order and as a vehicle for their own professional aspirations. New specialist societies were founded to foster and represent the aspirations of these new men. Campaigns were mounted to reform the Royal Society, removing from power the aristocratic amateurs who had dominated the society under the presidency of the autocratic Joseph Banks.[5] The key to the new sciences that these institutions fostered was discipline. Experiment and discovery were to be the provinces of a trained elite whose expert knowledge gave them privileged access to natural phenomena.

More recently some historians have also argued that other groups outside this small, middle-class elite also sought to engage in and fashion themselves around natural philosophical discourse and practice.[6] Not only did the gentlemen of science themselves quite often have startlingly different perceptions of the proper practice and use of experiment, others, marginal to elite society, could also forge themselves about notions of experiment very much opposed to those of the self-styled leaders of disciplined and expert science. Electricity could be an important resource in all of these debates. It held out the possibility of progress through new machines and processes. It also provided new ways of organizing nature and society and even of understanding the human body itself.

Looking at the uses of electricity is therefore a good way of approaching a range of issues surrounding science and society during the first half of the nineteenth century in Britain. Mary Shelley's *Frankenstein* provides an example of electricity's ramifications. Shelley focused her attention on the problematic identity of the experimenter and his role in public culture. The laboratory where the galvanic fluid flowed to make the Creature was an isolated and dangerous place outside society. It was a space that others could not (and would not wish to) enter and from which Frankenstein himself emerged only on furtive midnight raids of graveyards and charnel houses.[7] The Creature, as product of that secluded space, emerged from the laboratory with disastrous impact. The novel thus concerned itself with the relationships among the electrical experimenter, his product, and the world around him. Drawing on the resources of contemporary science, Shelley could convincingly portray the problematics of the laboratory experimenter's attempts to carve spaces for himself in early-nineteenth-century culture.[8]

This is one of the main issues with which this book concerns itself. The practice of experimental natural philosophy during this period was not a straightforward enterprise with clear-cut boundaries and protocols. In other words there was no obvious and self-evident way of going on that a hopeful experimenter might follow. Neither was it obvious to what audience the experimenter should address his work. Electrical experiments took place in a range of different places and aimed their productions at a corresponding variety of constituencies. Michael Faraday, for example, worked out of the prestigious setting of the Royal Institution, aiming his work at that establishment's fashionable clientele and at the Royal Society's vocationally minded and disciplined gentlemen. At the other end of the spectrum Joseph Saxton or William Sturgeon operated from the Adelaide Gallery of Practical Science for an audience anxious to witness spectacular electrical wonders.

These men used electricity in different ways to fashion their careers.[9] To carry out experiments, they needed to mobilize all kinds of resources, both material and social. If they did not have access to these resources and ways of transforming them into visible, tangible products, electrical experiments could not take place. The spaces where these men operated, and the media they used to make their products public bridged the gap between the experiments and their publics. They shaped the ways in which electricity could be seen and understood by providing appropriate contexts through which the imponderable fluid could be approached and assimilated into culture. They thus had an important role to play in defining what counted as experiment. They also defined the experimenter. In many ways the places where the experimenter appeared shaped the ways he acted as well.

Experimental natural philosophy was a contentious issue during the troubled decades following the Napoleonic Wars. The ability to define what might and what might not count as a legitimate science had important consequences. Electrical experiments could mold politics as much as politics did electricity. Electricity was being used to define the operations of the natural economy. In the early nineteenth century, this meant that it could define political economy as well.[10] The way that nature worked had consequences for the distribution of authority in the social world. As a result, it mattered who did what with electricity. It also mattered where and by whom electricity's operations could be seen.

This first section of the book aims to survey some of the places for electrical experiment during the first half of the nineteenth century and some of the men who constructed those spaces and operated within them. The survey does not aim to be exhaustive or systematic. The intention is, rather, to provide an indication of the variety of ways electrical experiments could be performed, and of the constituencies that could be forged around those different experimental practices. The emphasis of the analysis will be on these spaces rather than on biographical studies of individuals. One reason for this is the constitutive role accorded here to performance in the history of experiment. Another reason is that a focus on single individuals would obscure the massive changes that took place in the performance of experiment during this period. Looking at spaces helps make these transformations more explicit.

The first substantive chapter deals with the early career of Michael Faraday at the Royal Institution. Faraday clearly poses something of a problem for a book of this nature. The aim of this work will certainly not be to zero in on one such major figure. On the other hand one aim of this book is to provide materials for a revisionist account of Faraday's career.[11] Almost without exception treatments of the development of electricity in Britain during the first half of the nineteenth century consider almost exclusively Michael Faraday. He has been celebrated as the archetypal experimenter and the hero of electricity.

The historiographical roots of this tradition are almost as old as Faraday himself. On the one hand his many nineteenth-century biographers without exception sought to portray him as the genius figure of experimental natural philosophy. His combination of humility before nature and society with patient dexterity in the laboratory was applauded as the epitome of all that an experimenter should aspire toward.[12] Of this nineteenth-century genre, only Silvanus P. Thompson's biography suggests that Faraday might not have been the only significant British contributor to the science of experimental electricity during the period.[13] Faraday's reputation was also central to the new discipline of mathematical physics as founded by James Clerk Maxwell, William Thomson, and their disciples. Maxwell in particular presented his electromagnetic theories as being the articulation in mathematical language of Faraday's experimental results. The integrity of the new physics thus depended on the preservation and defense of Michael Faraday's reputation.

Any work dealing with the physical sciences during the first half of the nineteenth century must therefore deal with Faraday. His reputation towered over the period, and his preeminence was recognized by most, if not all, of his contemporaries. The strategy adopted in this book has been to attempt a deconstruction by focusing on the ways in which Faraday constructed spaces for himself, so that he could fashion a career through experiment. The chapter follows Faraday from his early days as Humphry Davy's laboratory assistant, through his efforts during the 1820s to reform the workings of the Royal Institution, to sample his major experimental work of the 1830s and 1840s. In particular it is suggested that his work to establish the Friday Evening Discourses and to strengthen the organization of the institution's lecture courses was crucial to the making of both his reputation and that of the Royal Institution as a place of knowledge-production. By making the Royal Institution a space for fashionable science, Faraday also made it into a space for his brand of experiment.

Others, however, had different notions concerning the practice and scope of electrical experimentation. Chapter 2 focuses on an electrician who was in many ways very similar to, but also in many important respects very different from, Michael Faraday. William Sturgeon came from a background similar to that of his more eminent contemporary, but his career developed in a strikingly different fashion. While Faraday became the doyen of electrical science, Sturgeon spent his career struggling to carve a niche for himself in the scientific world. His practice of electricity and the audiences at which he aimed his work contrasted sharply with those of the elite culture of the Royal Institution. His relative failure to establish himself and his particular brand of experimentation has resulted in his being relegated, by and large, to the footnotes of history. Much of his experimental work was, however, crucial to the development of electrical technologies during the second half of the nineteenth century.[14]

The chapter outlines, first of all, the electrical cosmology articulated by Sturgeon through his experimental practice, linking that cosmology both to the practice of his electrical experiments and to the audiences at which he aimed them. Sturgeon articulated a vision of nature in terms of electrical instruments and machines. One could best understand the universe by building electrical apparatus that reproduced its phenomena, and by making visible the workings of that apparatus. Experimentation was therefore a matter of display. Faraday remains a major figure in this chapter, in part at least since Sturgeon so explicitly used his contemporary as a foil against which to display his own experimental work and his views concerning the proper practice of experiment. The chapter focuses on two disputes between Faraday and Sturgeon in the pages of the *Annals of Electricity*, which Sturgeon edited, showing how Sturgeon tried to use his journal as a space where he could establish himself as an authoratative voice and final arbiter in electrical matters. The London Electrical Society provided a similar space where Sturgeon could articulate his views on electricity and experiment. Discussion of the society is reserved, however, for a subsequent chapter.

Chapter 3 examines some spaces for electrical experiment very different from the exclusive surroundings of the Royal Institution. During the 1830s in London, galleries of practical science were founded where spectacular experiments were performed as part of exhibitions displaying the products of invention to the metropolitan public. The workers at places like the Adelaide Gallery on the Strand or the Royal Polytechnic Institute on Regent's Street catered to the same audience that flocked to the city's many shows and panoramas.[15] The chapter examines the networks of resources surrounding these places and the ways those resources were put together to make electricity.

The world of these exhibitions was the world of the artisan and skilled mechanic rather than that of the fashionable gentlemen at the Royal Institution. The galleries of practical science needed London's networks of workshops and small factories to provide them with the resources to put their shows together. Exhibitors put their wares on display there to show off their inventive skills at producing ingenious contrivances. This was the context of electricity for a large segment of London's population. The chapter aims to show how electricity's operation at these exhibitions fitted into the world of invention and spectacular display. Electricity would be fashioned in the same ways as the machines around it as a commodity for public consumption.

The introduction of new machines and new forms of disciplinary organization into the workshop brought about changes in the ways in which skilled artisans fashioned themselves and organized their labor. It also brought about changes in the ways they regarded themselves in relationship to the inventions and machines that were the product of their labor. This had important consequences for the inventions on show at the Adelaide Gallery or the Royal Polytechnic Institute. Exhibitors sought to use the displays to redefine their status as inventors and to clarify their claims of property over the products of their labor. This was the case for electricity as well. Inventors of new magnetoelectric machines such as Edward Clarke or Joseph Saxton used the Adelaide Gallery as a battleground in their rival claims to priority in invention and discovery.

The chapter focuses therefore on the importance of work in the production of electrical experiments as well as on the labor required for successful public performances. This issue is also central to the concerns of chapter 4, which examines the activities of the short-lived London Electrical Society. The primary issue here is the organization of experimental labor and the definition of a community of electrical practitioners. The founders of the London Electrical Society sought to define themselves as the locus of experimental expertise and to build a network of electricians through which their craft skills could proliferate and replicate.[16]

Founding a scientific society to foster a particular discipline was, of course, no new thing during the late 1830s. The emergence of a number of such societies, notably the Astronomical and Geological Societies, during the first half of the century has been taken as one of the landmarks on the way to the professionalization of science during the nineteenth century.[17] The origins and founders of the London Electrical Society, however, were rather different from those of these large, prestigious organizations that dominated nineteenth-century British science. The society's collapse within barely five years of its establishment is not the only mark of difference. None of its members had the status of one of the scientific community's "big guns" such as Babbage or Herschel or Whewell. Instead, they were operative chemists, instrument-makers, and itinerant lecturers.[18]

The chapter gives an overview of the society's activities during its brief period of existence. In their meetings at the Adelaide Gallery for the first few years and then at the Royal Polytechnic Institute, the organization's members delivered papers and carried out experiments on all manner of electrical matters. Their emphasis was on the machines that made electricity and on the details of their working. Improvements in the economy of galvanic batteries or magnetoelectric devices were central to their concerns. Displaying electricity's universality in nature was another ongoing interest. By making the world an electrical machine, they could argue that they as practical electricians were its proper interpreters.

The final section of the chapter examines in detail a day of experiments conducted by some of the London Electrical Society's members to uncover how their aims and interests worked in practice. Their goal was to assemble and assess a massive battery of galvanic cells. The experiment, the labor required to make it work, and the controversy that soon surrounded it reveal the fragility of the electricians' culture. Making the source of electricity secure and reliable was a time-consuming and intensive process. Establishing the proper division of labor for experimentation was fraught with difficulty. They were committed to electrical experimentation as a collaborative activity but could not resolve the problem of proper allocation of credit to participating individuals in such circumstances. Following presentation of their results to the society, the former collaborators quarreled bitterly over the spoils of the experiment.

Chapter 5, concluding the first half of the book, looks at electricity from a slightly different perspective. Rather than focusing on electricity's uses in a particular space, it examines some of the range of uses to which electricity was put in the effort to account for nature. In other words it looks at some of the different ways in which an electrical universe could be put together. The emphasis here is on the heterogeneity of electricity during the early part of the nineteenth century. The fluid had a variety of different uses. It could explain the movements of the planets, the structure of the earth, the development of plants, and the organization of the human brain. It could bring back the dead, as well as challenging the hegemony of elite experts over scientific discourse.

The chapter's opening section looks at the way in which the science of galvanism was put together by William Sturgeon through his lectures. Two things are clear from these presentations. First, that the genealogy Sturgeon presented for the subject differed significantly from the usual accounts: Galvani's claims to discovery were strongly supported over Volta's and the existence of a specific animal electricity affirmed. Second, that galvanism was defined through its instrumentation: the subject's development was seen in terms of the invention of batteries and devices for its display—which included dead human and animal bodies—rather than through the development of theoretical apparatus. The result was a history that marginalized the usual heroes such as Davy and Faraday, celebrating Galvani, Aldini, and various British battery-makers instead.

The core of the chapter then examines some attempts to show the universal workings of electricity in nature. These range from the Owenite Thomas Simmons Mackintosh's comprehensive attack on Newtonianism, through an elaboration of the role of electricity in governing celestial motions, to Andrew Crosse's production of artificial crystals (and insects) through electricity and Robert Were Fox's experiments on the electricity of terrestrial mineral formations. The use of electricity by heterodox medical practitioners is examined, and finally the surgeon Alfred Smee's ambitious and controversial account of the electrical workings of the human body is presented. The chapter emphasizes the ways in which electricity could be used in quite different ways by various interested parties. Its reception, too, depended very much on where and by whom claims were made on its behalf. The social identity of spaces and spokesmen often mattered as much as what was said concerning electricity's powers.

This chapter—indeed, much of the book's first part—emphasizes the ways in which knowledge about electrical instruments and machines, the everyday apparatus of an electrician's laboratory, was coextensive and interchangeable with knowledge about nature. There was a direct link between the practical epistemology of the electricians' culture, their construction and display of a whole range of different instruments for making electricity visible, and their ontology, their view that nature was to be understood as being constituted of such machinery. The resources put to work in constructing and articulating this culture were diverse and heterogeneous. So too were the uses to which electricity could be adapted.

This first half of the book is very much an exercise in the historiography and sociology of experiment. These topics have received a great deal of attention over the last two decades following the pioneering work of Harry Collins and others during the 1970s.[19] This work in the sociology of scientific knowledge did much to question traditional accounts of experimentation as being simply a matter of theory testing and argued that more attention should be devoted to the practical details of experiments themselves. Case studies in a special issue of *Social Studies of Science* in 1981 established the viability of the program.[20] Steven Shapin and Simon Schaffer showed, despite the doubts of many sociologists, how fruitful this kind of sociological analysis could be when put to work in a historical account of experimentation.[21]

Collins's work in particular drew attention to the difficulty of replicating an experiment and to what he described as the experimenter's regress. This focus on replication and the problems encountered by experimenters when they attempt to reproduce another's work made explicit the tacit dimension of experimental practice.[22] In order to replicate an experiment, the experimenter needs more than a simple literary account of the apparatus and the means of manipulation. He or she must be situated in a context that shares with the original experimenter a whole repertoire of unarticulated skills, practices, and assumptions. They must share the same tacit knowledge. Similarly, the notion of the experimenter's regress drew attention to the problematic nature of closure and to the strategies adopted by experimenters in attempting to convince others of the validity of their work.

Many historians have argued recently that the performance of experiments cannot be understood independent of the surrounding material culture that provides the basic resources without which an experiment cannot take place.[23] Bringing such resources together and dominating them is seen as being a central, rather than a peripheral, part of performing an experiment. Attention to this material culture also foregrounds the problem of social order, both inside and outside the laboratory. Understanding how an experiment works means more than understanding how the apparatus works. Issues of social class, hierarchy, and the division of labor within the laboratory are crucial. The outcome of an experiment depends as much on the process of negotiation among participants as it does on the successful manipulation of apparatus.

Also at stake during the period examined in this section of the book was the question of what kind of activity might even legitimately be taken to count as being an experiment at all. *Experiment*, after all, is an attribution that endows a particular event or activity with significance and status. It is not a transcendental category. That a particular way of going on counts as doing an experiment is the outcome of a number of factors ranging from the social status of the performer, through the space where the performance took place, to the outcome of the activity. Different actors in different social and physical spaces may decide differently as to what kind of practice might legitimately be called experimental. As a result the description of a particular practice or set of practices as an experiment is always contestable. It was certainly contestable during the first half of the nineteenth century.

One advantage of the approach adopted in this book is that it does make clear the ways in which the practice of experiment varies across culture. The experimenters discussed here had different aims and goals and a range of accounts of what the proper practice of experiment might be. They worked in different spaces and aimed their productions at a variety of audiences. These spaces and audiences constructed the meaning of their experiments as much as did the experimenters themselves. Practices and strategies of experimentation that might count as successful in one place for a particular audience might not carry conviction in the same way in another context. This is one reason why audiences were so important. Practices had to change as they moved from laboratory to lecture theater and from one audience to another.

In many respects, the classic term *closure* as applied to the end of experimentation may be inappropriate unless taken to mean no more than the end of a dispute.[24] Experiments never end in that their meanings are never fully stabilized.[25] Any particular sequence of experimental practices is always available for reinterpretation by different groups and in different settings. Contingency never goes away entirely. In fact one conclusion of this book is that the *openness* of experimental practice is central to the resolution of discoveries and inventions and the settling of priority disputes. The boundaries around an experiment that mark it as a successful achievement are a matter of negotiation and are continually being redrawn and redefined.

One feature of this book is that little if any distinction is made between science and technology. Equally, knowledge and practice are taken to be largely interchangeable. In this, the analysis offered here owes much to the earlier work of Bruno Latour.[26] The book in a certain sense does simply follow its actors around looking at how different elements were brought together by different people in different places to forge an electricians' culture. The production of knowledge (usually experimental knowledge in this book) is regarded as a practical matter of bringing together a range of heterogeneous resources and binding them together, convincing an audience that this mixture then constitutes a coherent way of looking at the world. Knowledge is regarded as having a practical rather than an abstract quality, constituted of instruments, machines, routines, and ways of going on. In contrast with Latour's later work, however (and recent work by Pickering also), the focus of attention is on the actors. There are no actants or material agencies in this book.[27] The emphasis is on the human effort and labor required to capture resources and make them work. Making scientific knowledge is taken to be a labor process. This is not to suggest that the material culture of experiment is in any way peripheral. Without material resources there would be no electricity. Such material resources gather their significance and power, however, from their place in human culture. They are not autonomous agencies. Material culture is central to the story told in this book. But it is central because it was actively constructed and manipulated by human actors in pursuit of their own particular goals.

Useful lessons for the sociology of knowledge, and of experiments in particular, may still be drawn from Marx.[28] If the process of experimentation is to be regarded as a matter of human labor, then the experiment itself might usefully be regarded as a commodity. A feature of successful experiments that many commentators have noted is their apparent divorce from context. If an experiment works, it becomes a product of the natural world rather than of the labor required to forge it. As David Gooding puts it, "in nature's school the phenomena displayed are natural facts, free of human ingenuity or artifice."[29] If so, then a good experiment is one that has succeeded in making its author and its context invisible.

This closely resembles what Marx has to say about commodities:

> The mysterious character of the commodity-form consists therefore simply in the fact that the commodity reflects the social characteristics of men's own labour as objective characteristics of the products of labour themselves, as the socio-natural properties of these things . . . to find an analogy we must take flight into the misty realm of religion. There the products of the human brain appear as autonomous figures endowed with a life of their own, which enter into relations both with each other and with the human race. So it is in the world of commodities with the products of men's hands. I call this the fetishism which attaches itself to the products of labour as soon as they are produced as commodities, and is therefore inseparable from the production of commodities.[30]

So it is in the world of experiments as well, and, as Marx notes for commodities, this feature of experiments is inseparable from the way in which they are produced and put together. Understanding an experiment, like understanding a commodity, is therefore a matter of reconstructing the context and the labor that made it.

But again as with commodities, experiments must be understood in their consumption as well as in their production.[31] Electricity, for example, proliferated in the early nineteenth century. It appeared in batteries and machines, in lecture theaters and exhibition halls, in Cornish mines and Italian volcanoes. It appeared in animal and human bodies. In all of these places electricity was made sense of through its display as spectacle. These were the places where electricity was consumed. As different audiences encountered electricity, experimenters worked to control those audiences as a way of demonstrating their mastery of nature. Experiments acquired their significance through such showmanship. That was their mode of consumption.[32]

CHAPTER 1

# The Errors of a
# Fashionable Man: Michael Faraday
# and the Royal Institution

ANY ACCOUNT of the cultures of electrical experiment during the second quarter of the nineteenth century must find a space for Michael Faraday. By the middle of his career in about 1840 he was without question, for the metropolitan middle and upper classes at whom he aimed his science, the very image of the experimental natural philosopher. Too many histories of science have taken that monumental success as self-evident. Faraday was a genius, either for his philosophical insight,[1] or more commonly for his experimental skill. He is as a result unavoidable, providing a yardstick against which his electrical contemporaries cannot but be measured, as on occasion they measured themselves. Understanding Faraday and the position he carved out for himself is therefore central to understanding, if only by contrast, the wider context of electricity that this work seeks to explore.

The aim of this chapter is to deconstruct Faraday and provide some outline of how that image might have been put together. Crucial to this story is Faraday's position at the Royal Institution, which provided him with resources that few other experimenters at this period could match. Simply pointing to these resources is, however, clearly insufficient. Other contemporary professors at the institution, with access to just those resources, had no such impact. Faraday used his resources to make experiments, to make an audience for his work, and to make himself such that he could capture that audience's interest. This account will therefore highlight Faraday's self-fashioning, his efforts to build an audience at the Royal Institution, and his electrical experiments and the labor expended in putting them together and defending his claims to the discoveries they revealed. The story is social, rather than biographical, at every stage, since at every stage Faraday had to generate and mobilize social resources and interests in order to succeed.[2]

The Royal Institution, on its foundation in 1799, provided London and its publics with a new and different space for the production and display of natural knowledge. The incentives behind the establishment are well documented.[3] The institution was founded by a coterie of aristocratic gentlemen with landowning interests who sought to harness natural knowledge in the service of improving agriculture. Its founders included members of the Board of Agriculture and the Society for Bettering the Condition of the Poor. They hoped that the systematic application of chemistry and natural history to the production of food could alleviate deprivation and therefore social unrest. Under the banner of scientific philanthropy, the Royal Institution's promoters hoped to stave off revolution without changing the hierarchical structure of English society.

The key figure behind the foundation of the Royal Institution was the flamboyant American Benjamin Thompson, Count Rumford.[4] Thompson had fled the colonies as a royalist at the outbreak of the American Revolution, having been employed as a spy by the British. In England and on the Continent (where he was made a count by the Elector of Bavaria in 1791) he courted the aristocracy, promoting his campaign to improve society through practical science. During the '90s he advocated the foundation in London of a "House of Industry" to display and disseminate practical inventions for scientifically improving society. With the patronage of Sir Joseph Banks, Rumford's campaign was successful. The Royal Institution was founded in March 1799 and a mansion purchased in fashionable Albemarle Street to house the new establishment.

The institution was primarily governed by a Committee of Managers, selected from among the "Proprietors" who had subscribed the substantial sum of fifty guineas for the right to attend lectures and exhibitions. As Morris Berman argues, much of the institution's direction over the next decades was determined by the social and intellectual interests of this small group of men.[5] Rumford himself soon lost interest in the project, following substantial disagreements with his patrons. He left London for Paris in 1802. Notions of the new institution's role and audience were already changing. Educating the poor was no longer considered a safe and patriotic alternative as war hysteria and anti-French sentiment intensified. Early plans for the building had included an outside staircase so that artisans might attend lectures without mingling with their social superiors in the lobby. Thomas Webster, the architect, reported that he "was asked rudely what I meant by instructing the lower classes in science. I was told likewise that it was resolved upon that the plan must be dropped as quietly as possible, it was thought to have a political tendency."[6] It rapidly became the case that lectures at the Royal Institution were aimed primarily at a fashionable and elite audience.

Along with a lecture theater—the first in London designed for lectures in natural philosophy—the Royal Institution established a laboratory.[7] This laboratory was in the basement, occupying the space once taken by the kitchen and outhouses of the original mansion. The laboratory's role was to furnish a space where the institution's professors of chemistry and their staff of assistants could prepare materials for their lectures and carry out commercial consulting work for the institution's members and others. It was in effect the "backstage" portion of the small basement lecture theater used for lectures on practical chemistry. The laboratory was well stocked for its purpose. William Thomas Brande two decades later claimed that "in its completeness and convenience; it comprises all that is required in the pursuit of experimental chemistry."[8] This included a forge and a furnace. The professor had six workmen as well as a mathematical instrument-maker at his service.[9]

The first incumbent as professor was Thomas Garnett, an experienced performer from Glasgow's Anderson's Institution. He soon left, following disagreements with the managers concerning his salary, and was replaced by his assistant, Humphry Davy. Davy had only recently arrived at the institution, having been recruited from the radical doctor Thomas Beddoes's Pneumatic Institute in Bristol. In his new position, Davy was expected to work under the direction of the managers in applied chemistry and to lecture on the practical uses of his work to the institution's polite audience.[10] Over the next few decades, as Davy's success with his audience provided him with increasingly greater autonomy, he decisively directed the institution's resources toward spectacular discoveries and equally spectacular public expositions of those discoveries to fashionable society.

Jan Golinski has recently given a brilliant account of the strategies employed by Davy in carving out his career.[11] Combining rhetorical flair in public lecturing and decisive experimentation on a massive scale, he made himself the epitome of the fashionable philosopher. In transforming himself, he also transformed the study of galvanism from a dubious enterprise tainted by radical connections into an orthodox and highly successful science.[12] The voltaic pile in Davy's hands, utilizing the resources at his disposal in the Royal Institution's laboratory, became a powerful instrument of analysis and discovery. Those discoveries, translated into the lecture theater, established his reputation.

Davy's career was not without its problems, however. His background as a humble apothecary's apprentice from Penzance and his later association with the radical Thomas Beddoes did not guarantee him an easy entry into the higher echelons of London society. Many detractors held that Davy's preoccupation with social status obscured his chemical reputation.[13] Maintaining his reputation was a constant preoccupation for him, as exemplified by his careful dissociation of himself from his early involvement with John George Children's commercial venture to manufacture gunpowder.[14] Aspiring as he did to be a gentleman, he had to avoid contact strenuously with anything that hinted of trade. He was contemptuous of his successor as professor at the Royal Institution, William Brande, dismissing him as a "mercenary" who had "come from the counter" and "had no lofty views."[15]

Men with different views on "trade" could be just as contemptuous of Davy. The short-lived *Chemist*, edited by the radical socialist Thomas Hodgskin, lost no opportunity for sniping at the Royal Society's noveau-riche president.[16] Davy was accused of professing "a sort of royal science."

> If in its pursuit he makes any discoveries which are useful to the multitude, they may, and welcome, have the benefit of them, but he had no appearance of labouring for the people. . . . Amidst all the great efforts which have been lately made to promote scientific instruction among the working classes, and amidst all the patronage which those efforts have found among opulent and clever men, it has been with regret that we have sought in vain to trace one exertion or one smile of encouragement bestowed on such efforts by the President of the Royal Society.[17]

Davy's chemistry, said the *Chemist*, was divisive and elitist, deliberately designed to exclude the masses. Even when reluctantly defending him in his dispute with the Lords of the Admiralty over the protection of ships' copper linings, it was clearly the lesser of two evils. Hodgskin made it clear that he had "no respect for the manner in which the learned President proceeded."[18]

Despite his difficulties, and the Royal Institution's own financial problems that his lecturing success did much to alleviate, Davy's reputation by the middle of the century's second decade was secure. Even to his enemies he was Britain's greatest chemist. Similarly, the Royal Institution basked in a reputation as the country's premier site of chemical and natural philosophical discovery. Thomas Carlyle, with characteristically sarcastic insight, described the place as "a kind of sublime Mechanics' Institute for the upper classes."[19] Davy's lectures attracted crowds of fashionable admirers. When Michael Faraday, having himself been a less than fashionable auditor at one of Davy's lecture courses, was hired by the Royal Institution as the great man's assistant, he was entering a well-established and prestigious laboratory to work under an equally prestigious master.

### THE PHILOSOPHER'S APPRENTICE

Faraday's biography is relatively well known.[20] He was born the son of a blacksmith in 1791. His parents had recently arrived in London from the north of England in search of employment. He was brought up in London's backstreets and poorly educated; he recorded that "my education was of the most ordinary description, consisting of little more than the rudiments of reading, writing, and arithmetic ... hours out of school were passed at home and in the streets."[21] In 1804 at the age of thirteen, he was employed as an errand boy by a bookseller, George Ribeau. A year later, he was bound apprentice to Ribeau, to learn the trade of bookseller and bookbinder. He was embarked, therefore, on a course that would provide him with a good and respectable trade, with the prospect of one day becoming his own master in his craft.

As an apprentice bookbinder, Faraday lived under his master's roof, learning the skills of his trade. He also had access to books and developed an interest in natural philosophy. With money borrowed from his brother, he attended lectures by John Tatum, one of London's many public lecturers.[22] There he met other young men, similarly bent on self-improvement, and joined the City Philosophical Society, which met at Tatum's house to read and discuss papers. He started to conduct his own experiments in chemistry and electricity, constructing apparatus from available materials. Through a contact with one of Ribeau's customers, Faraday received a ticket to attend a course of Humphry Davy's lectures at the Royal Institution. He took careful notes as he had been taught to do at the City Philosophical Society.

On 7 October 1812, Faraday's seven-year apprenticeship came to an end. He was now a journeyman bookbinder, in search of employment. A few months later, he took a curious step. He bound the notes he had taken of Davy's lectures and sent them to him, along with a letter begging for employment at the Royal Institution. For a man of Faraday's background the step was less curious than it may seem. Apprentices were required at the end of their term to present their master with a "masterpiece"—an example of their work—to display the technical competence they had achieved. Similarly, journeymen seeking employment would present a prospective employer with such an example of their craft.[23] Faraday was sending Davy a "masterpiece" that combined both his skills as a bookbinder and his aspirations to natural philosophy.

Some months following Faraday's appeal, one of the Royal Institution's laboratory assistants, William Payne, was sacked. He had been accused by John Newman, the institution's instrument-maker, of failing in his duty of attending and assisting at William Brande's chemistry lectures. A brawl ensued during which Newman was injured. Payne was then dismissed after ten years' service, and on Davy's recommendation his post was offered to Michael Faraday. The managers resolved "that Michael Faraday be engaged to fill the situation lately occupied by Mr. Payne on the same terms."[24] The terms were quite generous: twenty-five shillings a week and his board. Faraday had commenced a new apprenticeship in experimental natural philosophy.

Faraday's duties as laboratory assistant had been laid down by the managers:

> To attend and assist the lecturers and professors in preparing for and during lectures. Where any instruments or apparatus may be required, to attend to their careful removal from the model-room and laboratory to the lecture-room, and to clean and replace them after being used, reporting to the managers such accidents as shall require repair, a constant diary being kept by him for that purpose. That in one day in each week he be employed in keeping clean the models in the repository, and that all the instruments in the glass cases be cleaned and dusted at least once within a month.[25]

Davy's colleague William Pepys had suggested that Faraday should be put to washing bottles. Davy had thought that too degrading, but Faraday's official list of duties was nonetheless largely menial.

In practice, however, Faraday was almost immediately given the opportunity to assist Davy in his own experiments and to prepare chemical compounds.[26] In this way over the next few months he gradually became familiar with the routines and practices of everyday laboratory life. He also became familiar with some of the laboratory's dangers. Davy was at the time experimenting with highly volatile compounds of chlorine and nitrogen. Both Faraday and his master came close to serious injury on several occasions.[27] He also had an opportunity to assist Davy and others at their lectures, proudly recording his participation in letters to his friends.

But the fragility of Faraday's social status became clear when after a few months' employment he was invited by Davy to join his entourage on a tour of the Continent. Faraday accepted the offer with some trepidation. He had never previously traveled more than twelve miles from London.[28] He would also be giving up his position as laboratory assistant to the Royal Institution. This was a difficult decision for a young man only just embarked on his career. The party, consisting of Davy, his wife, Faraday, and Lady Jane Davy's maidservant, left London on 14 October 1813 and embarked from Plymouth to Morlaix a few days later on the 17th. Faraday almost immediately encountered problems concerning his precise social status within the party from his master's wife, Lady Jane. The main cause of difficulty was that Davy's valet had at very short notice decided not to accompany them, with the result that Davy asked Faraday to perform some of the duties that a valet would normally perform.[29] As a result his status was unclear.

It seems probable that Faraday in many respects regarded himself as Davy's apprentice.[30] As such he did not regard himself as a servant and felt entitled to the rights and privileges normally accorded an apprentice by his master. Davy, as a former apprentice himself, probably regarded the relationship in a similar light. Lady Jane, however, being unfamiliar with workshop codes of behavior and conduct, saw Faraday as performing a valet's duties and therefore regarded him as a servant. This caused Faraday much discomfort:

> She is haughty & proud to an excessive degree and delights in making her inferiors feel her power. . . . When I first left England unused as I was to high life & to politeness unversed as I was in the art of expressing sentiments I did not feel I was little suited to come within the observation and under the power in some degree of one whose whole life consists of forms etiquette & manners. . . . This at first was a source of great uneasiness to me and often made me feel very dull & discontented and if I could have come home again at that time you would have seen me before I had left England six months.[31]

Faraday was slowly learning some of the skills he would need if he was to move in polite society and the care with which he needed to define his own position with regard to his social superiors.

Faraday was in a difficult position. Davy had resigned his professorship at the Royal Institution in order to embark on his Continental travels. Faraday, in order to accompany him, had likewise resigned his post as laboratory assistant. It was not clear what his position would be when he returned to England with Davy, though his master assured him that all would be well. He was also aware of rumors concerning the institution's precarious financial position. He even publicly contemplated abandoning philosophy and returning to his trade of bookselling and binding. On the other hand he still had the opportunity of assisting Davy with his experiments and of meeting the host of eminent philosophers whom Davy visited. Many of the contacts forged on this tour were to prove very useful to Faraday in his later career.

The philosopher's apprentice eventually returned to London on 23 April 1815. The party's tour had been cut short by the turmoil following Napoleon's escape from Elba and the recommencement of the European war. Davy did not resume his position as professor of chemistry at the Royal Institution. That post continued to be held by his successor, William Thomas Brande. Instead Davy took up a position as one of the institution's managers and also served as vice president. Faraday was soon able to return to his former position as chemical assistant, with a five-shilling increase in his weekly wages. Stephen Slatter, who had been appointed to replace Faraday, returned to his previous position as assistant porter, while Blackett Thomas Wallis, who had replaced him, was discharged.[32]

This European tour had brought home to Faraday the importance of carefully defining his relationships with others if he was to succeed in the genteel surroundings of the Royal Institution. Experimental dexterity in the laboratory would clearly not be enough to make a natural philosopher. Faraday had been interested in self-improvement since his days as a bookbinder's apprentice. He made use of Isaac Watts's *Improvement of the Mind*, which he read shortly after the publication of a new edition in 1809, to organize a concerted campaign of betterment.

Watts recommended five ways of learning: observation, reading, instruction by lectures, conversation, and meditation.[33] Interestingly, Watts particularly recommended learning by lecture:

> There is something more sprightly, more delightful and entertaining in the living discourse of a wise, learned, and well-qualified teacher, than there is in the silent and sedentary practice of reading. The very turn of voice, the good pronunciation, and the polite and alluring manner which some teachers have attained, will engage the attention, keep the soul fixed, and convey and insinuate into the mind the ideas of things in a more lively and forcible way, than the mere reading of books in the silence and retirement of the closet.[34]

Faraday in his future career was to pay particular attention to the proper performance of lectures. In 1810, presumably shortly after reading Watts, Faraday started attending John Tatum's lectures and, following Watts's prescriptions, took careful notes in preparation for the solitary meditation without which proper training of the mind was impossible.[35]

Attendance at Tatum's lectures had brought Faraday into contact with the City Philosophical Society, founded by Tatum in 1808 as a forum where attendees at his lectures could further discuss and improve their knowledge.[36] Members and their friends met every second Wednesday to hear a lecture by one of the members. Alternate Wednesdays were devoted to private discussion among the members. Following his return from the Continent, Faraday participated enthusiastically in the society's affairs. He was elected a member, thereby gaining the privilege of placing the letters M.C.P.S. after his name, and delivered his first lectures there between 1816 and 1818.

Faraday took advantage of his contacts at the City Philosophical Society to follow another of Watts's prescriptions. He commenced corresponding with some of his new acquaintances, particularly Benjamin Abbott and Thomas Huxtable, on matters of self-improvement. His letters to Abbott on lecturing show how self-conscious he was of the need for acquiring proper skills of self-presentation and performance. Being by now as a result of his duties a constant attendee of lectures at the Royal Institution, he was in a good position to observe what strategies were or were not successful, and to begin to formulate his own performance tactics.

As early as June 1813, only a few months after taking up his position as laboratory assistant, Faraday was discussing with Abbott the best ways of presenting a lecture. He fully realized that "Polite Company expect to be entertained not only by the subject of the Lecture but by the manner of the Lecturer, they look for respect, for language consonant to their dignity and ideas on a levell with their own."[37] Proper delivery was crucial:

> The utterance should not be rapid and hurried and consequently unintelligible but slow and deliberate conveying ideas with ease from the Lecturer and infusing them with clearness and readiness into the minds of the audience. A Lecturer should endeavour by all means to obtain a facility of utterance and the power of cloathing his thoughts and ideas in language smooth and harmonious and at the same time simple and easy his periods should be round not too long or unequal they should be complete & expressive conveying the whole of the ideas intended to be conveyed if they are long and obscure or incomplete they give rise to a degree of labour in the minds of the hearers which causes lassitude indifference and even disgust.[38]

Just as important was the way in which the lecturer carried himself: "he must by all means appear as a body distinct and separate from the things around him and must have some motion apart from that which they possess."[39]

Faraday was sufficiently concerned about improving his lecturing ability to set up a private group within the City Philosophical Society with his fellow member Edward Magrath to criticize and improve each other's use of language.[40] He also took the step of taking private lessons in elocution with one of London's most prominent teachers, Benjamin Smart.[41] This was not a light undertaking. Smart charged his pupils half a guinea for each lesson.[42] Faraday enouraged both Smart and Magrath to attend his lectures "for the sole purpose of noting down for him any faults of delivery or defective pronunciation that could be detected."[43]

Some indication of the strategies Faraday was taught by Smart may be gleaned from the latter's published works on elocution.[44] Smart took his pupils through three stages of learning: mechanical reading, significant reading, and impassioned reading, or speaking. Mechanical reading was designed simply to teach the utterance of words "justly, completely, and in smooth unbroken series between the written stops when they are joined into sentences." Significant reading took the additional step of making "the construction and meaning of every sentence plain by appropriate *tunes* (or inflections) of the voice."[45] Impassioned reading, which became "*distinct, significant, impressive* Speaking," was the final goal of elocution.[46] To enable students to reach this goal, Smart provided a series of graded exercises, the supervised reading of which gradually allowed the pupil to develop the proper skills.

Impassioned speaking was not, however, confined to using the voice: "the looks, the gesture, the whole deportment of the speaker, lend assistance; and it

1.1 Illustration from Benjamin Smart, *The Practice of Elocution*
(London, 1819), facing p. 63, demonstrating some of the body language
of public speaking as Faraday might have been taught it.

is the union of all of these that constitutes *expression*."[47] Smart gave detailed
instructions to his pupils on how they should stand and how they should hold
their bodies in order to achieve certain effects. He taught them how to use their
eyes and hands to address the audience, practices that gave the speaker the
"notion or feeling of *laying*, as it were, his facts or truths *before* his auditors."[48]
Constant practice and repetition of the various exercises, he urged, would convert these conventional gestures into natural motions of the body.

Faraday took considerable pains to acquire these skills of self-presentation,
which would be crucial to his success when he first had the opportunity of
lecturing before the Royal Institution's stylish audience. If he was to appear
in command of the knowledge he conveyed and the experiments he displayed before them, he had to appear to be in command of himself as well.
Public speaking was an art as difficult to master as was the art of successfully
manipulating the experimental apparatus in the Royal Institution's basement
laboratory.

Not long after his reinstatement at the Royal Institution, Faraday was given
his first opportunity to perform for an audience. This audience comprised not
the spectators in the institution's lecture theater but the readership of its *Quarterly Journal of Science*, edited by the professor of chemistry, William Brande.
Faraday's first effort at publication was modest. It was simply a note appended
to a communication by the marquis of Ridolfi, outlining a method of separating
platina from its ore and describing a sample of Tuscan caustic lime. Faraday's
note merely contained an analysis of a sample of that lime.[49] His analysis was
followed by "Observations on the Preceding Paper" by Humphry Davy, which
interpreted the findings in the light of volcanic activity and the geology of the
Italian states. Faraday was clearly still working under Davy's direction rather
than experimenting in his own right.

Subsequent papers showed the gradual development of Faraday's autonomy in the laboratory. His third publication, in early 1817, saw his first use of the personal pronoun in describing his work. He was still working under Davy's direction; he stated that he had been "requested . . . to make some experiments" by Davy, but was now drawing his own conclusions from his experimental work.[50] It was not long before his publications represented his experiments as his own autonomous and original productions.[51] Davy was gradually allowing his apprentice to practice his craft in his own right rather than simply as his master's assistant. Davy was also frequently absent from London during the late 1810s. At such times Faraday would work without supervision in the laboratory, though still largely following Davy's instructions.[52]

One such task allotted to him by Davy during one of his absences gave the apprentice an opportunity to stand up as an authority. A Viennese doctor, von Vest, had claimed the discovery of a new metallic element, which he christened "Sirium." Davy, visiting Rome, acquired a sample of the new substance and sent it to Faraday at the Royal Institution for analysis. Faraday went to work and soon published an account of his analysis, concluding: "It is evident that it has no claim to the character of a peculiar metal, but is a very impure mixed regulus. It contains sulphur, iron, nickel, and arsenic; and these make up very nearly, if not quite, its whole mass. I should guess the iron to be nearly equal to one third of the whole weight."[53] The apprentice was now of sufficient stature to enter into dispute, if only with an obscure and unknown foreigner.

He returned to the attack later in the year, having been sent a larger sample of the metal by Davy, and having access to a translation of von Vest's account of its production. He went through a careful series of experiments designed to separate out all the elements of which he suspected the metal to be made. Any residue remaining at the end of these processes could plausibly be regarded as a new element. He regretfully informed the readers of the *Quarterly Journal*, however, that when the series of analyses was concluded, absolutely no trace of the mysterious metal remained.[54] Davy was delighted with his protégé's success: "I find that you have found the parallax of Mr Wests Sirius & that as I expected He is mistaken."[55]

Davy was also now publicly acknowledging Faraday's assistance in his published work. In his work on phosphorus, for example, he emphasized that "[i]n these experiments, and in all the others detailed in this paper, I received much useful assistance from Mr. Faraday, of the Royal Institution; and much of their value, if they shall be found to possess any, will be owing to his accuracy and steadiness of manipulation."[56] This recognition was not without its perils. The Swedish chemist Berzelius, responding to attacks on his work in this paper, was withering about some of the experiments detailed and took Davy severely to task for depending on the skills of a mere technician: "If M. Davy would be so kind as to take the pains of repeating these experiments himself he should be convinced of the fact that when it comes to exact analyses, he should never entrust them into the care of another person."[57] While Faraday was now sufficiently authoritative to combat the claims of an obscure Austrian, he was still no match for the eminent Berzelius.

By 1820 Faraday was in an ambiguous position. On the one hand he was an autonomous experimenter in his own right, with a list of publications. On the other he remained Sir Humphry Davy's assistant, working on tasks allotted to him by his master. It was not always clear whether the products of his labor in the laboratory were his own or Davy's. Other patrons of the Royal Institution such as William Wollaston, as well as the institution's professors, could also take advantage of his services. During the early 1820s, Faraday, as he attempted to establish himself as an independent worker, found himself in difficulties over the question of his social status in the laboratory and the issue of whether his work was to be considered as his own property or that of his master.

In 1821, at the request of Richard Phillips, the editor of the *Annals of Philosophy*, Faraday wrote an anonymous piece, "Historical Sketch of Electromagnetism," describing the rapid development made in electromagnetic experimentation since Oersted's announcement of the movement of a magnetic needle near a wire carrying an electric current.[58] Following on from that work, in the course of preparation for which he had repeated all experiments published on the matter, Faraday during September 1821 continued to experiment and succeeded in making a current-carrying wire rotate about a magnet and vice versa. He rapidly published this result in the Royal Institution's in-house journal.[59] Within days of the paper's publication, however, Faraday became aware of rumors circulating concerning his rights of priority to the new experiment.

David Gooding has provided detailed analysis of Faraday's own labors leading up to the successful rotation experiment.[60] He shows how a range of material and social resources needed to be put in motion in the process of producing the first tentative laboratory results. Gooding also emphasizes the open-ended nature of Faraday's own interactions with the apparatus and the ways in which he was required to learn and direct his skills as he manipulated the material objects on his laboratory bench. The importance of the Royal Institution is also clear. Faraday could call on Newman, the institution's instrument-maker, to manufacture a more robust and compact version of his rotation apparatus, which he could use to convince others of his success. Gooding does not, however, sufficiently emphasize the social significance of Faraday's experiment. Just as important as the material negotiations in the laboratory were the delicate social negotiations that were required before Faraday could be regarded as a discoverer of anything at all.

Some months before Faraday's work on electromagnetism, William Wollaston had attempted at the Royal Institution to make a current-carrying wire rotate about its own axis. He had also made public his views concerning the possibility of electromagnetic rotation in general. It was now held that in

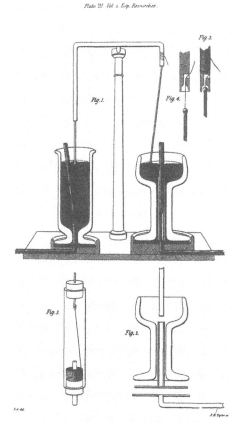

*Plate IV. Vol 2. Exp. Researches.*

1.2 Apparatus for Faraday's experiment demonstrating the rotation of a current-carrying wire around a magnet (and vice versa) in 1821. This was the experiment that led to accusations of ungentlemanly conduct. Michael Faraday, *Experimental Researches in Electricity*, 3 vols. (London, 1839–55), vol. 2, pl. 4.

choosing to work on establishing such rotations, Faraday had illegitimately and without acknowledgment taken an experiment that belonged to Wollaston. Faraday had to work hard to recover his integrity:

> I hear every day more and more of those sounds which though only whispers to me are I suspect spoken aloud amongst scientific men and which as they in part affect my honor and honesty I am anxious to do away with or at least to prove erroneous in those parts which are dishonorable to me. You know perfectly well what distress the very unexpected reception of my paper on Magnetism in public has caused me and you will not therefore be surprised at my anxiety to get out of it though I give trouble to you and other of my friends in doing so.[61]

Faraday was desperate for an interview with Wollaston that would allow him to clear his name.

It was clear that his difficulties revolved about the issue of his rights to the experiment and his social status. This is how he expressed it himself:

> If I understand aright I am charged 1 with not acknowledging the information I received in assisting Sir H. Davy in his experiments on this subject 2 with concealing the theory and views of Dr Wollaston 3 with taking the subject whilst Dr. Wollaston was at work on it and 4 with dishonorably taking Dr. Wollastons thoughts and pursuing them without acknowledgement to the results I have brought out.[62]

It was clear that Faraday himself regarded the last of these charges as the most serious. He had acted in a fashion inappropriate for a gentleman.

In early-nineteenth-century London, ungentlemanly behavior was unphilosophical behavior as well.[63] Wollaston clearly concurred in this view. Replying to Faraday's plea, he remarked,

> As to the opinions which others may have of your conduct, that is your concern not mine; and if you fully acquit yourself of making any incorrect use of the suggestions of others, it seems to me that you have no occasion to concern yourself much about the matter.[64]

At stake was the question of whether it was appropriate for a laboratory assistant to carry out experiments in a field that had already been appropriated by other, more exalted figures. Also at issue was the question of to what extent an assistant's labor could be considered his own rather than the property of his master.

The question arose again a few years later in the context of the liquefaction of chlorine. Faraday had been instructed by Davy to carry out a series of experiments, one of which was to heat chlorine gas in a sealed glass tube. The result of this particular experiment was that the chlorine liquefied under pressure. Again, when Faraday published this result in the *Philosophical Transactions*, it was not clear whose experiments they were. Davy added a note to Faraday's paper emphasizing that this result was one that he had anticipated when he instructed Faraday to carry out the experiment. He also insisted that Faraday point out in the paper that the experiment had been undertaken on his instructions.[65]

The uncertainty on this occasion had potentially disastrous consequences. Faraday had been nominated to become a fellow of the Royal Society. Davy was extremely unhappy about his candidacy, and a number of other fellows were reviving the question of his behavior over the electromagnetic rotation experiment in the light of the renewed uncertainty concerning his rights to the liquefaction of chlorine. Several fellows, notably Henry Warburton, threatened to blackball Faraday's nomination, and Davy himself attempted to force Faraday to withdraw.[66] Again Faraday had to work hard to overcome suspicions that he had acted in a manner unbecoming a gentleman and a fellow of the Royal Society.

With his election as a fellow of the Royal Society, Faraday's apprenticeship can be said to have come to an end. He was now an experimental natural philosopher in his own right. A little over a year later, at Davy's request, the managers of the Royal Institution marked this fact by changing Faraday's status. They resolved that "Mr. Faraday be appointed Director of the Laboratory under the superintendence of the Professor of Chemistry."[67] His autonomy had been achieved only with difficulty and the near alienation of his former master. As he admitted, "I was by no means in the same relation as to scientific communication with Sir Humphry Davy after I became a fellow of the Royal Society in 1824, as before that period."[68] Moving from artisan to gentleman was not an easy task.

FRIDAY EVENING FASHION

As director of the laboratory, Faraday had considerable freedom to follow his own experimental inquiries. At the same time, however, he had considerable responsibilities for the day-to-day running of the Royal Institution. These responsibilities increased during the second half of the 1820s as the institution once again found itself in a difficult financial position.[69] At some stage during the course of 1825, Faraday gradually introduced a new feature into the institution's range of offerings to its members and subscribers. On Friday evenings, members and their guests would be invited to attend the institution to examine exhibits laid out on display in the library and to hear an account of some new scientific discovery or curiosity from Faraday or an invited speaker. These Friday Evening Discourses soon became central to the Royal Institution's image. They also had a crucial role to play in the making of Michael Faraday as the metropolis's premier interpreter of nature for the leisured classes.

The institution's managers were not long in recognizing the Friday evening meetings' potential. They were noted at a meeting of 13 June 1825:

> The Board of Managers taking into consideration the advantages likely to result to Science from the meetings held in the Royal Institution on Friday Evenings, and being desirous of affording them all the support in their power, have
> *Resolved* That one of the Board of Managers do attend at each of the meetings in a rotation to be fixed amongst themselves, for the purpose of receiving the Visiters, and superintending the arrangements of the Evening.
> *Resolved* That every Member introducing Visiters to the Evening Meetings of the Royal Institution, do write the names of persons so introduced, in a book to be provided for the purpose.
> *Resolved* That Mrs. Greenwood [the housekeeper] be ordered to provide Tea and Coffee &c. at the Friday Evening Meetings.
> *Ordered* That the above resolutions be hung up in the Hall, and Library, and Reading Rooms, of the Institution.[70]

What had commenced as a private and largely informal event orchestrated by Faraday now became, under the managers' auspices, a crucial linchpin in the institution's activities. With typical Benthamite zeal, a committee was set up to oversee the proceedings.

The evening meetings required considerable organization, however. Much of Faraday's correspondence during the second half of the 1820s was taken up with inviting potential speakers and discouraging those who were thought inappropriate. Any hint of controversy was to be avoided at all costs at these gatherings, as Faraday sharply pointed out to William Hosking, one hopeful aspirant:

> Our Committee return you their best thanks for your kind offer but think the subject one which in its nature does not consort with the character of the Friday Evening meetings as they wish to preserve them.
> New points in philosophy—or new modes of experimental illustration—or new applications to useful purposes are what is wished for after these follow new matters in taste or literature. But mere matter of opinion which can be settled only by reference to taste and not by reference to natural facts is they think inadmissible at our Lecture table on the Friday Evenings.[71]

The sensibilities of the Royal Institution's genteel patrons and their guests were not to be disturbed. The managers were well aware of the fragility of their position. Any loss of members or subscribers could have disastrous consequences for the institution.

The establishment of the Friday Evening Discourses had a marked impact on the institution's own organization. It was soon clear that the meetings needed to be carefully managed to be successful, and regulations were accordingly drawn up:

> The Library is to be opened at half past Eight o'clock. All the Members of the Institution have the privilege of introducing their Friends either personally or by proper tickets signed by themselves.
> No persons, except Members of the Royal Institution or those especially invited by the Board of Managers can be admitted unless personally introduced by a Member or the production of a ticket signed by a Member. Gentlemen introduced by ticket are requested to deliver them to the Porter in attendance. . . . Members may have on application to the Assistant Secretary an unlimited supply of tickets.[72]

It soon became clear that these regulations were too generous. A few years later the Friday Evening Committee resolved to limit the privilege of bringing guests to no more than two per member.[73]

A committee was also established to consider the redeployment of the institution's servants in the light of the new pressures. The institution employed two porters to attend the hall of the building and admit members and their guests. They were also required to carry messages and ensure that fires were kept burning in the various rooms. The Friday evenings' popularity meant that they could no longer fulfill their duties and still be in attendance at the door when guests arrived. According to revised regulations drawn up by Faraday in 1828: "No manufacturing employment to be carried out in the Hall of the Institution. The Porters are to be clean and to be disengaged from any other duty whilst in the Hall, than that of attending to applications and to the convenience of the Members and Subscribers."[74] Less than a year later, the Hall Committee reported "[t]hat both servants agree in stating their inability to conform with the regulations more strictly than at present, and that much additional duties have accrued from the Friday Evening Meetings."[75] The upshot was that the managers resolved to reorganize the porters' duties, to hire a third servant, and to place all three directly under Faraday's superintendence.[76]

The Friday meetings also had a quite direct physical effect on the institution. Early in 1830, Faraday reported to the managers his observation that one of the bookcases in the library appeared to be sinking. Investigation determined that the library's floor was collapsing under the weight of the additional visitors passing through during the Friday evening meetings.[77] At the end of the season, an iron pillar was erected in the Newspaper Room below the library to support the subsiding floor. There could be no question any more of the Friday Evening Discourses' success. They were clearly quite literally bringing the house down.

As Faraday had suggested in his rebuff to Hosking, topics for the Friday Evening Discourses were carefully selected. All matter that might be considered offensive to the tastes of the Royal Institution's fashionable members was excluded. Early topics explained to the discourses' audience ranged from the new block machinery invented by Brunel and installed at the Naval Dockyards in Portsmouth to expositions of the latest developments in electricity. Faraday in 1832, for example, gave three discourses on his recent investigations of magnetoelectric induction.[78] His audience could hear an authoritative account of new discoveries, carefully tailored to meet their requirements.

Faraday did not restrict himself to his own specialties in his discourses. His second performance of 1837, for example, was an account of Marshall Hall's controversial theory of the reflex function of the spinal marrow. His last discourse of that year, "Early Arts: The Bow and Arrow," was attended by, among others, Mrs. Richard Owen and her father William Clift:

> In the ante-room I had some conversation with Mr. F., who said this lecture was the last of his course for the season. It was chiefly on arrows and weapons. Faraday showed us the various flints used in different times and different countries for arrow-heads, knives, &c. It was most interesting and amusing, and of course well delivered. Mr. F. shot or rather blew several small arrows through tubes—and with good aim—at a band-box with a centre mark. The place full, but the heat and draught dreadful.[79]

Mrs. Owen was a frequent attendee at Faraday's performances, and a great admirer. Commenting on his discourse titled "The Gymnotus and the Torpedo" in January 1839, she remarked that "Faraday is the *beau idéal* of a popular lecturer."[80]

These performances were not achieved without effort, however. This was where Smart's early training and Faraday's self-conscious self-fashioning came into play. The discourses were the product of hard labor:

For his Friday discourses, and for his other set lectures in the theatre, he always made ample preparation beforehand. His matter was always over-abundant, and, if his experiments were always successful, this was not solely attributable to his exceeding skill of hand. For, unrivalled as he was as a manipulator, in the cases in which he attempted to show complicated or difficult experiments, that which was to be shown was always well rehearsed beforehand in the laboratory.[81]

The result of all this was that "his manner was so natural, that the thought of any art in his lecturing never occurred to anyone."[82]

Publicity was also required to make the discourses a success and attract a prestigious audience. Faraday made sure that William Jerdan, editor of the *Literary Gazette*, received abstracts of the Friday evening performances. If necessary he wrote the accounts himself.[83] He also took more informal steps to ensure that appropriate guests were invited. Writing to the naturalist Walter Calverley Trevelyan, for example, he invited him to attend a meeting: "I have thought that perhaps you would like to see how we spend our Friday Evenings here and under that impression beg to enclose a ticket which, if you do not use, perhaps you will do me the favour to burn."[84] It was clear that the precious tickets should not be permitted to fall into the wrong hands.

The discourses were in any case immensely popular and successful. During the second half of the 1830s, attendance at one of Faraday's performances only once fell below 400. His "Early Arts: The Bow and Arrow," which so impressed Mrs. Owen, drew a crowd of 583 admirers.[85] By the late 1840s attendance at one of Faraday's discourses more than once exceeded 1,000. In terms of the more pragmatic goal of increasing the membership and thereby saving the Royal Institution from financial ruin, the discourses seem to have been successful as well. In the first five years of their existence, from 1826 to 1830, 325 new members were elected into the Royal Institution. This should be compared with the 53 members elected over the previous five years.[86]

The establishment of the Friday Evening Discourses was not the only step taken to improve the Royal Institution's finances and reputation. Early in 1826, the committee established to consider the institution's lectures reported the results of their deliberations to the managers.[87] They put forward recommendations aimed at using the lectures to improve both the Royal Institution's finances and its reputation as a premier site of scientific knowledge making. The institution's lectures were divided into two categories. On the one hand there were the private lectures, delivered by Brande and Faraday in the mornings and primarily aimed at medical students. On the other were the public lectures, delivered in the afternoons to the institution's members and subscribers. The institution's members were also permitted to attend the private lectures in the mornings.

The Lecture Committee suggested that both categories of lectures should be to some extent reorganized. The morning lectures were at that time held in the small lecture theater in the basement of the building, opening out from the chemical laboratory. It was suggested that they should be held instead in the large theater upstairs so as to increase the numbers who could attend and to bring them more prominently to the notice of members. They held that these lectures had already "been productive of great advantage to the Royal Institution and have greatly raised its character as a school of Chemistry."[88] They argued that their removal to the main theater would add to the members' benefits, and also suggested that they should be expanded to include other subjects. Other private lecturers should be encouraged to hold their lectures at the institution on condition that members be permitted to attend free of charge. This would add to the institution's range of services to the members, thus encouraging more subscribers to become full members, without any additional cost to the institution itself.

The public lectures held in the afternoons were one of the institution's chief sources of income, generated largely by annual subscribers. The committee suggested that this income could be increased if individuals were permitted to subscribe to attend particular courses rather than subscribing for the entire season. The reason for this was straightforward:

> According to the present arrangements of the Institution, the lecture Season usually commences in the beginning of February and continues till the month of June including a period of about four months. Of course all who subscribe to the lectures generally have the opportunity of attending the whole of the lectures during the Season, but the fashionable Season in London seldom commences till after Easter, so that the persons who come to town at that time although they pay the same subscription have only half the advantage of them who are in London for the whole period.[89]

Allowing subscribers to enroll for particular lecture courses would accommodate the institution's program to the constraints of the fashionable London season and therefore increase their revenue.

The reforms suggested by the Lecture Committee were highly successful. Large numbers took advantage of the opportunity to subscribe for particular courses rather than for the whole season. Interestingly, by far the majority of these subscribers to individual courses were female.[90] The institution's managers and Faraday himself as director of the laboratory had worked hard and successfully to integrate the institution's program into the rhythms of fashionable metropolitan society and to provide the services that society wanted. For Faraday's own program of experimental discovery in electricity, this was to be crucial. Establishing the Royal Institution as a secure space for his activities was central to his success as a natural philosopher. Just as important were his efforts to tailor his performances to the requirements and expectations of his audience.

Faraday was well aware of the central role the Royal Institution played for his career, and was jealous of its reputation. He responded furiously when William Upcott, assistant librarian at the rival London Institution, cast doubt upon its preeminence:

1.3 *Fraser's Magazine*'s impression of Faraday at work. *Fraser's Magazine* 13 (1836): facing p. 224.

> I am amused and a little offended at Upcots hypocrisy. He knows well enough that to the world an hours existence of our Institution is worth a years of the London and that though it were destroyed still the remembrance of it would live for years to come in places where the one he lives at has never been heard off [*sic*] . . . I think I could make the man wince if I were inclined and yet all in mere chat chat over a cup of tea.[91]

A slur on the institution for which he labored so hard was a slur on Faraday's own reputation as well.

By the mid-1830s Faraday was sufficiently well known among London's fashionable classes to be used by the Tory press as a stick with which to beat the Whig administration. The scurrilously Tory *Fraser's Magazine* made great play of the Whigs' initial reluctance to award Faraday a government pension, unfavorably comparing their miserly attitude toward science and the arts with the generosity of the previous administration under Peel.[92] The government's behavior and Faraday's response were characterized as "a striking monument of Whig liberality to men of science, and of the honourable and spirited behaviour of a gentleman who knows when to feel and how to resent an insult."[93]

*Fraser's* continued their attack in the next issue, lampooning Melbourne's government for having had to award Faraday his pension after all, under pressure from the king, and accusing them of having conducted a deliberate campaign to discredit Faraday.[94] The same year they published a literary sketch of their new hero, suggesting that his social success was a guarantee of knighthood, and that he would follow Davy's path in this respect as well: "he is now what Davy was when he first saw Davy—in all but *money*." *"Far-a-Day"* they joked, could easily be interpreted as *"Near-a-Knight."*[95] He had become the epitome of polite natural philosophy and London's main exponent of science to the fashionable.

*Fraser's* was right to compare Faraday's social success to the example of his predecessor. Like Davy before him, Faraday had fashioned himself carefully to meet the requirements of his polite audience. Unlike Davy, however, he was clearly aware of the boundaries between himself and that audience. Part of Faraday's social strategy was to deliberately distance himself from his patrons rather than attempt, as Davy had done, to become one of them. This was central to his success. Faraday was lauded as being one of "nature's gentlemen," whereas Davy had been condemned for attempting to ape his social superiors. Faraday never accepted a knighthood. He was aware that such an act might breach the delicate barrier he had built between himself and his audience.

READING THE BOOK OF NATURE?

In 1831 Faraday commenced the series *Experimental Researches*, which he read before the Royal Society and published in their *Philosophical Transactions*—papers that would make his name as the authority on electricity and the century's greatest experimenter.[96] The first series of these researches, presented to the Royal Society on 24 November 1831, contained the first full account of his discovery of voltaelectric and magnetoelectric induction. Over the next thirty years, until 1855, Faraday produced thirty of these series, detailing the discoveries made in the basement laboratory of the Royal Institution. This laboratory was a very private space. It was separated even from the institution's main chemical laboratory where most of the staff labored. It was a space completely devoted to Faraday's electromagnetism and the *Experimental Researches*.[97] This section will examine two of these series in particular: his first account of magnetoelectric induction and his announcement of the magneto-optic effect. The aim will be to uncover some of the strategies he adopted in transferring his experiments from the domain of his laboratory in the Royal Institution and into the pages of the *Philosophical Transactions*.

Faraday began the experiments that led to the first series of *Experimental Researches* on 29 August 1831. He had ordered the construction of a soft iron ring, wrapped around with separate coils of insulated copper wire. One coil was connected to a battery and the other passed over a magnetic needle. He immediately noted that at the moment the first coil was connected to the battery, there was a simultaneous "sensible effect" on the needle, which "oscillated and settled at last in original position." The same occurred when the connection with the battery was broken.[98] It seems clear that he already knew what to look for. He carried out a few more experiments that day, varying the conditions. He attempted unsuccessfully to produce a spark and concluded that the effect was "due to a wave of electricity caused at moments of breaking and completing contacts."[99]

No more experiments were done on the matter until the end of September. On the 12th he recorded the preparation of a number of coils but did no work with them until 24 September.[100] A number of experiments were carried out with these various coils, but no effect could be perceived. Not until 1 October did Faraday succeed in producing the effect again with two coaxial coils of copper wire wound around a block of wood. This was his first indication that the iron core was not essential to the phenomenon.[101] As before, he then attempted to produce other signs of the presence of electricity such as heat and chemical effects. He was unsuccessful but did succeed in getting a "very distinct though small" spark at the end of the inducing wires.[102]

The experiments continued over the next month as Faraday attempted to ascertain the exact conditions under which what he now described as the inductive effect appeared. He found that he could produce an effect from a single coil by quickly inserting a bar magnet into the coil, showing that "a wave of Electricity was so produced from *mere approximation of a magnet* and not from its formation in *situ*."[103] He visited Samuel Hunter Christie in Woolwich to experiment with the Royal Society's large magnet; his endeavors there culminated in an extensive series in which a copper plate was revolved between the poles of the magnet, producing powerful effects on the galvanometer. These experiments were repeated and extended again a few days later on 4 November.[104] This was the last day of experiments before Faraday delivered the first series of *Experimental Researches* before the Royal Society on 24 November.

Faraday's *Diary* demonstrates how he used the resources of both the Royal Institution and his contacts at Woolwich to produce his experiments.[105] His laboratory in the basement provided him with a private space where he could experiment undisturbed by the institution's other workers. Both Faraday and his apparatus needed to be isolated at first to work successfully. The tedious and laborious task of winding the wire coils could be delegated to his assistants.[106] Through Christie at Woolwich he had access to the Royal Society's massive magnet, which provided him with the power he needed to find the delicate effects for which he was searching. Mobilizing these different resources allowed Faraday to transform the first tentative twitchings of his magnetic needle into reliable, replicable, and therefore public phenomena.[107]

In the paper delivered before the Royal Society, Faraday presented his experiments in a very different order from that in which they were made according to his *Diary*. This he did in order to solidify the distinction he wanted between two types of induction: voltaelectric and magnetoelectric. The decision to describe the effect as one of "induction" was strategic. It invoked for the audience the familiar territory of common electricity. After brief general remarks concerning electrical induction, Faraday first gave an account of the experiments he had carried out on 24 September with helices of wire of various lengths connected to a battery and a galvanometer, respectively. He then recounted the successful attempt of 1 October:

> Two hundred and three feet of copper wire in one length were coiled round a large block of wood; other two hundred and three feet of similar wire were interposed as a spiral between the turns of the first coil, and metallic contact everywhere prevented by twine. One of these helices was connected with a galvanometer, and the other with a battery of one hundred pairs of plates four inches square, with double coppers, and well charged. When the contact was made, there was a sudden and very slight effect at the galvanometer, and there was also a similar slight effect when the contact with the battery was broken.[108]

He could render the effect more certain both by increasing the strength of the battery and by substituting a helix containing a needle for the galvanometer. When contact was made, the needle was found to be magnetized.[109]

He described further experiments which established that the inductive effect was instantaneous and temporary. The current, when contact was made, was always opposed to the original current, and, when contact was broken, the current was in the same direction. Faraday reported that he could find no other signs of electricity such as shock, spark, heating, or chemical effect, but attributed this to the brief duration and feebleness of the current. The effect in general he characterized as voltaelectric induction.

The next section of Faraday's account contained the first experiment he had actually performed, with the iron ring. He detailed how, with this contrivance and using charcoal points, he had been able to produce a spark. Experiments were then described in which the actions of the helices used to establish voltaelectric induction were augmented by the insertion of a soft iron cylinder. He also described how the insertion of a bar magnet into a helix was itself enough to produce the induced current. These various effects were classified as magnetoelectric induction. The main difference between the two kinds was that voltaelectric induction was instantaneous, while magnetoelectric induction took some time.[110]

In the remainder of the paper, these experiments were put to work in establishing what Faraday described as the electrotonic state, and in providing an explanation for Arago's phenomenon, in which a bar magnet suspended over a spinning copper plate tended to move with the spin. The electrotonic state was the "peculiar condition" of the wire while it was subject to volta- or magnetoelectric induction. Arago's phenomenon was explained as being the result of currents induced in the spinning plate as it moved past the poles of the suspended magnet. The movement of the magnet following the plate could then be understood as a rotation of the same kind as those established by Faraday himself in his 1821 experiments.[111]

Faraday clearly felt that his newest discoveries constituted a major coup. A few days after the reading of his paper at the Royal Society, he rushed off an enthusiastic letter to his friend Richard Phillips, describing his progress.[112] In this private communication he was (as he expressed it) "egotistical" that he had succeeded in explaining a phenomenon that had eluded "great names" such as Herschel, Babbage, and Arago himself. In December he also communicated his discoveries to Jean Nicolas Hachette in Paris, an old acquaintance whom he had met during his tour with Davy. Hachette promptly passed the account on to Arago, who as the permanent secretary of the Royal Academy of Sciences arranged to have it read at the next meeting on 26 December. This version was published, with the result that two Italians, Leopoldo Nobili and Vincenzo Antinori, repeated and published on the induction experiment before Faraday's own paper appeared in the *Philosophical Transactions*.[113] Faraday was furious.

As a result of his own experiences of the past decade, Faraday was acutely sensitive on matters of priority. Despite the fact that the Italian philosophers unambiguously stated that their work was a replication and extension of Faraday's discoveries, he regarded them as plagiarists who had stolen his work before he had been given due credit. He blamed Hachette for making a private communication public and thus making possible the Italian plagiarism. Hachette replied that the announcement of his discoveries at both the Royal Society in London and the Royal Academy of Sciences in Paris left Faraday's priority in no doubt:

> Les publications que vous avez faites à la Société Royale, confirmées par la Communication à l'académie Royale des Sciences de Paris, ne laissent aucun doute sur la priorité de vos découvertes, dont l'honneur vous appartient sans passage.[114]

For Hachette these announcements were quite sufficient.

Faraday refused to be mollified and continued to insist that the Italians' (and Hachette's) behavior was dishonorable. He had published a translation of the Nobili and Antinori paper in the *Philosophical Magazine* with a note by himself making public his dissatisfaction:

> I may perhaps be allowed to say, (more in reference however to what I think ought to be a general regulation than to the present case), that had I thought that that letter to M. Hachette would be considered as giving the subject to the philosophical world for general pursuit, I should not have written it; or at least not until after the publication of my *first paper*.[115]

Faraday regarded not only the discovery, but the right to develop new experiments in the field, as being his private property.

Hachette clearly maintained that this position was ludicrous:

> [V]ous annoncez une grande découverte à la Société Royale; elle est transmise de suite à Paris en votre nom; toute la gloire de l'inventeur vous est assurée; que pouviez-vous désirer de plus? que personne ne travaille dans la mine que vous aviez ouverte; cela est impossible.[116]

Faraday, however, did not see his "grande découverte" as an isolated instance. It was the first link in a chain of researches that he would pursue for the rest of his career. The title of his paper communicated to the Royal Society made that quite explicit. It was self-consciously titled as the "First Series" of his *Experimental Researches in Electricity*. In a private note, Faraday characterized the "discoverer of a fact" as the lesser man. The most exalted was "[h]e who refers all to still more general principles."[117] His complaint about the Italians was that they had interfered with his path to those general principles. As Tyndall noted, "Faraday entertained the opinion that the discoverer of a great law or principle had a right to the 'spoils'—this was his term, arising from its illustration."[118]

Almost one and a half decades after the appearance of his first series, Faraday presented to the Royal Society the results of his nineteenth series, in which he claimed to have succeeded in making a connection between the forces of magnetism and light.[119] This series of experiments was the result of his "conviction" that

> the various forms under which the forces of matter are made manifest have one common origin; or in other words, are so directly related and mutually dependent, that they are convertible, as it were, one into another, and possess equivalents of power in their action.[120]

He had attempted previously to discover such a relationship, observing the effect of passing a ray of polarized light through an electrolytic conductor, but had been unable to find any effect.[121]

On this occasion, he commenced experimenting on 30 August 1845, following much the same track that he had attempted unsuccessfully a decade earlier. He had constructed a glass trough in which electrolytes could be placed for decomposition while a ray of polarized light was passed through them. He used a Grove battery with five plates and a magnetic breaker and coils so that he could vary and alternate the current passing across the glass trough. A variety of electrolytes were then tried under different conditions. The light source was an Argand lamp, the light being polarized by reflection off a glass plate.[122] None of these experiments showed any effect on the polarized light when electricity was passed across the electrolytes.

These experiments continued through the first half of September as Faraday tried different types of electricity and different kinds of electrolyte. On 13 September he abandoned his attempt to show some effect with electricity and tried magnets instead, using two electromagnets and five Grove cells. He tried various forms of glass and crystal, placing the electromagnets in different positions as the polarized light was passed along the sample. One of the substances tried was the heavy glass that Faraday had labored so hard to produce during his work with the Royal Society's Optical Committee in the late 1820s.[123] With this specimen, when contrary magnetic poles were on the same side, there *was an effect produced on the polarized ray*, and thus magnetic force and light were proved to have relation to each other. This fact will most likely prove exceedingly fertile and of great value in the investigation of both conditions of natural force.[124]

Faraday had found the link he had been searching for. He spent the rest of the day varying the conditions and trying different substances. He finally concluded with satisfaction, "Have got enough for to day."[125]

Before continuing his experiments, Faraday borrowed a large electromagnet from Woolwich, to provide more power for his trials. With the new magnet at his disposal he proceeded on 18 September to establish the newly discovered effect's characteristics. He also established the best way of seeing the phenomenon:

> I find that it is easier for the eye to distinguish the effect of the new power conferred on the *dimagnetic* when the image is (by the Nichol eye piece revolution) rendered slightly visible upon one side or the other of utter darkness. The Magnetic curves then cause increase or diminution of the light of the image, and either is more sensible to the eye than the effect when one begins to observe with a dark field of view. Must observe bodies feeble in power in this way.[126]

He was training himself to see the phenomenon to best advantage so that he could then see it in cases where he had previously been unable to find any effect. He could now be certain that the effect took place when the light was passed in the direction of the magnetic curves, that its intensity depended on the intensity of the magnetic curves, and that the direction of the ray's rotation depended on the direction of the magnetic curves. At the end of the day he was clearly satisfied: *"An excellent day's work."*[127]

For the next few days he continued the experiments, using his new experience to see the phenomenon in a variety of different substances. On 30 September he returned to his original attempt to find a relationship between light and electricity. By 11 October he had constructed some apparatus that could produce an effect. This was a helix consisting of two glass tubes joined together and wound around with a coil of insulated copper. When a current passed through the coil, with distilled water in the tubes, the ray of light was circularly polarized.[128] Over the next few days he refined the effect, again ascertaining the best ways of seeing it and establishing that the direction of the rays' rotation was the same as that of the electric current in the helices.

Faraday's communication to the Royal Society on the results of these new experiments was received there on4 November 1845 and read on the 20th. Rumors that Faraday had made a new discovery did not take long to circulate. The *Athenaeum* announced the news in its weekly gossip column:

> Mr. Faraday, on Monday, announced at a meeting of the Council of the Royal Institution, a very remarkable discovery; which appears to connect the imponderable agencies yet closer together, if it does not indeed prove that Light, Heat, and Electricity are merely modifications of one great universal principle.[129]

From the outset, the new discovery was being framed as the latest confirmation of a grand general principle. Talk of the unity of imponderable agencies was popular at the time. William Robert Grove at the rival London Institution had been proclaiming his doctrine of the correlation of forces for several years.[130]

Faraday's nineteenth series was subtitled "On the Magnetization of Light and the Illumination of Magnetic Lines of Force." After defining these terms, he proceeded immediately to an account of the first successful experiment with the heavy glass of his own making. He then very briefly summarized the exhaustive experiments he had conducted to establish the circumstances under which the rotation phenomenon could be best observed, and in which substances it could be seen. The law regulating the phenomenon was that

> if a magnetic line of force be *going from* a north pole, or *coming* from a south pole, along the path of a polarized ray coming to the observer, it will rotate that ray to the right-hand; or, that if such a line of force be coming from a north pole, or going from a south pole, it will rotate such a ray to the left-hand.[131]

This law was invariant, regardless of the nature of the different substances in which the effect could be made visible.

Having established the existence of the rotation effect with magnets, Faraday then outlined the way in which it operated with electrical currents. He described the experiments in which diamagnetics were placed in helices through which currents were passed. He then expressed the law:

> When an electric current passes round a ray of polarized light in a plane perpendicular to the ray, it causes the ray to revolve on its axis, as long as it is under the influence of the current, in the *same direction* as that in which the current is passing.[132]

He emphasized the simplicity of the law and its identity with the previously enunciated law of the action of magnetism on light. He also emphasized the homology of both laws with Ampère's theory of magnetism. He suggested that if the diamagnetic in both setups were replaced by an iron core, then the lines representing the direction of the rays' rotations would be the directions of the currents in Ampère's theory.[133]

The paper concluded with some general considerations of Faraday's findings. He again emphasized that the discoveries

> established . . . for the first time, a direct relation and dependence between light and the magnetic and electric forces; and thus a great addition is made to the facts . . . which tend to prove that all natural forces are tied together, and have one common origin.[134]

The role of the diamagnetic in the phenomenon was carefully explained. Faraday asserted that the substance through which the polarized light passed played no direct role in producing the phenomenon: "it is the magnetic lines of force *only* which are effectual on the rays of light."[135] The medium through which the light passed played a role in facilitating and making visible the interaction:

> The magnetic forces do not act on the ray of light directly and without the intervention of matter, but through the mediation of the substance in which they and the ray have a simultaneous existence; the substances and the forces giving to and receiving from each other the power of acting on the light.[136]

This was shown by the absence of any effect in vacuum, air, or gases and also by the different degrees to which the phenomenon was visible in different substances.

In this way, Faraday was using the apparatus of his experiments as a means to an end. The material objects constituting his apparatus were to be regarded as tools for looking at something else, in this case the lines of magnetic force. The focus of attention was shifted therefore away from the artifacts that Faraday manipulated and toward an abstracted nature, which, while it could be approached only through the experiment, existed independently of the experiment's constituent materials.[137] Faraday's audience was directed to see the phenomenon as being divorced both from Faraday himself and from the apparatus through which it was constituted.

This was more than a rhetorical strategy. It was also a matter of ontology. Faraday directed his audience to look away from the material apparatus on his workbench because he wished them to see that nature lay elsewhere. In this instance he was quite emphatic that these experiments on the magnetic and electric relationships of light buttressed his view of matter as an artifact of disembodied power:

> Recognizing or perceiving *matter* only by its powers, and knowing nothing of any imaginary nucleus, abstracted from the idea of these powers, the phaenomena described in this paper much strengthen my inclination to trust in the views I have on a former occasion advanced in reference to its nature.[138]

He was referring to the views he had presented almost two years previously at a Friday Evening Discourse on the nature of matter and then had published in the *Philosophical Magazine*.[139] Here he had argued that the abstract notion of matter should be abandoned since we had no clear conception of matter other than through the various powers it manifested. This disembodied ontology went along well with his deliberate epistemological strategy of divorcing the phenomenon from the apparatus that produced it.[140]

As had happened thirteen years previously, news of Faraday's spectacular discovery spread before he could properly present it. The notice in the *Athenaeum* appeared on 8 November, almost two weeks before Faraday's paper was to be read before the Royal Society. It was noticed immediately. John Herschel wrote to Faraday the next day, requesting more information and reminding him of Herschel's own anticipation of such a result.[141] Herschel described the experiment he had made in expectation of just such an effect:

> It is now a great many years ago that I tried to bring this to the test of experiment (I think it was between 1822 and 1825) when on the occasion of a great magnetic display by Mr. Pepys at the London Institution I came prepared with a copper helix in an

earthen tube (as a non conductor) & a pair of black glass plates so arranged as that the 2nd reflexion should extinguish a ray polarised by the first after traversing the axis of the copper helix I expected to see light take the place of darkness—perhaps coloured bands—when contact was made. The effect was *nil*. But the battery was exhausted and the wire long and not thick and it was doubtful whether the full charge remaining in the battery *did* pass, being only a single couple of large plates.[142]

He then went on to suggest further lines of research to Faraday and to offer his own interpretation of the phenomena.

Faraday answered immediately, expressing ignorance of (and some annoyance at) the report in the *Athenaeum*. He made it clear that Herschel's experiment could never have worked, since he did not use a diamagnetic inside the helix, before briefly describing his findings and the directions in which his work was proceeding: "my next paper to the R.S. will contain an account of *new magnetic actions & conditions* of matter. In the mean time let me ask you to say nothing about this part for I want to make it out as quietly as I can."[143] Faraday had clearly not forgotten the Hachette affair.

A letter soon arrived from William Whewell, also alerted to the discovery by the *Athenaeum* notice. Faraday responded with a short account and again briefly mentioned that he was hot on the trail of new discoveries. Again also, however, he warned Whewell to be discreet: "But do not say, even, that, you are aware I am so engaged. I do not want men's minds to be turned to my present working until I am a little more advanced."[144] Faraday clearly felt that his discovery of the new phenomenon had opened up a new field ripe for his picking. He did not want anybody else poaching on his territory. This became clear as Herschel continued to pester him for more details and suggestions of new experiments. Faraday's response was to send him a sealed note, not to be opened without Faraday's permission. In it was an outline of his developing researches. He would not allow himself to be anticipated.[145]

The response to Faraday's new researches was ambivalent. Both Herschel and Whewell, for example, were clearly interested but unsure as to how the phenomenon should be interpreted. Herschel inclined to the view that the phenomenon was evidence of a relationship between the magnetic influence and the crystalline forces in the diamagnetic substance rather than of a direct relationship between magnetism and light.[146] It was evidence concerning the structure of matter rather than of the disposition of lines of force. He cited Faraday's failure to find the effect in the absence of a crystalline medium as proof of his view. Faraday was adamant that although the phenomenon could so far be manifested only in the presence of a crystalline medium, the experiment demonstrated a direct relationship between light and magnetism. This was a consequence of his view that all forces were manifestations of a single power. Herschel had been intermittently studying the relationship between polarized light and crystalline structure since 1820 and was not disposed to defer to Faraday's authority. He took the view that what Faraday's magneto-optic effect provided was a new tool to study the internal structure of crystals, and that the presence of a crystalline medium was necessary for the phenomenon to be produced.[147]

Whewell (in private at least) was inclined to be even more dismissive. In a letter to Forbes he complained about "the overcharged importance of Faraday's view of his recent discoveries."[148] Whewell, as Schaffer has argued, was concerned that Faraday's interpretation of the new phenomenon threatened the fundamental idea of polarity that Whewell held to be crucial to the science of electromagnetism in its relationship to chemistry.[149] Neither he nor Herschel was keen to allow Faraday the monopoly over further research and interpretation that he clearly saw as a prerequisite and a privilege of his status as discoverer. While Faraday was the archetype of the fashionable and successful discoverer and interpreter of nature in the Royal Institution's lecture theater, his views could be contested by those who had a different view of the proper conduct and role of experimental philosophy. Whewell might endorse Faraday's removal of "extraneous machinery" from his philosophy, but he remained cautious about the abstractions that Faraday sought to erect in its place.

THE GREAT EXPERIMENTER

The aim of this chapter has been to uncover the work and resources that went into the making of Michael Faraday. That the effort of fashioning was successful was testified to by Faraday's first biographer:

> Nature, not education, rendered Faraday strong and refined. . . . By some such natural process in the formation of this man, beauty and nobleness coalesced, to the exclusion of everything vulgar and low. He did not learn his gentleness in the world, for he withdrew himself from its culture; and still this land of England contained no truer gentleman than he.[150]

Tyndall, of course, had his own reasons for celebrating Michael Faraday, his patron and predecessor at the Royal Institution. He was aware also of the work that had been required to build the image:

> Underneath his sweetness and gentleness was the heat of a volcano. He was a man of excitable and fiery nature; but through high self-discipline he had converted the fire into a central glow and motive power of life, instead of permitting it to waste itself in useless passion. . . . Faraday was *not* slow to anger, but he completely ruled his own spirit, and thus, though he took no cities, he captivated all hearts.[151]

The key was discipline. Faraday had learned how to act in the manner appropriate to one who wished to be the purveyor of polite knowledge to London's genteel patrons of science.

Proper behavior encapsulated Faraday's success. Nothing that he did in the laboratory could be sufficient other than as part of a complex of tactics that allowed him to make that place a source of knowledge. In fact that private place of knowledge production in the basement of the Royal Institution could not function at all without the wall of public activity that surrounded it.[152] At the same time, however, the glittering facades of the Royal Institution's Friday Evening Discourses, for example, would be just as fragile without the constant labor going on out of sight to produce and maintain their integrity.

The difference, of course, was not a straightforward matter of public and private spaces.[153] Nothing about Faraday's basement laboratory made it essentially private, just as the Friday Evening Discourses were not wholly public. Faraday's success was at least partly the result of his having strategically secured different spaces with different degrees of seclusion and exclusivity. His tactics were not always successful. In both episodes of discovery described here, Faraday was unable to insulate himself completely. His letter to Hachette describing the induction experiments became a public document. Similarly, his announcement of the magneto-optic effect to a private meeting at the Royal Institution was rapidly translated into the public sphere. The fragility of these boundaries illustrates their contingency and the work required to keep them in place. Things said and done in one place did not necessarily work as well in other settings where other conventions might be in play.

Faraday's status was the product of careful management, both by himself and by others. All his contemporary biographers noted his careful construction of boundaries around several aspects of his activities. Many recorded with reverence their occasional glimpses of the "real" Faraday behind the scenes.[154] By being made into such an icon, Faraday could be made to serve a variety of interests other than his own. He came to stand for the humble and disinterested investigator of the natural world.[155] Even the ascerbically agnostic John Tyndall celebrated his Christian virtues. The resulting distance helped Faraday as well. It made it possible for the blacksmith's son to move easily in a society very different from the one he had left behind. It also helped experiment. Securing the "spoils" of his discoveries was easier if both the place of experiment and its author were at a remove.

Faraday made the Royal Institution as much as the institution made him. On a practical level, his activities in establishing the Friday Evening Discourses and the equally popular Juvenile Christmas Lectures secured the institution's financial future and its public reputation among fashionable London society. More important, by representing himself and being represented as the ideal type of disembodied genius, pursuing abstract knowledge of the natural world, he allowed the Royal Institution to represent itself as the ideal type of place where such an activity could take place.[156] Behind this facade were the laboratories and workshops where technicians laboring with forges, furnaces, and foul-smelling galvanic batteries supported and sustained the dazzling natural philosopher in the lecture theater above them.

# The Vast Laboratory of Nature: William Sturgeon and Popular Electricity

ONE OF THE FEW critical remarks aimed at Michael Faraday by one of his biographers was by Silvanus P. Thompson. The remark concerned Faraday's attitude toward the work of his near contemporary and rival, William Sturgeon. According to Thompson, Faraday had never given to Sturgeon the credit he deserved for his experimental work, and for his invention of the soft iron electromagnet in particular.[1] Thompson went so far as to identify Sturgeon's invention as being equal to Faraday's discovery of electromagnetic rotation in its significance for the history of electricity. There is no question but that Faraday's, rather than Thompson's, view of Sturgeon has prevailed. Very few people beyond a small number of historians of electricity will have heard of Sturgeon or his experimental work in electricity. While it is certainly the case that Sturgeon was never regarded, even by his contemporaries, as being equal to Faraday in significance as an interpreter of the electrical world, there is equally no question that he was regarded by many as a central figure. The aim of this chapter, in part at least, is to recover some of Sturgeon's significance in the interests of historical symmetry.

While this chapter focuses on William Sturgeon, Faraday will continue to play a key role. There are several reasons for using Faraday in a comparative fashion. In the first instance, of course, both men worked at the same time in the burgeoning field of electrical experimentation. Their productions in that field were, however, very different. Second, Faraday and Sturgeon came from relatively similar backgrounds. They were both sons of poor, artisan families from the north of England, although Faraday spent all of his early life in the east end of London. Both therefore shared characteristics that were uncommon in the increasingly elite world of London science. Most important, however, Sturgeon himself for much of his working career as an experimental philosopher explicitly used Faraday as a foil against which to position his own researches in electricity. Much of his work during the 1830s in particular took the form of direct responses to some of Faraday's *Experimental Researches*. This chapter looks in some detail at two of those episodes in particular. In many respects it is arguable that Sturgeon represented Faraday as an exemplar of everything that an experimental natural philosopher should not be. He was elitist and secretive, while the ideal experimenter should be populist and open.

Finally, the comparison between the two men and their careers shows how dissimilar accounts of authority and competence in experiment during this period could be. Their aims and means of experimenting were very different. Issues of authority and experimental competence are most explicit during episodes of controversy. At such stages, normal and taken-for-granted ways of proceeding become open to question in the absence of agreed protocols for the resolution of debate. This is the difference between closed and open experimental settings. In a closed setting, experiments are black boxes and their outcomes remain unproblematic. In open settings, on the other hand, no consensus exists concerning the proper outcome of experiment. In these situations, black boxes become transparent once more and are held up for scrutiny.[2]

This sociological account needs to be placed in a historical context. During the first half of the nineteenth century the criteria that determined the outcome of debates in electrical science were not those which operate in modern science. Concepts such as the "core set" cannot easily be used for periods when the kinds of disciplinary training that produces such groups of authoritative witnesses had not yet emerged.[3] Other resources were, however, available to bind communities of practitioners and their audiences together. Alliances might be constructed on the basis of patronage or of institutional affiliation. Notions of gentlemanly behavior might be appealed to as a resource to govern behavior when controversy threatened, and to regulate the course of controversies when they occurred. Constituencies might be built around a whole range of different institutions, from shared education and training to subscription and contribution to the same journal. Different protocols prevailed in different contexts, however.

The relationship between an experimenter and his work during a contested experiment can be ambiguous in a variety of ways. On the one hand a successful experimenter must convince his audiences that an experiment models or represents some feature of the natural world. In order to achieve this, he must represent the outcome of the experiment as being independent of his own actions. At the same time, however, he must also convince his audience that he is himself a competent practitioner. This requires that he make visible his own expertise and skill. In other words he must establish his own credentials as a reliable spokesman for Nature while simultaneously representing Nature as speaking for herself.

For Michael Faraday, the lecture room of the Royal Institution was a closed setting in which his authoritative presence was the final adjudicator. By the 1830s he had successfully constructed a carefully tailored image such that he could persuade his audience that the proper outcome of any experiment was not open to debate.[4] This was the culmination of careful management of human and material resources. Faraday had succeeded in making his apparatus transparent and had produced an influential persona that allowed him to speak with authority about invisible phenomena of which the productions of his experiments were only signs. In the lecture room of the Royal Institution, the facts of nature simply were as Faraday said they were.

There remained places, however, where Faraday's writ did not apply.[5] When his work was published in William Sturgeon's *Annals of Electricity*, for example, it appeared in an open setting. Faraday's carefully managed authority in this forum did not necessarily make him a reliable spokesman for nature. Here his experiments were open to careful and critical scrutiny in the course of which their competence and significance were matters of dispute. Black boxes that had remained securely closed in the lecture room of the Royal Institution were opened for inspection in the pages of Sturgeon's *Annals*. For Sturgeon the *Annals of Electricity*, which he edited, provided a different kind of experimental space. Its pages provided a forum where he could dictate the criteria and the protocols that governed the progress and the outcome of dispute. Thus a setting that was dangerously open from Faraday's perspective was securely closed for William Sturgeon.

This is an aspect of what Collins describes as the experimenter's regress. The experimenter's regress arises because the representativeness of an experiment can be ascertained only post hoc—one cannot know beforehand whether an experiment will work. An experiment is a good representative of the natural world if it produces the correct result. But the claims of an experiment to represent nature are contingent upon the competence of the performer who conducted the experiment.[6] Resolving the regress can therefore be understood in terms of treading the tightrope between nature and society. The successful experimenter is one who has persuasively demonstrated that the phenomena, rather than his own interventions, produced the result. This typically requires at least some degree of control over the place of experiment, its organization, and its protocols. Relinquishing that control increases the danger of relinquishing the phenomena.[7] Sturgeon's attacks on Faraday's competence and integrity as an experimenter in the pages of his journal were efforts to establish himself as an authority instead.

There are no book-length biographies of Sturgeon.[8] The few obituaries and biographical notices written by his contemporaries portray him, as Faraday was portrayed, as a self-educated, self-made philosopher. He was born in 1783, the son of John Sturgeon, a shoemaker, and Betsy Adcock, in Whittington, Lancashire. At the age of thirteen he was apprenticed to another shoemaker in a nearby village, where he was "doomed to be even more cruelly treated than at home."[9] Following his qualification as a journeyman shoemaker, Sturgeon enlisted in 1802 in the Westmoreland Militia and two years later volunteered into the Second Battalion of the Royal Artillery. According to his biographers, his interest in electricity arose out of his experiences in foreign service, particularly from his encounter with "a terrific thunderstorm" in Newfoundland. He taught himself reading and writing and, having been befriended by an artillery sergeant who owned a small library of books, set to mastering the elements of languages, mathematics, and natural philosophy at night in his spare time after guard duty.

Sturgeon retired from the military in 1820, settling in Woolwich with his first wife. He resumed his trade as a shoemaker and at the same time continued his philosophical studies, making his own instruments with an old lathe. Woolwich was well suited for his pursuits. Sturgeon's skills as an instrument-maker and electrical experimenter brought him to the notice of local practitioners such as Peter Barlow, who was professor of mathematics at the nearby Royal Military Academy and a well-known electrical experimenter in his own right. With Barlow's patronage as well as that of Olinthus Gregory and Samuel Hunter Christie, Sturgeon was appointed a lecturer in experimental philosophy at the East India Company's Royal Military Academy at Addiscombe. While hardly providing him with financial security, such a position did at least provide Sturgeon with a recognized place on London's philosophical map.

Throughout the second half of the 1820s and the 1830s Sturgeon was active as a lecturer and demonstrator of electrical experiments in London and the surrounding area. According to Joule, his lectures were popular, "distinguished by his power of impressing the truths of science clearly and accurately on the minds of his auditory, and especially by the uniform success of his experimental illustrations."[10] Not everyone agreed, however, with this assessment of Sturgeon's performances. Joseph Henry, visiting London in the late 1830s, thought him "a very good experimenter but an indifferent lecturer," complaining that he was "very obscure in his theoretical notions."[11] Elsewhere he described him as being "at the head of the second rate philosophers of London."[12] From the mid-1830s, Sturgeon embarked on a major venture, editing a new journal, the *Annals of Electricity*, devoted to the electrical sciences. At about the same time, he was instrumental in the foundation of the short-lived London Electrical Society. These, arguably, were both efforts on Sturgeon's part to forge a constituency for himself. Both institutions would provide, if only briefly, spaces where Sturgeon to at least some degree had a voice in the organization and resolution of debate. The London Electrical Society collapsed after a few years, although Sturgeon had parted company with it even before then.[13] The *Annals of Electricity* foundered after ten volumes owing to a paucity of subscribers.[14]

About 1838, Sturgeon left London for Manchester, where he had been invited to become the superintendent of the Royal Victoria Gallery of Practical Science, an institution modeled on London's popular exhibitions. He befriended the young James Prescott Joule at this time, as Joule was just embarking on his experimental career.[15] The gallery failed after a few years, and until his death in 1850 Sturgeon lived a precarious existence as an itinerant popular lecturer in Manchester and the surrounding district. Joule was instrumental in securing him an annual government pension of fifty pounds shortly before he died.

By examining some of the interactions between Michael Faraday and William Sturgeon as they sought to construct their experimental lives, this chapter seeks to illuminate some of the different strategies and resources that were available to early-nineteenth-century experimentalists. In particular it focuses on the ways in which the experimentalist's own role in the experiment was both effaced and displayed in order to guarantee the integrity of nature. The first section discusses William Sturgeon's early experiments during the 1820s, showing how his public presentations of his experiments, the technical details of his apparatus, and his electrical ontology meshed together. The next two sections then examine in some detail two disputes concerning the interpretation and practice of experiment between Sturgeon and Michael Faraday, showing how Sturgeon's notions of the proprieties of experimentation and the constitution of electricity impacted on the details of his experimental practice. Both these disputes took place within the pages of Sturgeon's *Annals of Electricity*. Sturgeon reprinted the relevant sections of Faraday's *Researches* in his journal as he sought to reinterpret them in the light of his own work. The disputes highlight, therefore, the important role the *Annals of Electricity* played in providing Sturgeon with a forum in which he could dictate the terms of debate.

## TABLETOP PHILOSOPHY

It is by now commonplace to assert that successful experimentation requires the "death of the author." David Gooding, for example, has skillfully shown how Faraday labored in order to make his own role in the production of phenomena invisible.[16] This, however, was not the only way of going on in convincing others of the reliability of experimental productions, their correspondence with nature, and the authority of the experimenter. William Sturgeon aimed to make his own role and that of his apparatus visible to his audiences. This is not to suggest that Sturgeon in any way produced transparent accounts of his proceedings. He was engaged in producing himself as an "ideal author" whose virtues as an experienced and skilled manipulator underwrote his productions, just as Faraday aimed to efface himself from the experiment. The construction of such an "ideal author" makes explicit the ways in which issues of ontology, epistemology, practice, and presentation are intimately linked.

Sturgeon's early work provides a good example. His first experiments on electromagnetism were published in Tilloch's *Philosophical Magazine* during 1824. They were his first publications in experimental natural philosophy, appearing four years after he had retired from the Royal Artillery and commenced his career as a lecturer and instrument-maker. By this time Faraday's rotation phenomenon was relatively well established, and other devices demonstrating electromagnetic rotation, such as Barlow's Wheel, had appeared on the philosophical market.[17] Details of Barlow's Wheel appeared in Tilloch's *Philosophical Magazine* early in 1822. As Barlow put it, it followed from attempts to replicate Faraday's rotation experiment when "the young man who was assisting me wished to try the effect of the horse-shoe magnet upon the freely suspended galvanic wire, as it hung with its lower end in the mercury."[18] The result was that the wire oscillated rapidly to and fro between the poles of the magnet. When a serrated wheel was used instead of a single wire, the result was rotation between the poles. The "young man" in question was James Marsh, who produced a variety of such devices demonstrating rotation in the early

1820s, and whom Barlow more than once commended to the *Philosophical Magazine*'s readers' attention.[19] Sturgeon's own early experiments were aimed at producing just these kinds of demonstration devices to exhibit rotation as well, being refinements of Marsh's experiments.[20] Like Marsh, he was under Barlow's patronage.

Sturgeon's aim in the first of his independent publications, as he expressed it, was "to understand the relation that subsists between the chemico- and thermo-electric phenomena, as influenced by the magnet, so as to form a comparison of the widely different apparatus for exhibiting those phenomena."[21] Simply speaking, the experiments described in the paper were designed to compare the ways in which chemically and thermally generated electric currents, respectively, responded to the presence of a magnet. It culminated in the detailed description of a "Comparing Galvanoscope," which could be used "for the purpose of exhibiting and comparing chemico and thermo phenomena as influenced by the magnet."[22]

Sturgeon's subsequent papers detailed refinements to the experiments described in his first publication, culminating in a description of a "rotative thermo-magnetical" instrument.[23] This last communication described an apparatus for producing electromagnetic rotation using a thermocouple as the source of electricity. A bimetallic circuit of silver and platinum was arranged around a magnet so that when it was heated, it would rotate freely. In a postscript to this paper, Sturgeon added that he had since succeeded "in forming a sphere of galvanized wires, to rotate by the influence of both poles of an internal magnet." He noted that this variation of the apparatus "was suggested on reading the late Dr. Halley on the theory of the earth; and although it may not be considered as a proof of that philosopher's notion of terrestrial magnetic variation, yet perhaps it may tend in some measure to strengthen the hypothesis."[24]

Later in 1824, Sturgeon published a detailed description of his rotating sphere, together with further considerations regarding its relevance to Halley's theory of terrestrial magnetism.[25] The main experimental and technical innovation embodied in the new apparatus, as Sturgeon presented it, was that the rotation took place under the influence of both poles of a straight bar magnet. As he pointed out, in previous apparatus only one magnetic pole was used at a time, and different poles caused the rotations to take place in different directions.

It was this new technical innovation, Sturgeon asserted, that made possible the application of the concept of electromagnetic rotation to the earth's rotation:

> It is by no means intended from this experiment to assert, that the rotatory motion of the earth and planets is really the effect of electro-magnetism . . . but merely to detail an experimental fact as exhibited by the apparatus in its present imperfect state, the mechanism of which confines the experiment greatly within the limits it might have been extended to. However, it already proves that a galvanized sphere, when free to move and containing within it a magnetic nucleus or kernel, *will rotate* by the influ-

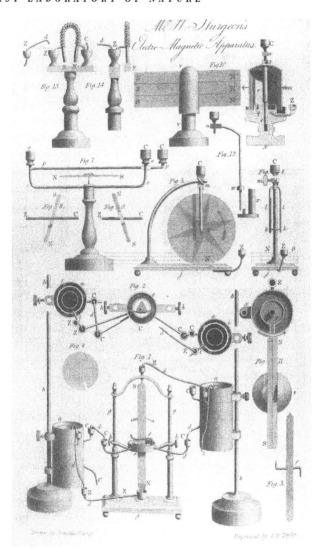

2.1 "Mr. W. Sturgeon's Electro-magnetic Apparatus." The
tabletop apparatus Sturgeon designed for the Royal Society of
Arts in 1824. *Annals of Electricity* 8 (1842): facing p. 225.

ence of that central magnet; and that also, were the magnet free to move, a probability
is manifested that it would rotate at the same time, i.e. they would both rotate at the
same time by the reciprocal action of each other.[26]

In other words the technical innovation embodied in the apparatus was presented as allowing the electrician to literally represent nature with his apparatus. The apparatus did not become invisible but rather stood for a natural
system.[27]

Sturgeon proceeded to detail the various ways in which terrestrial features conformed to the model provided by his rotating apparatus. He drew attention, for example, to the increased incidence of thunder and lightning in the tropics, which would be expected if the sun's heat in those regions produced more thermoelectric current than it would at the poles. In this way, by making explicit the electrical ontology that his apparatus revealed, Sturgeon could validate the practices that made his apparatus possible. At the same time the success of his practices in making the apparatus work validated his electrical ontology of the universe.

This electrical ontology was itself also closely connected to the practice of experiment and the putative role of the lecturer in conveying experimental knowledge to his public. In describing his new apparatus, Sturgeon emphasized its advantages as a demonstration device for use in lectures. His main point was that in the case of this instrument, unlike those previously devised for displaying electromagnetic rotations, the detailed mechanism of the apparatus would be visible to the audience.

> Another advantage in the manner of making this experiment is, that the glass vessel and mercury for floating the magnet are here not used; therefore the whole of the magnet is in view; whereas in the old mode only a small portion of the magnet, about the thickness of a quill, was visible above the surface of the mercury. It can be no small gratification to those who are in the habit of giving public lectures, to be enabled to exhibit this experiment to the satisfaction of a large audience; for as the lecturer can now have his rotating magnet of almost any size he pleases, and likewise of any figure, this interesting experiment may be viewed from the remotest part of the lecture-room.[28]

For Sturgeon, witnessing the actual technical details of an experiment's operation was an integral part of understanding phenomena.

This interest in making the new electromagnetic phenomena, and the apparatus through which they were manifested, visible to his audience culminated in Sturgeon's work of 1824–25 in constructing an array of tabletop apparatus for the economical display of electromagnetism. He was awarded a silver medal and thirty guineas for this work by the Royal Society of Arts.[29] The set of apparatus consisted of about a dozen different instruments, designed to display the full variety of rotation and other electromagnetic effects. One of these new instruments, consisting of varnished copper wire coiled around a horseshoe-shaped bar of soft iron, later became known as the electromagnet.

Sturgeon was again following in James Marsh's footsteps. A year previously Marsh had submitted a "portable electro-magnetic apparatus" to the Society of Arts and had also been awarded a large silver medal and thirty guineas.[30] His apparatus consisted of a voltaic battery fitted into a specially designed wooden box, along with a variety of devices, including his own electromagnetic pendulum, a set of Ampère cylinders to demonstrate rotation around a magnet, and a spiral wire designed to act like a compass. His apparatus was hailed by the society as being "peculiarly well adapted for service on board ship, in some of those naval expeditions for the promotion of science and general knowledge which so honourably characterize the present Board of Admiralty."[31] The apparatus was designed to simultaneously exhibit "all the known facts of electromagnetism" and to allow its owner "to prosecute further researches in this interesting and important branch of natural philosophy."[32]

In submitting his own tabletop apparatus to the society's inspection, Sturgeon emphasized both that his new instruments were specifically designed to make electromagnetic effects visible to a large audience, and that this in itself implied that they were peculiarly suited for the conduct of new experimental research. He repeated his claim that it was essential that a lecture audience be able to see clearly how an experiment worked:

> Another . . . obstacle to the advancement of this interesting science, is, that the experiments being hitherto exhibited on so small a scale, are by no means calculated to illustrate the subject in public lectures; for when the experimenter succeeds even to his wishes (which is not frequently the case), the experiment can only be seen by a very near observer, and the more distant part of the auditory are obliged to take for granted what they hear reported (from those persons who are more favourably situated) of some of the most interesting facts, which they, from their distance, are unable to witness.[33]

He emphasized that it was important that experimental details should be visible.

The key innovation that had led to his ability to reproduce electromagnetic phenomena on a large scale was, Sturgeon suggested, that he had focused his efforts on attempting to augment the magnetic, rather than the galvanic, power of his apparatus. His experiments had shown that no particular ratio of galvanic to magnetic power was required for successful experimentation, and that as a result one might compensate for using small and relatively feeble voltaic batteries by using large and powerful magnets:

> This discovery led me to the use of powerful magnets and small Galvanic batteries, for with small magnets the experiments can never be made on a large scale, although the Galvanic force be ever so powerful; and, as minute and delicate experiments are not calculated for sufficiently conspicuous illustration in public lectures, I considered that an apparatus for exhibiting the experiments on a large scale, and with easy management, would not only be well adapted to the lecture room, but absolutely valuable to the advancement of the science.[34]

Making experimental details and the workings of the apparatus highly visible and public was crucial, Sturgeon suggested, not only for display but also for further experimentation.

According to his own account, Sturgeon produced this apparatus by treating his experiments as if they were machines. By varying the ratio of Galvanic and magnetic power employed in his apparatus, he aimed to discover the most efficient and economic combination for the production and display of electromagnetic phenomena. Certain combinations and machines were especially efficient for particular kinds of display. Faraday's simple rotation device, which stood for the whole array of electromagnetic phenomena, was therefore replaced by Sturgeon with a variety of instruments, each particularly suited for the production of a given phenomenon.

Sturgeon's interests lay in ensuring that his apparatus remained visible and closely associated with himself. This was at least partly a matter of trade. Part of Sturgeon's livelihood was derived from the sale of his experimental instruments. It was also, however, a matter of epistemology. Unlike Faraday, who wished to draw his audience's attention away from the experiment and toward nature, Sturgeon insisted that his audience be able to focus on the experiment and see clearly the technical details of its operation. His mode of experimentation was to build new instruments and to progress by manipulating their components in order to maximize their efficiency.

Sturgeon held that the universe itself operated like an item of electrical apparatus: "The electric fluid is so universally diffused throughout every part of nature's productions, that every particle of created matter, both animate and inanimate, which has hitherto been contemplated by the philosopher, is full of this surprizingly animated elemental fire."[35] Experiments such as his attempts to reproduce the earth's rotation by means of electricity displayed the viability of such an ontology.[36] At the same time the ontology constructed an important role for Sturgeon himself. If nature was made up of electrical machines, then a skilled maker of such apparatus was the appropriate authority in matters of natural fact.

This approach to experimentation was very different from that of Faraday, who went out of his way to present his results as disembodied facts, abstracted from the machines and the labor that made them. He directed attention away from the material artifacts on his lecturing table because he argued that nature lay elsewhere. The world for Faraday consisted of lines of force in space, so that facts about nature would be facts about those lines of force in the space surrounding the apparatus rather than the details of the apparatus itself.[37] Making visible the details of his experiments would, for Faraday, be irrelevant to the project of making visible Nature and Nature's God.

For Sturgeon, electric currents were to be understood as "the effects of a series of distinct discharges, in such rapid succession as not to be individually distinguished by the senses."[38] He called these discharges "electro-pulsations." Electrical phenomena were to be understood, therefore, as the result of these pulsations' behavior as modified by the characteristics of a particular item of apparatus. They were the result of the ways in which the electric fluid flowed in particular circumstances. In a coil of copper wire excited by magnetism, for example, the pulsations would flow at high velocity. If, however, a part of the circuit had a high resistance, the fluid on meeting this resistance would need to build up momentum to overcome it. The result would be "a sudden disturbance of electric fluid, previously at rest . . . and a violent rush of the current would as suddenly follow."[39] This was the origin of the shocks and sparks experienced in such circumstances, which Sturgeon compared to the blow given by a Montgolfier hydraulic ram. In this sense, for Sturgeon, explaining electrical phenomena was very much a matter of understanding the technical details of the apparatus in which they were produced.

## THE TRIBULATIONS OF MEASUREMENT

The differences between two such different ways of going on can best be understood during moments of controversy. Faraday, famously, went out of his way to avoid engaging in debate with those who disputed his work. Historians therefore have not typically examined his activities as being controversial. Faraday's refusal to engage with his opponents can, however, be read as a strategic move.[40] By not responding, he was literally placing his work beyond dispute. Silence was a mark of Faraday's own confidence in his authority as the spokesman for nature to the gentlemanly elite. Refusal to stoop to the level of dispute was, of course, also the mark of a gentleman.[41]

Faraday's reluctance to respond to challenges does not, however, mean that his work remained unchallenged or was in any way uncontroversial. William Sturgeon on several occasions during the 1830s mounted determined attacks on Faraday's experimental researches, seeking to cast doubt both on the validity of Faraday's results and on the authority upon which they rested. Examining Sturgeon's attacks can provide useful insights into the ways in which his and Faraday's interests interacted and into some of the ways in which natural philosophical authority could be challenged in the early nineteenth century. Crucially, Sturgeon's interventions render explicit the ways in which his emphasis on public display and demonstration was embedded in his experimental practice and his ontology. His response to Faraday's work shows not only how he thought experiments should be conducted, but also whom he held to be competent practitioners and what the proper aim of experiment should be. The means by which the debate was conducted in the pages of the *Annals of Electricity* demonstrate as well, as has been suggested, the ways in which Sturgeon could utilize his control over that forum to represent himself as an authoritative spokesman and arbiter on matters of electrical experimentation.

Their differences are clear in Faraday's and Sturgeon's respective claims concerning the proper role and significance of electrical measuring apparatus. Efforts to measure electricity were widespread during the 1830s as rival battery-makers sought to establish criteria that could be effectively used to establish their own apparatus's superiority.[42] A whole range of different devices and techniques, from the length of sparks produced by a battery to its capacity to deflect a magnetized needle, were used in efforts to find a reliable means of calibrating different batteries and determining their respective powers. In experiments such as these, the issue of what precisely it was that instruments measured, and to what extent measurements by different techniques could

THE ANNALS     Q357.C.4.1

OF

ELECTRICITY,

MAGNETISM, & CHEMISTRY;

AND

Guardian of Experimental Science.

CONDUCTED BY

WILLIAM STURGEON,

Lecturer on Experimental Philosophy, at the Honourable East India Company's
Military Seminary, Addiscombe, &c. &c.

AND ASSISTED BY GENTLEMEN EMINENT IN THESE DEPARTMENTS
OF PHILOSOPHY.

VOL. I.—OCTOBER, 1836, TO OCTOBER, 1837.

London:

Published by Sherwood, Gilbert, and Piper, Paternoster Row; and W.
Annan, 12, Gracechurch Street.
Sold also by Messrs. Hodges and Smith, and Fannin and Co. Dublin;
Maclachlan and Stewart, and Carfrae and Son, Edinburgh; Mr.
Robertson, Glasgow: Mr. Smith, Aberdeen; and Mr. Dobson, No.
108, Chestnut Street, Philadelphia.

1837.

2.2 The title page of the first volume of Sturgeon's *Annals of
Electricity*.

even be counted as measuring the same thing, was highly contentious. Faraday's researches had a central role in these debates. In particular he was responsible for the important role played by the galvanometer and the voltaelectrometer as means of establishing assessments of battery performance in
particular and carrying out electrical measurements in general.

A crucial part of Faraday's work on the galvanometer as "an actual measurer
of the electricity passing through it" appeared in his discussion of the "relation
by measure of common and voltaic electricity" in the third series of his *Experimental Researches*.[43] The sequence of experiments laid out there were primarily intended to provide a detailed empirical confirmation of the claims that "if
the same absolute quantity of electricity passes through the galvanometer,

whatever may be its intensity, the deflecting force upon the needle is the same," and consequently that "the deflecting force of an electric current is directly proportional to the absolute quantity of electricity passed, at whatever intensity that electricity may be."[44] Effectively, these assertions amounted to the claim that the galvanometer could be reliably utilized to comparatively measure the electricity from different sources, regardless of the particular characteristics of the source itself.

These claims were established through the use of a battery of Leyden jars charged by an electrical machine. First Faraday showed that regardless of the variations in intensity he introduced into the circuit by varying its components, a given number of revolutions of the electrical machine always produced the same transient deflection of a galvanometer needle. If he doubled the charge on the Leyden jar by doubling the number of revolutions of the electrical machine, then the transient deflection of the galvanometer would similarly be increased in proportion. The deflecting force was therefore directly proportional to the absolute quantity of electricity passing through the circuit. The quantity of electricity produced by thirty revolutions of the electrical machine, for example, gave a consistent deflection of 5½ divisions on Faraday's galvanometer.[45]

To render his claims general, Faraday needed to show that his results applied also to electricity from sources other than an electrical machine. In particular, in the early 1830s, he needed to show that his results could be extended to apply to the voltaic battery as well. He therefore needed to find a voltaic arrangement that would produce the same readings on his galvanometer as did his electrical machine. Faraday found such an arrangement that consisted of parallel platinum and zinc wires separated by 5/16 of an inch. These electrodes were "plunged" into dilute sulfuric acid (one drop of strong sulfuric acid in four ounces of distilled water) to a depth of 5/8 of an inch and retained there for eight beats of Faraday's watch (8/150 of a minute) before being quickly withdrawn. When the apparatus was manipulated in this way the galvanometer needle was deflected "and continued to advance in the same direction some time after the voltaic apparatus was removed from the acid."[46] The needle arrived at 5½ divisions before swinging back to the same distance on the other side of the scale. Faraday concluded that this voltaic arrangement produced the same amount of electricity in 8/150 of a minute as the battery of Leyden jars charged by thirty turns of the electrical machine, and that the electricity from both sources produced identical magnetic effects on the galvanometer needle.[47]

To generalize his results, Faraday then aimed to show that electricity from both sources could produce identical chemical as well as magnetic effects. To accomplish this, Faraday first needed to find a way of relating the transient current from the Leyden jars to a constant current from the voltaic arrangement. When his voltaic arrangement was immersed permanently in the solution, it gave a galvanometer deflection of 5⅓ divisions. Faraday assumed that "a permanent deflection to that extent might be considered as indicating a constant voltaic current which in eight beats of my watch could supply as much electricity as the electrical battery [of Leyden jars] charged by thirty turns of the machine."[48] Armed with this relationship between transient and permanent sources of electricity, Faraday could determine whether the electricity from these two sources had the same chemical effect. He conducted a series of experiments using paper soaked in a solution of hydriodate of potassa. When four thicknesses of this paper were placed in circuit with the electrical machine, ten turns of the machine produced a pale spot of iodine on the paper, twenty turns produced a darker spot, and thirty turns produced "a dark brown spot penetrating to the second thickness of the paper."[49]

These experiments were then repeated with the voltaic apparatus. However, for this set of experiments, Faraday immersed the wires in a solution of nitric acid and water rather than in dilute sulfuric acid, adjusting the apparatus so that the galvanometer still gave a reading of 5⅓ divisions. Faraday found that if the current was passed through the paper for eight beats of his watch, it produced a spot of iodine on the paper of the same depth and tint as that produced by thirty turns of the electrical machine. He concluded that "both in magnetic deflection and in chemical force, the current of electricity of the standard voltaic battery for eight beats of a watch was equal to that of the machine evolved by thirty revolutions."[50] More generally, Faraday asserted on the basis of these experiments that "the chemical power, like the magnetic force, is in direct proportion to the absolute quantity of electricity which passes."[51] The results also led him to suggest that both magnetic and chemical effects could be used as parameters to measure the quantity of electricity passing through the circuit.

Faraday did not attempt to embody these conclusions in a standard instrument until his seventh series of *Experimental Researches*, some years later. Here he set out to describe "a new measurer of electricity" founded on the basic principle

> that the chemical action of a current *is constant for a constant quantity of electricity*, notwithstanding the greatest variations in its sources, in its intensity, in the size of the electrodes used, in the nature of the conductors . . . through which it is passed, or in any other circumstance.[52]

Faraday's electrical measuring device consisted of a glass tube filled with acidulated water, inverted over a trough also filled with water. Platinum electrodes from the source of electricity were inserted into the tube. When electricity passed through the electrodes, gases were given off, the quantity of which could be used as an indicator of the exact quantity of electricity that had passed through. He provided careful experiments demonstrating that variations in parameters such as electrode size, current intensity, and the strength of the electrolyte had no effect on the instrument's action.[53]

To show, for example, that electrode size was not a significant criterion, he placed two of his "volta-electrometers" with different sized electrodes, in series. When a common current passed through the instruments, "very nearly the same quantity of gas was evolved in all. The difference was sometimes in favour of one, and sometimes on the side of another, but the general result was that the largest quantity of gas was evolved at the smallest electrode."[54] Faraday attributed this to the fact that the gases (hydrogen and oxygen) were soluble in water, and that when they were more rapidly evolved per unit surface area, as in the case of the smaller surface area, they had less time to dissolve. He suggested that the difficulty could be avoided if the electrodes were arranged vertically so that the gases might rise more quickly, and if only the hydrogen were collected, since it was less soluble in water than the oxygen.[55]

In this way Faraday presented his instrument as having the capacity for absolute measurement of electrical quantity.

> The instrument affords us the only actual measurer of voltaic electricity which we at present posses. For without being affected by variations in time or intensity, or alterations in the current itself, of any kind, or from any cause, or even intermissions of action, it takes note with accuracy of the quantity of electricity which has passed through it, and reveals that quantity by inspection. I have therefore named it a VOLTA-ELECTROMETER.[56]

The instrument offered Faraday a means of measuring and calibrating electricity, regardless of the source from which it had been derived. The electricity produced by a voltaic battery could be absolutely compared with that deriving from a magnetoelectric machine or even a common electrical machine.

William Sturgeon by no means shared Faraday's confidence that these experiments had established the voltaelectrometer as an absolute measurer of electricity. His attacks on Faraday's experiments and their interpretation serve, on the one hand, to make explicit the amount of interpretive work required to make such measurements self-evident, and, on the other, more importantly here, to cast light on Sturgeon's own views of the proprieties of experimentation. In the first volume of his *Annals of Electricity*, Sturgeon inserted an open letter to Faraday questioning the conclusions of his researches "on the relation by measure of common and voltaic electricity" in the third series of his *Experimental Researches*. The relevant sections of Faraday's third series were reprinted by way of preface to Sturgeon's attack.[57]

Sturgeon's strategy in this letter was to cast doubt both on Faraday's interpretation of his results and, more crucially, on his skills as an experimenter. In response to Faraday's assertion that the intensity of an electric current had no effect on the galvanometer, Sturgeon pointed out that "the intensity of any given quantity will be inversely as the time it occupies.—Therefore if the intensity be of no consequence, neither can the time be of any."[58] This implied that regardless of whether a given quantity of electricity took a second, a minute, or a year to pass through the galvanometer, the reading would be the same in all cases. Sturgeon presented this as a reductio ad absurdum, maintaining that Faraday's conclusions in this case were patently ridiculous and therefore untrue. The invalidity of the interpretation was the outcome of Faraday's ineptitude in the choice of an appropriate measuring instrument. Sturgeon claimed that the galvanometer was "unsuitable for ascertaining the exact amount of deflection due to discharge of electric jars, and more particularly so when those jars are charged to high densities."[59] This was the result of the galvanometer coils' being tightly packed so that in such cases "the fluid invariably springs from one to another," spoiling the reading.[60]

These few remarks in themselves tell us a great deal concerning Sturgeon's views on electricity and the practice of experiment. Unlike Faraday, Sturgeon was here again deploying a straightforward fluid ontology of electricity. Electrical intensity was simply to be understood as the rate at which the electrical fluid passed a certain point, just like the intensity of a flow of water. On this view Faraday's assertions did indeed appear absurd. Different ontologies had clear consequences even for such an apparently mundane activity as measurement. It mattered what exactly the experimenter held the measurement was *of*. The physical characteristics of the measuring device were also central to Sturgeon's argument. On his account, the entire project of trying to compare common and voltaic electricity using a galvanometer was misguided. By its nature, according to Sturgeon, the galvanometer simply was not the kind of instrument that could be used for such a purpose. Its construction precluded its being used as an accurate measurer of common electricity. In fact, according to Sturgeon, the very precision of Faraday's results was a matter for suspicion:

> I am very well aware that the precautions against error which you have described, and the correspondency which appears in the results of those few experiments which you have chosen as favourable to your views, are well calculated to captivate the credulity of the unexperienced in electricity; but that apparent exactness which would prevailingly command *their* ascent [*sic*], might possibly be the very means of suscitating the electrician to a closer examination of the process by which such precision was obtained.[61]

Sturgeon was representing himself as the kind of experienced electrical practitioner whose tacit skills allowed him to see through the rhetoric of precision in Faraday's presentation.

Discussing Faraday's attempts to show that his voltaic arrangement could be manipulated to produce the same results as the electrical machine, Sturgeon adopted the same strategy again. According to Sturgeon, the figures Faraday had given for the permanent and transient deflections of his galvanometer needle could be accurate only if he had actually adopted an experimental procedure different from the one he had outlined. Faraday had neglected to take into account the fact that the transient deflections of the galvanometer needle depended on the initial intensity of the current. In the case of the Leyden jar battery, the initial intensity was the maximum, while for the voltaic arrangement, it was the minimum. Since Faraday had ignored this feature of the apparatus, any efforts at producing comparative results on his part were useless.[62] Similarly, Sturgeon objected to Faraday's experiments on the chemical effects of electricity. Since Faraday had used a different electrolyte for his voltaic arrangement in this case, nitric acid instead of dilute sulfuric acid, the experiment could not be legitimately related to the previous ones with the galvanometer. Sturgeon insisted that using a different electrolyte broke the similarity relationship, particularly since the aim of the experiments was to show that a particular voltaic arrangement having the same magnetic effect as an electrical machine had the same chemical effect as well. As far as Sturgeon was concerned, the voltaic arrangements in question simply were not the same.[63]

In any case, Sturgeon claimed that the experiments were again badly performed: the solution of hydriodate of potassa that Faraday used was clearly too strong, since it could maintain a current of 5⅓ divisions on the galvanometer and would therefore decompose too easily. Sturgeon also pointed out that the depth of brown coloring on a piece of paper was hardly a sufficiently accurate parameter to permit Faraday to claim the relationship between common and voltaic electricity to be demonstrated by measurement. In conclusion, Sturgeon totally rejected Faraday's assertion that chemical power and magnetic force were in direct proportion to the absolute quantity of electricity passing through the circuit, and in particular he rejected the implication that "magnetic deflections and chemical decompositions go hand in hand, or are indicative of the extent of each other."[64] As far as Sturgeon was concerned, Faraday's experiments warranted no such conclusion.

Sturgeon's response to Faraday's experiments establishing the voltaelectrometer as an absolute measuring instrument adopted a similar approach and again took the form of "remarks" in his *Annals of Electricity*.[65] His main counterexperiment showed that contrary to Faraday's claims, there was a significant variation in the chemical action with different electrode sizes. Furthermore, while

> the same current . . . is productive of different degrees of chemical action inversely as some function of the transverse sections of the fluid part of the circuit undergoing decomposition . . . the deflections of the magnetic needle . . . are in the reverse order; or, when a fluid conductor (not in the battery) is in the circuit, the deflections are directly as some functions of the transverse sections of the fluid conductor, the distance between the terminals being constant.[66]

Sturgeon concluded from this "that a considerable portion of the electric currents, which, between small terminals, was employed in the decomposing process, traversed the larger fluid section as a mere conducting channel, being totally unoccupied as a decomposing agent."[67] As a measurer of the quantity of electricity passing through the circuit therefore, the voltaelectrometer was useless, since it appeared that in some, if not all, cases an indeterminate quantity of electricity passed through the instrument without causing any chemical action.

The consequences of Sturgeon's rejection of Faraday's experimental approach were made explicit in his "Fifth Memoir on Experimental and Theoretical Researches in Electricity, Magnetism &c.," containing a discussion of the various methods of assessing batteries and their relative usefulness. The paper described several series of experiments, primarily concerned with assessing the extent to which different types and arrangements of voltaic batteries produced chemical and magnetic effects. In the first sequence of experiments, Sturgeon used a voltaelectrometer (or "electro-gasometer," as he called it) to assess the decomposing effects of different types of cell arranged in different series and combinations. His conclusion was that different types of cell needed to be combined in different ways to produce their maximum decomposing power.

For example, five pairs of Daniell cells connected in series would produce more decomposition than any other combination—so that if ten pairs were to be used, the maximum decomposing power would be achieved through the connection of the ten pairs in two (parallel) series of five rather than as a single series. With the Grove cell, on the other hand, it appeared that ten pairs in a single series would produce the maximum effect, and in this case, therefore, connecting them in two parallel series of five would diminish rather than enhance the decomposing effect. Similar results were achieved when magnetic effects were assessed. In this sequence of experiments, Sturgeon compared the power of different arrangements of batteries to sustain weights on an electromagnet, drawing the conclusion once more that different arrangements of the same number of voltaic pairs were necessary to sustain different weights, and that the arrangements required for maximum magnetic power varied from battery to battery. In both cases the number of cells in series did not necessarily lead to an increased effect but rather could, after a certain point, lead to a considerable diminution in power.[68]

The implications of experiments such as these were for Sturgeon quite clear. They demonstrated that there was no underlying quantifiable relationship among the various phenomenal effects of the electric current such as magnetism, heat, or chemical action. Equally, the experiments demonstrated that there was no direct proportion between any of these phenomenal effects and the absolute quantity of electricity passing through the circuit, or at least that no such relationship was measurable. As a result, insofar as any of these techniques could be used to assess the power of a battery, that technique could at best provide information concerning only the battery's power to produce the class of phenomena in question, since the battery's capacity to produce one kind of phenomenon provided no indication of its capacity to produce another. Even the battery's capacity to produce a particular phenomenon such as the decomposition of water could not be assessed absolutely, since different combinations of the same battery could produce different results.

For Sturgeon, at least, Faraday's efforts to quantify electrical measurements were a misguided failure and an indication of his ignorance of the basic operation of electrical apparatus. The instruments he had developed could not be regarded as providing reliable techniques for measuring the quantity of electricity. Instead they could be regarded only as instruments for the display of battery power.

> However, since the decomposition of water is a phenomenon of one particular class only; and that other classes of phenomena, quite as important as the electro-chemical, are displayed to the greatest advantage by very different arrangements of voltaic batteries to that which gives a maximum of chemical action, it would be absurd in the

extreme to continue the term Voltameter, to any piece of apparatus which does not indicate the powers of voltaic batteries, in the production even of one class of phenomena; and which gives no idea whatever respecting the powers of batteries in the production of other classes. It is well-known in the scientific world, that I have given to the water-decomposing apparatus the name of Electro-gasometer, because it shows the absolute quantity of gas liberated by each battery employed, and presumes nothing more. . . . The instrument which we continue to call a galvanometer, is in precisely the same predicament as the voltameter; because it indicates nothing more than electro-magnetic deflections. The proper name for this instrument would be the electromagnetometer.[69]

The different nomenclature can almost be taken to indicate literally different instrumentation. In Sturgeon's experimental world, these devices simply did not measure the same things as they did in Faraday's. They had a different purpose.

Faraday made no public reply to Sturgeon's criticisms. The debate as it was played out in the pages of Sturgeon's *Annals* took place without any direct input on Faraday's part. He was, however, well aware of Sturgeon's views, and his lack of response was deliberate. As he communicated to Edward Solly:

> I had seen Mr Sturgeon's criticism but it did not disturb me simply because I was glad to find that the arguments he uses are of no force. It might have been that he had found a good objection & then I should have been very sorry.
>
> I have however no intention of taking any notice of it. It would be some fun to send him Baron Humboldts letter to me in which he selects that very paper as the foundation of compliment and praise only I cannot consent to use the letter for such a small purpose or quote Humboldts name in sport.[70]

Sturgeon's and Faraday's worlds did not intersect sufficiently for Faraday to consider a response to be either necessary or in his interest. Sturgeon's attacks were simply an occasion for private "fun."

## SELF-INDUCTION

Similar points may be made concerning Sturgeon's attacks on Faraday's work on self-inductance during the mid-1830s. These are of particular interest as well since in this case Sturgeon had carried out his own research on the phenomenon independently of Faraday, before he became aware of his adversary's published researches. Again it is worth noting that Sturgeon's editorship of the *Annals of Electricity* was a crucial resource. The journal provided a space where Sturgeon could put together his own account of electrical experimentation, casting himself as the authoritative voice, commenting upon and challenging both Faraday's practice and his interpretation of experiment.

Both in his *Diary* and in the published paper, Faraday noted that his work on the "induction of a current on itself" was instigated by the observation of a

Mr. William Jenkin that an electric shock could be produced at the moment when contact was broken, from a single pair of voltaic plates if a helix were placed in the circuit.[71] In a series of experiments commencing on 15 October 1834 and continuing until 9 January 1835, Faraday attempted to analyze and elucidate this new phenomenon, communicating the results to the Royal Society on 18 December 1834, where they were read out on 29 January 1835 as the "Ninth Series of Experimental Researches in Electricity."

Both the *Diary* entry and the published researches seem to indicate that Faraday very quickly decided to treat the new phenomenon as an example of induction.[72] The first stage of his investigation was to vary the circumstances under which the phenomenon was produced by using coils of different lengths and wire of different thicknesses, both with and without an iron core to the helix. It is interesting to note that while in the published researches the results of these preliminary experiments were expressed in terms of both the appearance of a spark and the sensation of a shock, the *Diary* seems to indicate that Faraday in all but two of the trials produced only a spark. He obviously assumed that the appearance of a bright spark and the electric shock were interchangeable surrogates for electrical action.[73]

Having proceeded to ascertain in some detail the circumstances under which the phenomenon could be produced, and concluding that it was best produced when the coil had a soft iron core, Faraday proceeded to a detailed analysis of the phenomenon itself:

> The bright spark at the electromotor, and the shock in the arms, appeared evidently to be due to *one* current in the long wire, divided into two parts by the double channel afforded through the body and through the electromotor. . . . It followed, therefore, that by using a better conductor in place of the human body, the *whole* of this extra current might be made to pass at that place; and thus be separated from that which the electromotor could produce by its immediate action, and its *direction* be examined apart from any interference of the original and originating current. . . . The *current* thus separated was examined by galvanometers and decomposing apparatus introduced into the course of this wire.[74]

The nature of the phenomenon had, however, been transformed. It was no longer the production of shocks and sparks upon the breaking of contact but the direction of electric current that had become the object of inquiry. The readings of the galvanometer and the decomposing apparatus were to be taken as surrogates for the original effects.[75]

Using a chemical decomposition apparatus consisting of paper soaked in a solution of iodide of potassium, Faraday showed that while the circuit was complete, no decomposition took place, indicating that all the current passed through the coil. When contact was broken, however, chemical decomposition immediately took place. Faraday noted that the "iodide appeared against the wire N[egative], and not against the wire P[ositive]; thus demonstrating that the current through the cross-wires, when contact was broken, was in the *reverse direction* to that . . . which the electromotor would have sent through it."[76]

Faraday presented the results produced with the galvanometer, however, as providing the crucial evidence of reversal in the direction of the extra current generated when contact was broken. In this case, while the battery was connected to the circuit, the galvanometer needle indicated a steady deflection in one direction. The needle was forced back to its usual position, and then, when the circuit was broken, it immediately indicated a strong instantaneous deflection in the opposite direction. On the basis of these results Faraday suggested that differences in the intensity, quantity, and direction of the extra current, as indicated by the decomposing apparatus and the galvanometer, indicated clearly that this extra current was identical to the induced currents he had described in the first series of his *Experimental Researches*.

In order to render this conclusion inescapable, Faraday then conducted some experiments using a double, rather than a single, helix in the circuit—in other words, an *induction coil*. He noted that when the circuit of the secondary coil was complete, such that a current could flow through it, no extra current was detected in the primary circuit when contact was broken. He concluded from these experiments that the "strong spark in the single long wire or helix, at the moment of disjunction, is therefore the equivalent of the current which would be produced in a neighbouring wire if such second current were permitted."[77]

The first point that may be made concerning Faraday's strategy here is that he was positing a similarity relationship. By representing the experiment described to him by Jenkin as an instance of induction, he was asserting that it was in essence the same phenomenon as the one he had established and analyzed in his first series of *Experimental Researches*. He was making it part of his research program. This was an attempt to shift the evidential context of the experiment.[78] What had been for Jenkin a means of enhancing the capacity of a battery for producing electric shocks became for Faraday further evidence for the existence of the electrotonic state.

It was unusual for Faraday to investigate isolated phenomena presented to him by other experimenters. As he noted himself,

> The number of suggestions, hints for discovery, and propositions of various kinds, offered to me very freely, and with perfect goodwill and simplicity on the part of the proposers for my exclusive investigation and final honour, is remarkably great, and it is no less remarkable that but for one exception—that of Mr. Jenkin—they have all been worthless.[79]

Jenkin's information was different since Faraday could reinterpret it as part of his own program of experimental researches. It was not an isolated fact but evidence of induction.

This was again a strategy that involved drawing the reader's attention away from the experimental apparatus. As Faraday presented the case, the enhanced shock was a sign that something was happening away from the experiment. In the same way that the rotation of a current-carrying wire about the pole of a magnet had become a sign of the disposition of electric and magnetic forces in space, the electric shock had become evidence for something other than itself. The experimental apparatus had ceased to play a role in the production of the phenomenon. It was simply the place where Nature made herself visible.

William Sturgeon initially conducted his experiments on the electric shock from a single pair of voltaic plates following a conversation with a Mr. Peaboddy at the Adelaide Gallery, in the course of which he was informed of some recent experiments on the matter by Joseph Henry.[80] The aim of these experiments, as Sturgeon understood it, was "to convert *quantity* of the electric fluid into *intensity*, by means of a single voltaic pair; the indication of intensity being that of producing a shock."[81] He attempted to replicate Henry's results using the coils from his magnetoelectric machine.

After having succeeded in replicating the shock effect, Sturgeon varied the experiment with the aim of ascertaining the best and most efficient form of apparatus for producing the phenomenon. In particular he focused on the question of "how far the experiment would permit the battery to be diminished in size."[82] He found that the battery could be reduced to the "size of a lady's thimble" while still producing "considerable shocks." In a postscript to the paper he mentioned that he had also tried the effect of a soft iron core in his coils but had soon been convinced that "the iron core had not much influence, either one way or another."[83]

Sturgeon interpreted the enhanced shock from the apparatus as being due to the momentum of the electric fluid in the circuit. He suggested that in a straight wire, a certain amount of the fluid's force was used up in maintaining the wire in a state of magnetic polarity. In a coil, however, the "polar magnetic energies once developed would be arranged in the best possible order for mutual polar attraction, and the magnetic resistance thereby very much abated."[84] The coil would therefore be a better conductor, and as a result the electric fluid would have greater momentum than it would develop in a straight wire. Since the electric shock was a consequence of the fluid's momentum, Sturgeon suggested, it ought therefore to be more intense when a coil was present in the circuit.

Sturgeon's fluid ontology was part and parcel of his insistence on paying close attention to the technical details of experimentation. The electric fluid was a feature of the apparatus rather than of the surrounding space. Changing the details of the experimental setup would have material consequences for the ways in which the fluid behaved. The aim of experimentation, for Sturgeon, was to discover that combination of components in his apparatus which would allow the fluid to produce its effects most visibly and most efficiently. In this way, the maximization of effect could be regarded as the best way of theorizing about the nature of the electric fluid.

In keeping with his insistence on foregrounding the technical details of his apparatus and their role in producing the phenomenon, Sturgeon was also keen to emphasize the labor required for successful experimentation. In concluding his paper, he drew the reader's attention to this feature of his work:

"It is now eleven o'clock at night, and I have been at work almost without intermission since seven in the morning, and every other part of this number is now in the press."[85] Most of this time had been taken up in the tedious business of his winding and unwinding long pieces of wire into coils in order to compare the shock-enhancing properties of the same length of wire in coiled and uncoiled form.

Sturgeon returned to the issue of the production of electric shocks from a single pair of plates in a subsequent number of his *Annals*. He apologized to his readers (and to Faraday) for having failed to mention Faraday's experiments in his previous paper and printed the ninth series in full so that his readers would be aware of Faraday's claims concerning the production of the phenomenon. However, he also published a detailed critique of Faraday's experiments, casting doubt on all the main points that he had established, and suggesting that "Dr. Faraday does not appear to have arrived at any distinct views respecting the nature of the action in the display of these curious phenomena."[86] The *Annals of Electricity* was playing its role as an experimental space—a virtual laboratory, so to speak—where Sturgeon could dictate the terms of debate and impose his own notions of experimental propriety.

In particular, Sturgeon denied Faraday's assertions that the effect was best produced when the coils used had a soft iron core and that the extra current produced at the breaking of contact was different in direction from the battery current. He suggested that "[w]ith regard to Dr. Faraday's experiments on the effects of soft iron in the process, they were obviously too limited, and not varied with that caution, and to that extent, to be productive of any other than unconnected results; which, though highly interesting as insulated facts, were not calculated to establish any general law."[87] In short, he suggested that Faraday had not performed the experiments competently and had prejudged the issue.

In the case of current direction, Sturgeon asserted that on the one hand, Faraday had misinterpreted the results, and that on the other, he had not in any case used the apparatus legitimately. In particular, Sturgeon objected strongly to Faraday's use of the galvanometer: "The inferences which Dr. Faraday has drawn from his experiment [with the galvanometer] . . . are the most unfortunate that he could possibly have hit upon, resting principally, if not solely, upon the idea of *permanent* and *transient* deflections being of the same extent, from the same source of excitation."[88] He described experiments illustrating that the reverse deflection described by Faraday also occurred at the breaking of contact without the presence of the coil in the circuit.

He suggested that this was a normal feature of galvanometer behavior from which no inferences could be drawn concerning the direction of the current.

These results lead to conclusions the very reverse of those which Dr. Faraday has drawn from the *partial*, and consequently *inconclusive* character of his experimental data; but they are precisely such as might have been expected by any one conversant with the nature of electro-dynamic action.[89]

He similarly suggested that Faraday's decomposition experiments were equally dubious. In short he asserted that none of the surrogate phenomena utilized by Faraday were adequate substitutes for the core phenomenon.

Sturgeon's strategy in this attack was to draw the reader's attention back to Faraday's own constitutive role in producing his experimental results. By suggesting that the apparatus's behavior as described by Faraday differed from that which would be expected under normal circumstances, he was suggesting that the results produced were of Faraday's own manufacture. In this way, Faraday's incompetent tinkering with his instruments could be represented as a reason for distrusting his results. At the same time, however, Sturgeon's own skills as shared with those "conversant with the nature of electro-dynamic action" were deployed to portray Sturgeon himself as a legitimate critic. It was Sturgeon's own practical knowledge of the ways in which electrical apparatus behaved that allowed him to perceive the manufactured nature of Faraday's results.

This suggests some of the ways in which categories such as skill and practical knowledge could be used polemically. On the one hand, by drawing attention back to the work that Faraday had done to produce his results, Sturgeon could dismiss his conclusions as invalid. On the other hand, Sturgeon could portray his own practical knowledge of the vagaries of electrical instruments as providing good reasons for treating him as an authoritative expert. Sturgeon had used precisely this strategy previously against Faraday's work on the relationship of common and voltaic electricity. He had suggested there that Faraday's results could have been achieved only had Faraday used a different procedure from the one actually described.[90] The point was both to display Faraday's own interventions and to draw attention to Sturgeon's practical expertise.

Sturgeon's appeal toward the group of competent practitioners and his exclusion of Faraday from that community provide some insight into his own social epistemology and the ways in which his experimental practice was embedded in that epistemology. Sturgeon saw himself as speaking to an audience of technically literate and skilled practitioners—the readers of the *Annals of Electricity* of which he was editor. Making a fact real was a matter of making it visible to such an audience. Making a fact public by displaying its mode of production was an essential part of making a fact. Faraday, by failing to make the workings of his apparatus visible, was excluding himself from the community of practitioners. This was tantamount to excluding himself from the forum in which facts were best made.

Having dismissed Faraday's claims, Sturgeon presented his own analysis of the phenomenon in a separate paper.[91] The differences between his results and Faraday's, he asserted, were a consequence of the fact that they had each employed a different type of coil. He then presented a series of experiments showing how the use of different forms of apparatus produced significant variations in the nature and strength of the shock produced. He concluded that his experiments showed that the presence of a soft iron bar did not invariably enhance the shock, but that nevertheless "magnetism of the iron under some circumstances becomes efficient."[92]

Sturgeon then proceeded to examine the circumstances under which the presence of an iron core did enhance the shock. He described a number of comparative experiments in which wires of the same length were coiled around straight and horseshoe-shaped iron bars, respectively. Using the results of these experiments, he concluded that "[i]n every trial the shocks were much the strongest from the coil on the straight bar; showing again that the shocks do not depend on the *quantity* of magnetism displayed, but upon a proper application of it."[93] He pointed out that these results showed clearly that the "form of the iron most suitable for magnetic display, is the least so, in the production of shocks."[94] In this he was reminding his readers of the dangers of blithely assuming that one form of electromagnetic display could be used as a surrogate for another.

The aim of this series of experiments, as Sturgeon made clear, was "to enhance the display by a proper application of the laws which govern the action."[95] This was achieved through a focus on the technical details of the apparatus that produced the phenomenon, its components varied and manipulated in order to reveal the most effective and efficient combination. This illustrates once again how Sturgeon's experimental practice was centered on the apparatus as the object of inquiry. The aim of his experiments was to elucidate the workings of the apparatus as a way of elucidating the workings of nature.

CONCLUSION

Questions of practice cannot be discussed in isolation from the context and aims of experimentation.[96] Both Faraday and Sturgeon experimented with particular goals in mind, and their perceptions of what constituted good experimental practice could not be dissociated from either the world they wished to represent or the public to which they spoke. This observation confirms that accounts of experimental practice and the resolution of experimental debates must pay attention to the historical context within which such debates took place. The society within which Faraday and Sturgeon lived and worked provided its own resources and its own barriers to success. They could not, for example, appeal to a "core set" of qualified experts to validate their work.[97] The social structures of shared education, training, and research that produce such a group had not yet emerged. In the early nineteenth century the resources that could be drawn upon for the resolution of debate were those of polite society. Actors were obliged either to position the presentation of their work in such a way that it would comply with the hidden codes of proper behavior or to find new ways of representing themselves and new audiences for their work.

From this perspective polar categories such as public and private or tacit and explicit knowledge must be reexamined. The laboratory, for example, is not an essentially private space. Faraday's decision to keep his laboratory private was strategic. It was a tactic designed to further his own account of how the natural philosopher should represent the world in a way acceptable to his genteel audience.[98] In the same way Sturgeon's decision to make public the technical details of his experimental apparatus was designed to support his claim that the world was an electrical machine, and that only those skilled in the workings of machinery could understand such a world. Similarly, the division between what was and was not said in the presentation of experiments was strategic. It was in Faraday's interest to be silent about the sort of skill and labor required to make his experiments work. It was in Sturgeon's interest, on the other hand, to draw attention once more to Faraday's own role in producing his results. By displaying Faraday's labor, he was deconstructing the carefully crafted image of disembodied genius that Faraday had made for himself. Sturgeon himself chose to portray his relationship to his experiments in a different way, designed to capture a different audience.

It is obvious that Sturgeon held a grudge against Faraday and the Royal Institution. Joseph Henry recorded that Sturgeon and Faraday were "not on good terms."[99] He recorded also that "Mr. Sturgeon first became dissatisfied with the Royal Institution by a harshe remarke of Sir H Davey who when Mr S shewed him some exp. on magnetism said he had better mind his last than be dabling in science."[100] Sturgeon clearly had a sense of his deliberate exclusion from the elite circles of London science. He pointedly remarked to Henry that "any Society which will not give ear to, or not notice, any scientific discovery, or project, but such as emanates from its own members, can have little claim to the character of being a promoter, or even a well-wisher, of the general interests of Science."[101] This was in all probability a reference to the Royal Society, which had refused to print one of Sturgeon's papers in the *Philosophical Transactions*. Attacking Faraday in the pages of his newly established *Annals of Electricity* quite clearly had polemic value for Sturgeon as well. It was a means of establishing a niche for himself and of claiming for himself a status as a philosopher who could legitimately oppose the pronouncements of London's leading exponent of electricity. Sturgeon complained constantly of the difficulties he encountered in accessing British natural philosophical institutions, the British Association for the Advancement of Science as much as the Royal Society.[102] He clearly felt the need to forge other constituencies. Given the limits of his resources, this was not to be an easy task.

This chapter has outlined the electrical cosmology within which Sturgeon embedded his experiments, as well as some of the strategies he adopted and the resources he drew upon to find a constituency for those experiments and for himself as an experimenter. The suggestion is that all of these things went hand in hand. Sturgeon's ontology of electrical fluid was made real through the experiments and instruments he designed to make that fluid and its operations visible. The particularity of his approach is emphasized by its contrast to Faraday's, who chose a very different way of linking himself to his experiments and those experiments to Nature. Sturgeon's insistence on the detail of experiments and the visibility of their operation was part and parcel of his emphasis that his instruments simulated natural phenomena and processes and that understanding the details of the apparatus was therefore central to understanding nature. Through institutions such as his *Annals of Electricity* and the London Electrical Society, he tried to forge a constituency of electricians whose shared skills and understanding of the detail of electrical experiments and instruments would lead to a proper appreciation of electricity's place in Nature.

Sturgeon in the end failed, of course. He died a penniless, itinerant lecturer trying to eke out a living pushing his electrical apparatus on a handcart from village to village in the Manchester area. There was, however, nothing intrinsically inevitable about that failure. During his early philosophical career in the 1820s he rapidly succeeded in attracting the patronage of prominent and influential natural philosophers such as Peter Barlow and Samuel Hunter Christie, both of whom were leading fellows of the Royal Society. By the 1840s he was himself the patron of James Prescott Joule, soon to be the rising star of nineteenth-century English physics. In many ways his career, pursued as it was outside elite, established natural philosophical institutions, was until his final years a resounding success. Decades after his death correspondents in the *Electrician* and the *Electrical Review* still drew attention to his contributions. He failed ultimately, not because his experiments failed, but because the audience he cultivated could not be sustained.

# Blending Instruction with Amusement: London's Galleries of Practical Science

As THE OPENING chapters have suggested, place was important for electrical experiment in early-nineteenth-century London. Electricians needed to position themselves on the metropolis's cultural map, both to provide themselves with an audience and to acquire the resources they needed for their work. The places where electricity was produced and displayed in many ways defined its significance. Cultural connections were put together in terms of the geography of experiments. Prestigious establishments such as the Royal Institution were not, of course, the only places where London's publics could encounter electricians and their experiments. A variety of venues existed where those barred from entry into elite social institutions could witness, and even participate in, the productions of experiment. During the 1830s in particular, galleries of practical science were founded where experiments were publicly performed by popular lecturers such as William Sturgeon and George Bachhoffner before audiences eager for entertainment and instruction.

A contemporary described these institutions as

> the places in fact which furnish the great mass of the public with demonstrations of science. The other institutions are in a manner exclusive: membership or the introduction of a member is, with a few exceptions, imperative on those who would be present. The exceptions are in the institutions where single tickets are sold for single lectures. But the exhibitions now before us . . . are accessible to all who proffer their shilling at the door; and we venture to say that these two institutions have played an important part in diffusing among the middle, and even the working classes, that better knowledge of natural phenomena which is fast spreading among us, and that taste for physical philosophy which is becoming so general.[1]

Electricity was, of course, a key component in the repertoire of such public exhibitions. Shocks and sparks were well calculated to impress a paying audience.

These exhibitions and the men who worked in them were integrated into a particular network of places and practices in early-nineteenth-century London. Just as the Royal Institution and its professors such as Faraday were adapted to conform to the codes of fashionable society, popular exhibitions also had to manage themselves such that they could have a place in London's social geography. Their position on the map was not that of high society. London's shows and exhibitions ranged from classical art and Shakespearean theater to peep shows and street entertainment. Machines were becoming increasingly ubiquitous as objects of display. Electricity in these exhibitions therefore also fitted into a world of machinery. Experiments and apparatus took their place surrounded by displays of the products and adjuncts of labor. This chapter aims to explore how electricity worked in this culture, and how its practitioners used their material and social resources to manage their activities. Central to this was the notion of work, and how workers were related to the machines they used.

Exhibitions were crucial to the process whereby experimenters, instrument-makers, and machine-makers established their claims of priority and novelty. They were more than simply places where inventors showed off their wares. They were central to the production of the inventor's self-image and public image. They provided spaces where the machine-maker as showman could demonstrate his control over the machine and over the audience for his display. Those who worked and displayed the results of their labors at such places were skilled mechanics who needed a space where by demonstrating their control over their machines and experiments, they could also demonstrate their ownership of these technologies and the phenomena they produced. Invention was very much about controlling the products of labor. It turned out that the inventor's need to control his product often required the showman's skill in controlling an audience as well. Understanding electricity's role in these exhibitions also requires an understanding of how the exhibitions fitted into their cultural networks.

The first half of the nineteenth century was witness to massive transformations in the everyday life of the skilled workers upon whom such exhibitions depended. In particular the very notions of work and of skill were being reformulated in the wake of basic changes in the economic structure of industrial culture. The social organization of work in the workplace and the relationship of master and worker were in a state of flux as new machines and new forms of discipline were introduced.[2] In this context artisans and mechanics sought to construct new forms of organization themselves, both to preserve what they regarded as traditional rights and privileges and to refashion the image of the skilled worker. Part of this process included an articulation of the practical mechanic's role as an interpreter of the natural world.

London during this period retained its position as a center of trade and manufacture. The metropolis was still the "Athens of the artisan," sustaining a large network of small workshops in a variety of trades. The majority of London trades still operated out of small shops, employing no more than a handful of journeymen and apprentices. The status of the skilled artisan or mechanic continued to be maintained by the system of apprenticeship, which functioned to transmit craft practices and traditions while at the same time serving to limit access to those resources.[3] To a large extent the economic and cultural boundary between artisans and the lower-middle class was fluid. The gap dividing the journeyman from a small master, shopkeeper, or tradesman, or even a journalist, surgeon, or dissenting clergyman, was often quite small.[4]

A crucial feature of artisans' self-image was their awareness of possessing a privileged craft skill that distinguished them from the mass of laborers. These skills they regarded as being their property in as strong a sense as a capitalist might regard his capital. During the first few decades of the nineteenth century, artisans saw many of the privileges associated with their craft as being under threat with the introduction of new machinery and an increased division of labor that allowed masters to use cheaper labor, particularly that of women and children. The artisanal concepts of labor and the property of skill therefore became an important ingredient in political radicalism as some artisans sought to redefine their self-image and public image in relation to new labor practices.[5] New institutions were advocated to nurture the new artisan.[6]

London was a major focus for these activities. New journals were founded to represent and educate the skilled mechanic. Some, such as the *Technical Repository*, edited by Thomas Gill, sought to inform its readers by compiling descriptions of the most recent patents for new devices and processes. Others, such as the *Mechanics' Magazine*, founded in 1823, provided a broader repertoire. The *Mechanics' Magazine* was founded and edited by the patent agent Joseph C. Robertson and the journalist Thomas Hodgskin. Both were active in radical politics. The explicit aim of the journal was to educate and defend the artisan:

> The object proposed by this publication at its outset, was one of entire novelty, and no inconsiderable importance. A numerous and valuable portion of the community, including all who are manually employed in our different trades and manufactures, had begun, for the first time, to feel the want of a periodical work, which, at a price suited to their humble means, would diffuse among them a better acquaintance with the history and principles of the arts they practice, convey to them earlier information than they had hitherto been able to procure of new discoveries, inventions, and improvements, and attend generally to their peculiar interests, as affected by passing events.[7]

The journal was designed to define the mechanic and fashion for him a particular role in society.

The *Mechanics' Magazine* almost immediately took a leading role in the campaign to establish the London Mechanics' Institute.[8] Both Hodgskin and Robertson saw education as a crucial means of defending the artisan and articulating for him a new social role. The reform of manners, the advancement of technical skills, and eventually the reform of government were all to be achieved as mechanics were enlightened through chemistry, mechanical philosophy, and economics. Major differences quickly emerged, however, between the *Mechanics' Magazine* and others of the London Mechanics' Institute's supporters, notably George Birkbeck and Francis Place. Birkbeck and Place aimed for a large and prestigious establishment, requiring the aid of wealthy sympathizers who would expect to play a major role in managing the institute. Hodgskin and Robertson, on the other hand, proposed a more modest establishment, conforming to the artisan's customary organizations and on a scale that would allow artisans themselves to be in control. The dispute was about the role and autonomy of the artisan.[9]

The question of how spaces should be organized and by whom was explicitly a political question. Disputes between masters and their workforces during these years hinged precisely on these issues. At stake was the question of who should control the space, and therefore the rate, of labor.[10] The workforce claimed the right to organize the nature and pace of work in their shops as theirs by tradition. Artisans and mechanics regulated their practices and their workshop organizations through a range of often informal and tacit agreements that gave priority to community rather than to individual rights. Work committees were democratic, and executive positions were frequently rotated through the workshop to ensure maximum participation. This was the type of organization that many wished to maintain in new institutions such as the mechanics' institutes.

Ideology was therefore at stake in these organizational quarrels. Hodgskin in particular aimed to use the London Mechanics' Institute as a forum for the promulgation of a new political economy articulating the relationship of the artisan to machinery. In 1826 he was successful, against Place's opposition, in delivering a series of lectures on political economy. Here he argued that the artisan's property of skill, embodied in the machines he manufactured, gave the maker property rights to the machine:

> The enlightened skill of the different classes of workmen alluded to, comes to be substituted in the natural progress of society for less skilful labour; and this enlightened skill produces an almost infinitely greater quantity of useful commodities, than the rude labour it has gradually displaced ... the productive power of this skill is attributed to its visible products, the instruments, the mere owners of which, who neither make nor use them, imagine themselves to be very productive persons.[11]

A politically loaded image was being produced of the artisan as inventor, whose understanding and use of the machines he made gave him a right of property.

The newly articulated role of the mechanic-inventor was, however, fraught. Craft skills were typically practiced and transmitted in private. Artisans worked hard to protect the secrets of their trade from public scrutiny. Many argued that secrecy was vital for the maintenance of their social status. This was not simply a matter of preventing outsiders from discovering the mysteries of the workshop. It was also a matter of maintaining autonomy and organizational control of labor against the encroachments of middle-class masters.[12] Others, however, suggested that particularly as machines became commodities as well as means of production, they needed to be displayed in a public market. One such initiative was attempted by the leaders of the London Mechanics' Institute, who called for the establishment of a publicly sponsored National Repository to display the products of industry and invention, which opened to the public in 1828.[13]

The National Repository did not meet with unmitigated approval. Robertson in the *Mechanics' Magazine* waged war against it from its inception. This was at

least partly the result of his increasing hostility toward the London Mechanics'
Institute and its promoters, but even so he was able to draw on a powerful
rhetoric to support his animosity. Such an exhibition was "at variance with the
established tastes and habits of the English people."

> Is the manufacturer so degraded as to require to be raised in his own estimation by
> seeing crowds of curious idlers, or fashionable loungers, assembled to admire his
> productions? Does he need the criticism of the public, who must be less skilful than
> himself, to improve those arts on which his existence, reputation, and fortune de-
> pend? Is his competition with his rivals in trade not sufficiently stimulated by the
> desertion, or the increase of his customers, unless he likewise sends samples of his
> handicraft to stand a comparison with theirs in the same gallery?

The taste of the public for manufactured products did not need to be "cultivated or purified."[14] A public exhibition violated artisanal norms of secrecy
and privacy that were essential for the maintenance of their economic and social status.

Accounts of the secrecy inherent in the maintenance of artisanal craft practices could, however, also be used to dismiss any claims they might make to
natural knowledge. According to John Herschel, for example, accessibility
and openness were hallmarks of the scientific. Science "should be divested, as
far as possible, of artificial difficulties, and stripped of all such technicalities as
tend to place it in the light of a craft and a mystery, inaccessible without a kind
of apprenticeship." This was what distinguished science from art, which tended
to "bury itself in technicalities, and to place its pride in particular short cuts
and mysteries known only to adepts; to surprise and astonish by results, but
conceal processes."[15] To be an inventor and, crucially, to lay claim to scientific
status, the mechanic had to straddle the awkward boundary between private
and public.

Such boundaries were, of course, contingent and mediated. They were delin-
eated in the course of negotiation over the question of where knowledge could
legitimately be made. Herschel's distinctions were polemical rather than simply descriptive. Other boundaries could be negotiated in much the same way.
Artisans could argue that their productive labor was a private matter within
the workshop while the master's role was the public business of selling the
product.[16] In the same way an inventor's experiments might be differentiated
from his public demonstrations. These boundaries between private and public,
production and consumption, experiment and display were all open to and
constructed by negotiation. As new spaces emerged and artisan radicals argued
for a new articulation of their social relations, these categories changed as well.

The metropolis's large population of instrument-makers was largely embedded in this changing artisanal culture. They were one of the crucial resources
without which the fashionable science of London's elite institutions simply
could not take place. These men inhabited a very different social sphere from
that of self-assured gentility and polite reform. During the eighteenth century,
London's instrument-makers had been the foremost in the world. Family

ties and the inheritance of workshops were crucial means of acquiring skills and equipment. Instrument making had been a respectable profession during the Georgian era, and a route to fellowship in the Royal Society. Practitioners were regarded as legitimate producers of knowledge as well as of mechanical artifacts.[17]

By the 1830s, however, London makers were losing out to French and German competitors. The intellectual status of artisan practitioners and instrument-makers was changing as the spaces where they plied their trade were transformed.[18] That trade was no longer an easy route to the Royal Society's prestigious fellowships. Their work was not likely to appear in the august pages of the society's *Philosophical Transactions*. They were not usually to be found among the Royal Institution's well-heeled clientele, though their productions were often on display there.[19] They, too, had a stake in establishing new spaces where they could display their wares and legitimate their claims as knowledge-makers.

## GALLERIES OF PRACTICAL SCIENCE

Such a venue was established in 1832 by the American Jacob Perkins. Perkins had arrived in London in 1819, having led a checkered career as an inventor and entrepreneur in Philadelphia.[20] He had gained some eminence in American circles as the designer of a device to engrave banknotes. He came to London hoping to win a prize offered by the Bank of England and gain a commission to engrave forgery-proof money. Having failed in that project, he moved on to other schemes, notably his attempts to construct high-pressure steam boilers. One offshoot of this work was Perkins's steam gun, which he constructed and attempted to market during the mid-1820s. Perkins made several attempts to interest the military in his steam gun, even persuading the French to conduct their own tests in 1828, but with no success.

During the early 1830s he conceived a new project to open an exhibition to display his own inventions to the London public. The scheme was soon expanded to include new inventions and curiosities of all kinds. On 4 June 1832 the National Gallery of Practical Science, Blending Instruction with Amusement, opened its doors on Adelaide Street and the Lowther Arcade near the Strand. The building, and indeed the surrounding streets, were new. They were part of the new development designed by John Nash that included the construction of Trafalgar Square, completed only the previous year. It was an ideal site for an exhibition. The arcade and its surrounding streets were filled with fashionable shops to attract the attention of the populace.[21]

A prime site was a prerequisite for a successful exhibition. Popular shows of all kinds proliferated in late Regency London. The curious could choose among dissolving views, panoramas and dioramas, waxworks, and exhibitions of sculpture and painting. Dedicated buildings such as the Egyptian Hall or the Great Globe in Leicester Square featured an almost infinite variety of attractions. The

Colosseum in Regent's Park, for example, had been built originally to house a massive and spectacular panorama of London as seen from the top of St. Paul's Cathedral. A good location and a careful massaging of the popular press were essential if a new exhibition was to be successful.[22] Perkins and his associates at the Adelaide Gallery, as it was popularly known, were well aware of this. Perkins was already a veteran of London showmanship, having been exhibiting his steam gun for several years to enthusiastic crowds.

As the *Literary Gazette* recorded:

> Of this institution we had the satisfaction to participate in the private view on the evening of Monday week, when a numerous company of artists, men of science &c., was invited. The exhibition consisted of models, and other subjects connected with arts and science. Mr. Perkins's steam-gun was one of the most curious articles. . . . Some very interesting models of steam-boats were also shewn, in a well-contrived reservoir, to explain Mr. Perkins's improvement in paddle-wheels, which render steam-vessels applicable to canal navigation. In this Gallery mechanics and artists are invited to exhibit and compare their inventions and improvements, which the public can also inspect; and we have no doubt it will prove a means of facilitating improvement in many objects of practical science.[23]

The Adelaide Gallery became a venue where London's instrument-makers and mechanics could appear before the public in their own right as men of science. Their inventions were on display there unmediated by the pronouncements of elite philosophers.

During its early days, as the *Gazette*'s correspondent hinted, the gallery's exhibits were dominated by Perkins's own productions. This feature of the gallery gave rise to some criticism. In its review of the opening night, the *Mechanics' Magazine* commented that

> [t]he notion of erecting, at an expense of many thousands of pounds, a hall to exhibit two or three inventions of a particular individual—not all original and some of them mere abortions—was, to say the least, exceedingly preposterous; but to build a general exhibition-room for new inventions was at once to confer a great benefit on the mechanical world, and to lay out their money with every prospect of a profitable return to themselves.[24]

The commentator urged the gallery's proprietors that Perkins's own productions should hold pride of place for no longer than was absolutely necessary.

Perkins's strenuous exertions to promote his various inventions had earned him some enemies among London's mechanics' community. The *Mechanics' Magazine*, after initial support for Perkins's efforts to publicize his high-pressure steam boiler, became more suspicious as his claims grew more extravagant. Many of the magazine's correspondents voiced doubts concerning both the viability of Perkins's design and its claim to novelty. Robertson, however, clearly preferred Perkins's venture on Adelaide Street to the despised National Repository:

The National Repository . . . has taken the improvement of the national manufactures in hand once more, and opened a room in Leicester Square with a collection of specimens of British ingenuity, and invites contributions to its stores from all quarters most movingly. The importance of this unfortunate concern as an exhibition, however, seems to be quite eclipsed by the superior attractions of the private speculation, the "National Gallery of Practical Science," in the Strand.[25]

Despite what the *Mechanics' Magazine* clearly regarded as the dubious nature of some of Perkins's claims to novelty, he was at least a working, practical mechanic rather than one of the bourgeois supporters of the London Mechanics' Institute.

The Adelaide Gallery's star exhibit was, however, without question Perkins's steam gun. At frequent intervals during the day, the gun would be fired up to discharge rounds of seventy balls against a target in four seconds. Also popular was the electric eel. The gallery's promotional literature invited their visitors to attend at feeding time at 3:00 P.M. in the afternoon and at 8:00 P.M. in the evening to watch the eel electrically stun its prey.[26] When the eel died, the event was considered of sufficient national importance to be reported in the *Times*.[27] It was, however, replaced by two new living specimens. The old eel's dissected remains were also put on display, to aid in the explanation of the species' remarkable electrical powers.

Entry into the gallery cost visitors a shilling. This was the usual price for entry into one of London's many public exhibitions. Men of science were admitted free of charge, although it is not recorded how they were to establish their credentials. Once inside they were ready to witness wonders. The proprietors announced their willingness to put on display the inventions of any ingenious mechanic, free of charge. They promised to "promote . . . the adoption of whatever may be found to be comparatively superior, or relatively perfect in the arts, sciences or manufactures [and to display] specimens and models of inventions and other works &c. of interest, for public exhibition, free from charge . . . thereby gratuitously offering every possible facility for the practical demonstration of discoveries in Natural Philosophy, and for the exhibition of any new application of known principles of mechanical contrivances of general utility."[28] The exhibits they received ranged from paintings and sculptures, through models of famous buildings, to patent fire engines and an improved ear tube. The gallery soon acquired an oxyhydrogen microscope, through which the monsters inhabiting the filthy waters of the Thames could be viewed, magnified three million times. A variety of electrical apparatus was on display. The instrument-maker E. M. Clarke exhibited a variety of batteries. William Sturgeon, the inventor of the electromagnet, exhibited a ferroelectromagnetic globe to illustrate the origins of the earth's magnetism. James Marsh of Woolwich had an electromagnet on display.[29]

The gallery was divided into a number of different rooms. Visitors entered through the entrance lobby and the anteroom where they were surrounded by

3.1 The Long Room at the Adelaide Gallery. Perkins's
steam gun is in the right foreground.

reproductions of classical sculpture. Going through into the salon, they were
faced with more classical reproductions of sculpture and paintings, along with
models and plans of ingenious devices. The gallery's center was the Long Room
with its seventy-foot canal for demonstrating and experimenting with models of
paddle-driven steam boats. Here were Perkins's steam gun and high-pressure
steam boilers along with a plethora of models and inventions. This was also
where the gallery's electrical machines and apparatus were on display. Surrounding the Long Room on the second floor was the Gallery, where yet more
models and inventions were displayed, along with curiosities from all over the
globe, including "weapons taken from the natives of Owhyhee, who were engaged in the murder of Captain Cook."[30] Other rooms on this floor contained a
similar selection, leading finally to the Microscope Room, where the "Grand
Oxy-Hydrogen Microscope" could be inspected.

One of these rooms, the Loom and Lithographic Press Room, prominently
displayed the Jacquard loom in action. This machine was celebrated as the

inspiration of Charles Babbage's Analytical Engine, and Ada Lovelace enthusiastically recommended that those interested in Babbage's work should view it at the gallery: "Those who desire to study the principles of the Jacquard-loom in the most effectual manner, viz. that of practical observation, have only to step into the Adelaide Gallery . . . a weaver is constantly working at a Jacquard-loom, and is ready to give any information that may be desired as to the construction and modes of acting of his apparatus."[31] For many, the Jacquard was the ideal of self-acting and self-regulating machinery.[32] It was an excellent choice for exhibition.

Exhibits were not the only form of entertainment and instruction available at the gallery. In January 1844, for example, visitors admiring the Carbon Battery, the Aerial Machine, or the "magnificent transparent dissolving views" could also be charmed by the Infant Thalia and her performances of comic impersonations.[33] When the American impresario and showman P. T. Barnum visited London that year, his star performer, the diminutive General Tom Thumb, performed at the gallery.[34] At the same time, lectures on natural philosophy and the mechanical arts were in progress. Demonstrators and assistants were constantly available to explain the workings of the various examples of ingenious invention. Natural philosophical experiments and instruments, examples of mechanical art, and human and nonhuman curiosities of all kinds competed with and complemented each other in the same space.

The combination of natural philosophy and musical entertainment was not a new feature. As early as 1834, the *Literary Gazette* recorded that its correspondent had visited "this excellent Institution" to attend the first installment of a series of chemical lectures delivered by the gallery's lecturer in chemistry, William Maugham. This first lecture was on bleaching, a subject that was treated, according to the *Gazette*'s correspondent, with "perspicuity and simplicity." He also noted that "after the lecture there was a conversazione, agreeably enlivened by music, which was kept up till a late hour; and we acknowledge having spent a very pleasant evening in the Gallery of Practical Science."[35]

The gallery's proprietors continued to host occasional private gatherings along the lines of their opening exhibition. On 5 November 1835, for example, "a number of scientific individuals assembled . . . to witness the splendid effects produced by Mr. Cary's oxyhydrogen microscope, the magnifying powers of which have just been increased to the surprising extent of 3,000,000 of times." A variety of experiments were conducted: "the flea, by Mr. Cary's ingenuity, appeared much larger than the elephant; a portion of the body of this little insect covered a circle eighteen feet in diameter." The highlight of the evening was, however, the voltaic decomposition of water, demonstrated by William Maugham. Seen through the microscope, the effect was "exceedingly distinct and beautiful."[36]

Recorded reactions by visitors to the gallery were predictably varied. On his first visit to the Adelaide Gallery in 1837, the American electrician Joseph Henry paid much attention, of course, to the various electromagnetic devices on display: "One of the first objects which attracted my attention when I entered the gallery was one of my magnets by March [*sic*] of Woolwich. It is formed of a pice of square iron about the diameter of my first magnet but the legs are too long and the whole arrangement not proper to produce the greatest effect. It is supported on a Tripod and has a hollow lifter which one of the attendants informed appeared to work better than a solid pice of iron of the same size."[37] He went on to describe apparatus for delivering shocks and sparks by means of electromagnetism. Most visitors were not as critical as Henry and were amazed by the electromagnet's prodigious powers.

William Fothergill Cooke visited the gallery several times to study its electromagnetic devices, searching for hints in his attempts to construct a viable electromagnetic telegraph.[38] Even Henry was impressed by one nonelectrical device: a rapidly revolving wheel of soft iron which could cut hard steel. This he thought was "truely astonishing."[39] Another visitor was the duke of Wellington, who reportedly electrocuted himself by accident with one of the batteries, so that "the hero of a hundred fights, the conqueror of Europe, was as helpless as an infant under the control of that mighty agency."[40]

The Adelaide Gallery soon acquired a competitor. On 6 August 1838, the Polytechnic Institution opened its doors at 5 Cavendish Square, off upper Regent Street. The institution was granted a royal charter a year later. The main inspiration for the new Polytechnic came from Sir George Cayley, a rich and eccentric Yorkshireman with a passion for the mechanical arts. He was later to become known as a pioneer of aviation.[41] He was joined by Charles Payne, who had until recently been the superintendent of the Adelaide Gallery. Payne was offered the position of manager at the Polytechnic. Like the Adelaide Gallery, the Polytechnic charged a shilling for admission, parties of schoolchildren were admitted for half that price, and an annual subscription of one guinea could also be taken out.

The new institution was launched, as its precursor had been, with a private exhibition for selected guests and the press. As the *Athenaeum* remarked, "[t]he directors, anxious, we presume, to make the Institution known, while London yet retained a handful of its fashionables, submitted the whole to what is called a 'private view,' on Thursday last."[42] The Polytechnic's prospectus committed it to "the advancement of practical science in connexion with agriculture, the arts, and manufactures; and the demonstration, by the most simple and interesting methods of illustration, of those new principles upon which every science is based, and the processes employed in the most useful arts and manufactures effected."[43]

Like the Adelaide Gallery, the Polytechnic Institution was divided into a number of different spaces. The Hall of Manufactures had on display the tools and machines of a variety of different industries. In the basement beneath the hall was a laboratory, where under the direction of the chemist J. T. Cooper hopeful experimenters and patentees could conduct their own experiments. Above the Hall of Manufactures was the Lecture Theatre, with seating for five hundred. Beyond that was the Great Hall, 120 feet long and 40 feet high, where the institution's main exhibits were on display.[44]

3.2 The Main Hall at the Royal
Polytechnic Institution. From
the frontispiece, *Year-Book of
Facts in Science and Arts* 3
(1841).

Where the Adelaide Gallery had its steam gun, the Polytechnic Institution
also had its own main attraction in the form of a diving bell. Four or five people
could fit into the bell at a time, and for an extra fee they could descend into the
depths of a large tank of water. This attraction, according to *Punch* at least, was
very popular with young men seeking to impress their sweethearts.[45] Again like
the Adelaide Gallery, the Polytechnic had an oxyhydrogen microscope, also
constructed by Cary. Their promotional literature described it as a "magnificent Hydro-oxygen Microscope" and "the largest ever seen."[46] Another attraction was a large electrical machine, worked by a steam engine. During the
1840s this exhibit was replaced by a still more powerful source of electricity in
the form of the recently invented Armstrong hydroelectric machine. This machine produced electricity of very high tension by means of friction, as steam
was forced at high pressure through a number of small nozzles. This was described as "at least six times more powerful than any other electrical machine
ever constructed."[47]

The Polytechnic soon eclipsed the Adelaide Gallery, which by the 1840s was
in considerable financial difficulty. By the mid-1840s it seemed that the public's enthusiasm for exhibitions of scientific and mechanical ingenuity was on

the wane. Both the Adelaide and the Polytechnic introduced new features to try to maintain their clientele, but in the Adelaide Gallery's case, such efforts were in vain. In 1845 the Adelaide Gallery closed its doors for the last time, and the building was converted into Laurent's Casino.[48] The Polytechnic, on the other hand, survived into the second half of the century. For a short period at least, the display of spectacular electrical experiments seemed sufficiently lucrative to inspire the proprietors of other London exhibition rooms to add such productions to their repertoire. During the late '30s, for example, the Colosseum boasted a "Department of Natural Magic" under the superintendence of the operative chemist William Leithead. The department featured what was described as the largest electrical machine in the world.[49]

Despite the comparatively short period for which they flourished, barely more than a decade in the case of the Adelaide Gallery, these institutions performed an important function for London's instrument-making and lecturing population. They were also the places where the majority of the metropolis's population had contact with experimenters and their experiments.[50] At the Adelaide Gallery, William Maugham gave regular lectures on chemistry while William Sturgeon lectured on natural philosophy and electricity until he departed for Manchester in 1840 to become superintendent of the Royal Victoria Gallery of Practical Science.[51] At the Royal Polytechnic Institution, George Bachhoffner lectured on chemistry and natural philosophy. He and the Polytechnic were certainly remembered with affection by those who attended there:

> Ah me! The Polytechnic with its diving-bell, the descent in which was so pleasantly productive of immanent head-splitting; its diver, who rapped his helmet playfully with the coppers which had been thrown at him; its half-globes, brass pillars, and water-troughs so charged with electricity as nearly to dislocate the arms of those that touched them; with its microscope, wherein the infinitesimal creatures in a drop of Thames water appeared like antediluvian animals engaged in combat; with its lectures in which Professor Bachhoffner was always exhibiting chemistry to "the tyro"; with its dissolving views of "A ship," afterwards "on fi-er," and an illustration of—as explained by the unseen chorus—"The Hall of Waters—at Constant-nopull-where an unfort-nate Englishman-lost his life-attempting-to discover the passage!"—with all these attractions and a hundred more which I have forgotten, no wonder that the Polytechnic cast the old Adelaide Gallery in the shade, and that the proprietors of the latter were fain to welcome an entire and sweeping change of programme.[52]

Such institutions provided these men with an opportunity to appear before the public as natural philosophers. They were provided with a place to experiment and the resources needed to make and to publicize their experiments. In turn, London's public came to see electrical experiments, for example, as taking place in the context of a vast and spectacular array of natural and man-made curiosities.

These exhibitions were themselves on a particular map. Scientific tourists keen to acquaint themselves with the new factory system and the workings of machinery visited the Adelaide and the Polytechnic just as they visited factories and workshops (where they could). The public exhibitions made accessible and simultaneously constructed the hidden world of labor. They were included in tourist guides to London's landmarks.[53] Some of the scientific tourists came from very far afield, notably two shipbuilders from Bombay who visited during the late 1830s to study the construction of steamboats. They waxed lyrical about the exhibitions:

> The effect of looking down from this gallery [in the Polytechnic] upon the several things in constant motion, is quite enchanting, and we do not hesitate to say, that if we had seen nothing else in England besides the Adelaide Gallery and the Polytechnic Institution, we should have thought ourselves amply repaid for our voyage from India to England.[54]

It was through such tours of factories and of machine exhibitions, and the expanding number of accounts of such tours, that the middle classes constructed and made sense to themselves of the factory system. A link was being forged between display and the progress of industry. Great engineering feats were just as important as tourist attractions as they were as utilities. The best example was the Thames Tunnel: its own construction was a matter for display on-site, while models of its design were on show at the galleries.[55]

Machines in such exhibitions became commodities. The Adelaide Gallery and its competitors were marketplaces where, as the *Mechanics' Magazine* had predicted, manufacturers vied for the attention of "curious idlers." The magazine had argued that

> [i]n seeing a manufactured article . . . to appreciate its excellence the taste of the purchaser or consumer does not require to be cultivated or purified, as in the case of the fine arts, by the repeated exhibition of masterpieces.[56]

This, however, was precisely what the galleries of practical science aimed to do. Their audiences were being taught to see machines as consumer goods, to be admired, fetishized, and desired.[57] The exhibition was a passage point in which production and consumption occupied the same space. The distinction between the relatively private space of the laboratory and the public space of the lecture theater was not as visible in such places. Similarly, such exhibitions straddled the divide between workshop and shop window. Electricity in the galleries was as much a commodity as any other exhibit on display, to be regarded in the same way. Spectacle was an integral feature of its production as well as its display.

## MAKING ELECTRICITY

One good example of the way in which institutions such as the Adelaide Gallery could help to forge a career is that of the young American mechanic Joseph Saxton. Saxton, like Jacob Perkins, was a Philadelphian. He had been assistant to Isaiah Lukens in his clockmaking trade in that city before traveling to London in the late '20s or early '30s to act as Perkins's assistant. His main task in the months leading up to the opening of the Adelaide Gallery had been to prepare large magnets for display. He was well qualified for the task. His teacher Isaiah Lukens was known in Philadelphia not only as a clockmaker but as one of the best makers of permanent steel magnets.[58]

It is recorded that Lukens was somewhat distressed when his assistant left him to work with Perkins. Perkins had left rather a dubious reputation behind him in Philadelphia. When George Escol Sellars, whose father had been associated with Perkins in a business partnership, visited him at the Adelaide Gallery, he felt that Perkins had "found his true level, the typical showman of the period."[59] He remained sufficiently prestigious, however, for Joseph Henry, when he visited London in 1837, to obtain a letter of introduction from Robert Hare to visit him. In fact the Adelaide Gallery was one of the first places Henry visited in London after he arrived there.[60]

Saxton's activities, as recorded in the diary he kept for the early part of his sojourn in London, provide an interesting insight into the kinds of opportunity made available to a young mechanic by the Adelaide Gallery. The gallery for Saxton formed the center of a network of resources and contacts. It was also the place where his electrical experiments were performed and where his reputation was made.[61] The public's perception of Saxton and his work was mediated by their experience of the Adelaide Gallery and its reputation. His experiments and instruments took their place in an exhibition designed to display nature's curiosities and man's ingenuity to a paying audience of London's lower-middle classes.

The exhibition and its products were deeply embedded in the working culture of London's artisans. The gallery could not exist without the supporting network of small workshops and factories that provided raw materials and skills. As its principal mechanic, Saxton attended the gallery on an almost daily basis. He also had his own workshop at 22 Sussex Street where he produced models and artifacts for the exhibition and also worked on his own projects, such as an ambitious scheme for propelling railway carriages and canal boats through the use of stationary steam engines.[62] He also spent much time visiting the metropolis's workplaces: borrowing tools, buying exhibits and equipment, and placing orders for finishing work.

Saxton's main priority in the month or so leading up to the opening of the gallery was to prepare an entirely novel form of magnet. A few months previously, Michael Faraday had announced the possibility of producing electricity by means of magnetism.[63] Saxton had been struck by the potential of this discovery for the production of spectacular display and had set about transforming Faraday's apparatus into a suitable form for demonstration at the Adelaide Gallery. In its essentials, the apparatus was quite straightforward. It consisted of a permanent magnet, a soft iron keeper, and a coil of wire wound around the keeper. When the keeper was removed from the magnet, electricity would be induced in the coil and would produce a spark at the point where contact was broken.[64]

The apparent simplicity of Saxton's apparatus concealed the amount of labor that had been involved in its construction. As his diary reveals, producing the spark and ascertaining the conditions under which the spark was best produced required weeks of experimentation. All kinds of factors could affect the apparatus's performance: the size of the magnet, the length of the wire coil, the shape of the soft iron keeper. Saxton experimented repeatedly in order to find the combination that would produce the most spectacular display when the apparatus was put into action at the Adelaide Gallery. He tried, for example, the effect of placing charcoal points and brass balls on the ends of the wire to improve the spark, but with no success.[65]

By the evening of 4 June, when a large gathering of distinguished guests congregated to celebrate the opening of the gallery, Saxton's efforts had paid off. His apparatus was on display, and the spark was produced successfully. John George Children, the secretary of the Royal Society, was present and invited Saxton to display his apparatus at one of the "Conversation Parties" regularly held at Kensington Palace by the duke of Sussex, president of the Royal Society and younger brother of the king.[66] This was precisely the kind of publicity that could make the career of an aspiring young instrument-maker. Saxton's diary also records that in the weeks following the demonstration, he constructed and sold a number of magnets for displaying the spark.

It is worth remembering, however, that, just as the display of electrical devices and phenomena was only a part of the Adelaide Gallery's activities, working on his magnets was only one of the many activities in which Saxton was engaged. At his own workshop at 22 Sussex Street, he built models and other exhibits for the gallery, including the Paradox Head, designed so that a sword could be passed through its neck without leaving a mark.[67] He also engraved medals, repaired clocks and watches, and worked on a range of inventions including the fountain pen with which he wrote much of his diary.[68] Aspiring young mechanics in Saxton's position could not afford to limit themselves to a single speciality. He was also keeping a close eye on politics, as England tottered on the verge of revolution during the spring and early summer of 1832. On the very night of the Adelaide Gallery's opening exhibition, the Reform Bill was passed through the House of Lords.[69]

Saxton was also making allies among his fellows in London. A few days after the exhibition of 4 June, he recorded a visit by the instrument-maker Francis Watkins:

> at work at the magnet was called on by Mr Watkins had some conversation on the subject of magnetism he seemed to take a great interest in my getting the credit of first getting the spark from the magnet in England from what motive I do not know however time will show motive it is love of justice or private enmity to other parties.[70]

Others tried to persuade him to attend the annual meeting of the recently founded British Association for the Advancement of Science, being held that year in Oxford, so that he could demonstrate his apparatus to the distinguished gathering that would assemble there. Saxton declined, being unable to afford the expense of travel to Oxford.[71]

Once he had succeeded in producing the electric spark reliably, Saxton worked at expanding the repertoire of effects obtainable from his apparatus. On the day following his first exhibition at the Adelaide Gallery, he demonstrated a method of producing electric shocks by applying the wires from the coil to each side of the victim's tongue as the keeper was detached from the magnet.[72] The next day he succeeded in igniting gunpowder with the spark and proudly recorded this demonstration in his diary as "the first time gunpowder was ever fired by the magneto electric spark." He continued to work at this effect, devising a "gun" that could be fired electrically, and attempting to use the spark to shoot a gum-elastic ball across the length of the gallery's exhibition room. He was inspired, presumably, by the popular success of Perkins's steam gun.[73]

One very important feature of the gallery's public spaces was the way in which they served a variety of different functions. They were not simply spaces where the public passively viewed the productions of nature and of human mechanical ingenuity. The gallery was also a laboratory. Saxton experimented there with his electrical apparatus and often tried effects in public that had not been produced before. The canal in the Long Room was the site of experiments where Saxton assisted the engineer Thomas Telford, one of the gallery's patrons, in testing the efficiency of different shapes and sizes of paddles for steamboats.[74] The gallery's audience could see and participate in knowledge in the making. In cases like the first production of electrical shocks from Saxton's magnet, they became part of the experimental apparatus. Unlike the basement laboratory of the Royal Institution where Faraday and his staff of assistants labored in private, the experimental spaces of the Adelaide Gallery were public.[75]

The gallery provided Saxton with a place to work and a place to display his experiments in public. It was not common for instrument-makers to publish papers describing their work in the philosophical press. When James Marsh, for example, as mentioned in the previous chapter, developed his electrical swinging pendulum in 1821, the first account of it to appear was by Peter Barlow, his employer, rather than by Marsh himself, in a postscript to a paper by Barlow, in which he commended his ingenious workman to public notice.[76] Venues such as the Adelaide Gallery were therefore essential as means of allowing instrument-makers to lay claim to novelty and establish themselves as reputable producers of apparatus and of natural phenomena. In these places they could attempt to overcome the distinction between instrument-makers and knowledge-makers that was in the process of being constructed by the gentlemen of science. This became very clear to Saxton a year later when a vastly improved version of his spark-producing apparatus was put on display.

PRIORITY AND INTELLECTUAL PROPERTY

Saxton did little or no work on electromagnetism for most of the remainder of 1832. He occasionally noted in his diary that he had sent off a magnet to another customer, but he was mostly engaged in other matters for the Adelaide Gallery such as acquiring equipment and building models.[77] In December of that year, however, he began work on a machine that would be able to produce continuous currents of electricity, rather than single sparks. The machine was to be based on his already successful device, the main innovation being that the soft iron keeper would be made to revolve in front of the permanent magnet's poles to produce a continuous succession of sparks.[78]

The new machine was not completed until about 20 June 1833. On that day, as Saxton recorded the achievement in his diary:

> Made a tryal of the new arrangement of the Magneto-Electrical Machine, found it to surpass expectations. it produces a continued spark and so mutch a shock to the tong and lips that it is impossible to bear it for any length of time.[79]

After spending a few days unsuccessfully trying to achieve the decomposition of water using the new machine, Saxton set off for that year's meeting of the British Association for the Advancement of Science at Cambridge.[80] The instrument was exhibited to the public for the first time, and "the ingenious invention of Mr. Saxton, by which the transient electrical currents might exhibit their effects in so brilliant and so powerful a manner," was hailed by no less than Samuel Taylor Coleridge in the same breath as Faraday's discovery of electromagnetic induction.[81]

On his return from Cambridge, Saxton continued his efforts to decompose water with his new invention and was soon successful. The novel electromagnetic apparatus was put on display at the Adelaide Gallery in August. At the beginning of the month, however, Saxton unexpectedly found himself being roundly condemned in the weekly *Literary Gazette* for attempting to lay claim to the discoveries and inventions of others. According to the *Gazette*, a report was making the rounds of the newspapers to the effect that Saxton on the basis of his invention claimed priority for the first production of an electric spark from the magnet and for the first successful decomposition of water by magnetism.[82]

The *Literary Gazette* provided an alternative history of these discoveries. According to that account, Faraday had the credit for being the first to so produce the electric spark. His experiments had been repeated with an ordinary magnet, rather than an electromagnet, by Nobili and Antinori in Italy and Forbes in Scotland. Only after these successes did Saxton manage to produce the spark. The decomposition of water by electromagnetism had first been achieved by the French instrument-maker Hippolyte Pixii. Saxton, it was alleged, had simply repeated Pixii's experiments without attribution. With heavy irony, the *Gazette* emphasized its willingness in a future issue "to insert any communication from Mr. Saxton, vindicating his claims to these brilliant discoveries."[83]

Saxton, in a letter published in the next issue of the *Gazette*, indignantly denied that he had ever laid claim to being the first to produce a spark from the magnet. He did not mention his claim to having been the first in England to produce the spark. His claims concerning the decomposition of water he limited to his having been the first in London to demonstrate the effect. He denied

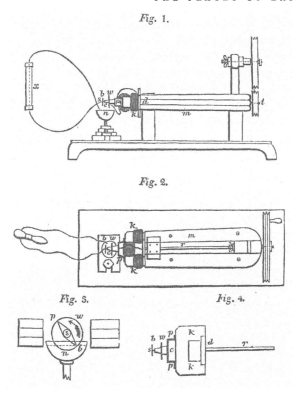

3.3 Saxton's magnetoelectric machine. *Journal of the Franklin Institute* 13 (1834): 155.

any prior knowledge of Pixii's work and asserted that since he regarded his own machine as being far superior to the Frenchman's, he would continue to claim it as a novelty until proved mistaken.[84] What was at stake here was the proper attribution of novelty and similarity. Saxton's machine could be either celebrated as an original invention or dismissed as a mere copy of Pixii's. Deciding such issues was no easy matter. It was not neccessarily obvious how different a device ought to be before it might count as a novelty. Neither was it clear what precisely such a claim to invention entailed. The *Gazette* apparently read Saxton's claim to invention as subsuming the discovery of the phenomena he could demonstrate as well. This was an example of the problematics of display. It was not always clear just what was being claimed by exhibition.

The affair left Saxton in an awkward position. Being primarily employed by the Adelaide Gallery as an instrument-maker and mechanic, he had used that institution to establish and promulgate his claims to novelty. He had no printed publication that might serve to establish the time or extent of his claims. He could, however, use the gallery's resources to fight for the superiority and originality of his invention. The gallery had facilities where a trial could take place and the resources to make such a trial a highly public affair. Accordingly, with the help of Francis Watkins, who had already announced his interest in establishing Saxton's priority and originality and who had access to a Pixii machine, moves were made to arrange a spectacular public contest between the two inventions and their inventors.[85]

The trial took place on 14 November and was reported at length in the *Literary Gazette*, the very journal that had led the way in casting doubt on Saxton's integrity and reputation a few months previously. Pixii's machine, operated by Francis Watkins,

> very rapidly decomposed water, first in a single tube, hydrogen gas being evolved from one wire connected with one pole of the armature while oxygen gas was given off at the other, precisely as when the elements of water are disunited by galvanic agency. The mixed gases were then reconverted into water by the electric spark after the usual manner. Water was next decomposed, and the elements received in two tubes; and it was observed that the proportions were, as near as possible, two to one, affording another proof of polar decomposition. We were next favoured with an experiment quite new in this country, namely, that of charging a Leyden jar with magneto-electricity; the truth of which was made evident by the aid of a delicate electroscope, the gold leaves of which very sensibly diverged.[86]

Saxton then proceeded to operate his own apparatus:

> It gave powerful shocks, brilliant sparks, heated a platinum wire red hot, and decomposed water; but the experimenter was not so fortunate in charging the jar as in the former instance, although there seems no doubt that it is capable of being affected by this magnet.[87]

The trial, on the whole, had been successful.

The journal described both machines in enthusiastic terms but was unequivocal concerning the superiority of Saxton's model: "This splendid apparatus attracted the universal admiration of the scientific company present, not only from the beautiful and extraordinary effects produced by it, but also from its very superior mechanical arrangement."[88] The "scientific company" assembled at the gallery had included Faraday, John Frederic Daniell (professor of chemistry at King's College, London), and Dionysus Lardner, popular author and lecturer. Through the display, their authority as witnesses had been thrown behind Saxton's claims to originality and invention. The exploitation of the Adelaide Gallery's resources had been successful.

The presence of prestigious witnesses was still an important component of establishing priority and authenticity during this period, particularly for individuals such as Saxton who aimed to establish their credentials by other than literary means.[89] Reports in the popular press, describing the spectacular phenomena associated with a new device or experiment and listing the prominent individuals who had been impressed by the display were crucial to the career of those who made their living producing such devices and displays. They brought their productions and the places where such productions could be witnessed at first hand to a far wider audience than would have been the case through a mere publication in a scientific journal, whose circulation in most cases was a few hundred rather than several thousand.[90] Ultimately, the relevant audience for Saxton was the public that paid its shillings at the doors of the Adelaide Gallery. The approval of the elite's natural philosophers such as Faraday was useful, but primarily only insofar as it certified the display, the device, and the discoverer for that wider audience.[91]

Even after this apparently convincing display of superiority, however, Saxton was still not completely safe in his intellectual property rights. While his machine had been accorded the laurels in most of the trials to which it was submitted, Pixii's machine had performed better at producing some of the relevant phenomena. In particular while Pixii's machine had successfully charged a Leyden jar, according to the *Literary Gazette* "an experiment quite new in this country," Saxton had been unable to produce the same effect with his apparatus. More than a year later, Saxton was still concerned to rectify this anomaly.

While his machine was superior to Pixii's in producing the effects associated with electrical quantity, it did not perform as well in producing effects for which intensity was required. "It appears that in constructing a magneto-electric machine that an important condition exists with regard to the object in view. That is if the object is to construct a machine to produce a violent shock it is necessary to have the helical wire in a few pieces, and as long as possible. But if the object is to produce a large spark the helix must be composed of a number of pieces of moderate length."[92] He noted that while the machine he had constructed for the Adelaide Gallery, whose coil consisted of twenty lengths of short wire connected at the ends, gave only a "moderate" shock, it "provides a very large spark with loud crackling noise, decomposes readily, & will melt a platinum wire of the 100th of an inch in diameter."

Another machine, with far less wire but with that wire in only two long strands, on the other hand, "gives so powerful a shock that but few persons could beare it for an instant." J. F. Daniell and the instrument-maker John Newman, who had constructed this other machine to Saxton's design, had attributed the difference in effect to the fact that the silk cloth insulating the wire was of different colors in the two machines. Saxton, however, felt that this explanation was inadequate.[93] A few days later he conducted some experiments connecting different coils of wire to his machine to demonstrate his supposition. While a coil of wire fifty-seven feet long produced a strong shock and a feeble spark, a coil of the same wire divided into four pieces did exactly the opposite, producing "a shock so feeble that it is scarcely sensible to the tongue, but a very brilliant spark."[94] These private experiments were to acquire a new significance twenty months later.

In the meantime, a new threat to Saxton's claims of priority and property appeared in the form of some experiments conducted by William Sturgeon in August 1834. Ironically, Sturgeon had carried out these experiments at the Adelaide Gallery, using Saxton's own magnetoelectric machine that was on display there. Sturgeon had been given permission by the gallery's proprietors, through the good offices of the superintendent, Charles Payne, to "employ their large magnet in any new experiments" he might wish to undertake.[95] Accordingly, Sturgeon had carried out a number of experiments in which a variety of substances—hydriodate of potassa, sulfate of copper, and water—were submitted to the action of the machine and decomposed. He also magnetized hard steel and soft iron and succeeded in producing "a great variety of electro-magnetic rotations, and some other rather novel motions, with electric currents by magnetic excitation." He achieved these effects, Sturgeon noted, "by changing and reversing the current; and the results were exhibited with as much promptitude as they could have been by the employment of a voltaic battery."[96]

A few months later, Francis Watkins sprang to Saxton's defense in the pages of the *Philosophical Magazine*. The experiments described by Sturgeon, he asserted, were unremarkable: "what Mr. Sturgeon brings forward as new in magneto-electricity in August 1834, had been noticed by Mr. Saxton, myself, and others, many months previously."[97] Referring to the demonstration in the Adelaide Gallery of the respective powers of Pixii's and Saxton's machines (at which the *Philosophical Magazine*'s editor, Richard Phillips, had been present) Watkins claimed that on that occasion, using Saxton's "large and splendid magnet," they had demonstrated the "mechanical, physical, and chemical effects of electricity developed by steel magnets." Sturgeon's experiments added nothing to those already established results.[98]

Sturgeon's reply, published in the *Philosophical Magazine* a few months later, was characteristically vitriolic: "a more disagreeable task could hardly fall to the lot of any individual than that of having to rescue from the unwarrantable attack of another, that which is justly and honestly his own right."[99] He castigated Watkins in uncompromising terms for attempting to gain credit for himself and Saxton for work that they had not done, and clarified his own claims to innovation. His originality, he asserted, lay in his having been successful in

> producing electro-dynamic phaenomena, of various classes, by giving to magnetically excited electric currents one uniform direction through the terminal conducting wires, by means of a certain contrivance which may very properly be called the Unio-directive Discharger; because it has the power of uniting, and discharging, in any one direction those currents which, in consequence of the mode of excitation, are originally urged, alternately, in directions opposite to each other.[100]

Any contrivance for producing magnetoelectricity that did not contain such a mechanism he dismissed as "comparatively useless."

Sturgeon's point was that the electricity produced by magnetoelectric action was alternating, or reciprocal, in character. While this was not a problem for the exhibition of sparks or shocks, or in the heating of wires, it was impossible to produce chemical decomposition, "with exact polar arrangement of the liberated constituents," without a uniform electric current. He admitted that more often than not in most magnetoelectric devices the current in one direction tended to predominate, but suggested that "it would be exceedingly unscientific, in cases where power is wanting, to employ a part only, when the whole is available; or, as in the present instance, to employ the difference only, instead of the sum of the reciprocating electric forces." His new unio-directive discharger, Sturgeon claimed, "gives to the magnetic electrometer a degree of importance which it could never have possessed without it."[101]

Sturgeon denied that any contrivance such as his for "collecting and giving a proper direction to the excited currents" had ever been used with Saxton's magnetoelectric machine at the Adelaide Gallery. He appealed to the testimony of the Adelaide's chemical lecturer, William Maugham, who had assisted him in his own experiments there, as a witness that nothing like the results produced by his experiments had been achieved before at the Adelaide Gallery. He appealed also to the physical appearance of the machine itself, as it could be inspected at the gallery. It did not include any contrivance "in the capacity of an unio-directive discharger." Finally, he drew attention to a recent paper by Watkins himself which showed that he had "not made even the slightest alteration, consequently no improvement, in the original apparatus." Watkins admitted that he had never produced "TRUE POLAR decomposition."[102]

His own results differed from those of Saxton and Watkins in that they could be controlled: "The results of all such experiments as those Mr. Watkins claims must ever be fortuitous, as it would be impossible to predict the direction in which the predominating force, by any new machine, would be exerted."[103] His own results, however, attained through the use of his unio-directive discharger, could be depended upon "with the same promptitude, uniformity, and precision as by any other source hitherto placed in the hands of the philosopher." Using his improvement in the apparatus, the experimenter could "confide in his predictions, and vary his exhibits in any way he pleases." He would be spared "all those corroding apprehensions and mortifying disappointments, which must ever molest his efforts, agonize his feelings, and chill the ardour of his inquiries, whilst operating with an apparatus over the powers of which he has not the slightest control."[104]

Control was central to the construction of the inventor as showman. It was only by controlling his machine, predicting how it would respond in different situations, that the inventor could also aim to control the audience for his showmanship. By claiming to be able to know what the machine would do and showing that it did so, the showman could also claim it as being his own invention. It was also a matter of self-control. A showman who could not command his apparatus and "vary his exhibits" at will could not confidently predict his own behavior either. Sturgeon was suggesting that by being able to know what the machine would do, he had made it his own. He, rather than Saxton, was the confident inventor-showman who could use the machine to control nature and command his audience. The self-control needed for a successful and compelling performance came through control over the machine.

Sturgeon's attack was—potentially, at any rate—extremely damaging since it was tantamount to an attack on the integrity of Saxton's machine. He was making a claim of property rights over the apparatus and its attendant phenomena that was in conflict with Saxton's own property rights. He was suggesting that at least some of the phenomena which Saxton claimed could be produced with his magnetoelectric machine could, in fact, be produced only through the additional use of his own unio-directive discharger. Watkins also chose to interpret Sturgeon's attack as a challenge to his own good name. He responded furiously in the *Philosophical Magazine*:

> In the last number of your Journal, page 231, I observe in the first paragraph of a long letter from Mr. William Sturgeon, that he has been pleased to indulge himself in using rather strong language against someone. I have accordingly inquired of this gentleman whether in that paragraph he alluded to me, and he has very *politely* answered me "Yes."[105]

Disputes concerning the novelty and value of an invention could very easily become matters of personal integrity.

A more serious challenge to Saxton's claims to novelty was soon mounted by a rival instrument-maker, Edward Clarke.[106] Clarke had moved from Dublin, where he had been associated with the optician Richard Spear, to London, where he had been employed in the workshops of Watkins & Hill, in whose firm Francis Watkins was a partner.[107] He also had links with the National Repository and may have displayed some of his productions there. Clarke left Watkins & Hill after a short time to set up his own instrument-making concern and specialized in electrical apparatus. He soon clashed with his former employers. By the late 1830s his instrument shop was on the Lowther Arcade, a stone's throw from the Adelaide Gallery.

In 1835, Clarke communicated two notices to the *Philosophical Magazine* describing improvements he had made to the magnetoelectric machine. Neither account mentioned Saxton's name.[108] A few months later, a veiled rebuke from Watkins appeared in the *Philosophical Magazine*, in the guise of a description of a magnetoelectric machine devised by Saxton:

> Having lately observed in the Philosophical Magazine some descriptions of slight modifications of apparatus for the development of magneto-electric phaenomena, and presuming from their insertion that you think such contributions interest your readers, I venture to offer to your notice an account of philosophical apparatus, or toy, which I have contrived, and which I believe to be novel.[109]

He proceeded with his account, emphasizing that Saxton was the inventor, that he had his express permission to build a modification of the instrument, and that the device could be seen on display at the Adelaide Gallery.

Clarke was not long in responding. To a brief note discussing voltaic magnetism, he appended some sharp remarks concerning Saxton's claims to invention, emphasizing that he too had been working on electromagnetism for several years. "This rotary motion was applied lately (and described in your valuable Journal for July) to an apparatus copied from that constructed by Mr. Saxton, which construction is merely a modification of one invented by

Professor Richie, and exhibited two years ago by his assistant."[110] If Watkins could insinuate that his former employee had contributed nothing more than slight modifications of philosophical toys, Clarke could respond that Watkins's new protégé Saxton had no great claim to originality and novelty either.

In the October 1836 issue of the *Philosophical Magazine*, Clarke published an account of his own new Magnetic Electrical Machine.[111] This new invention, he claimed, was the result of many years' intensive research:

> From the time Dr. Faraday first discovered magnetic electricity to the present, my attention, as a philosophical instrument maker, has been entirely devoted to that important branch of science, more especially to the construction of an efficacious magnetic electrical machine, which after much anxious thought, labour and expense, I now submit to your notice.[112]

He proceeded to give a detailed account of his machine's construction and a long list of its many advantages.

Prominent among these advantages was the inclusion of two different armatures, one for intensity and the other for quantity effects, that could be attached to the apparatus. Clarke noted that in November 1835 he had experimented with wire of different diameters in the armatures: "I found that the thick copper bell-wire gave brilliant sparks, but no perceptible shock, whilst, on the contrary, very fine wire gave powerful shocks, but very indifferent sparks."[113] As a result he had furnished his new machine with armatures of both kinds so as to "give the separate effects of quantity and intensity to the fullest extent of power that my magnetic battery was capable of supplying."

With his intensity armature, on which was wound fifteen hundred yards of fine insulated copper wire, a variety of spectacular experiments could be performed:

> The effect this armature produces on the nervous and muscular system is such, that no person out of the hundreds who have tried it could possibly endure the intense agony it is capable of producing; it is capable also of electrifying the most nervous person without giving them the least uneasiness; it shows the decomposition of water . . . and also of the neutral salts; it deflects the gold leaves of the electroscope, charges the Leyden jar, and by an arrangement of wires . . . the electricity is made distinctly visible, passing from the magnetic battery to the armature, and by the same arrangement not only shocks, sparks, but brilliant scintillations of steel can be obtained.[114]

A similarly spectacular array of effects, including "large and brilliant sparks, sufficiently so that a person can read small print from the light it produces," was obtainable from the quantity armature.

Another novelty claimed by Clarke was the inclusion of several attachments, such as a small electromagnet and decomposition apparatus, that could conveniently be added to the machine to show off its various effects. Clarke also recommended his machine for the use of medical practitioners wishing to utilize the therapeutic effects of electricity. He suggested that in the past it had been the "uncertainty of action of all electrical machines" which had prevented their extensive use by medical practitioners. He pointed in particular to the ease with which his machine could be operated and to its convenience and portability.[115]

Saxton responded swiftly to what he clearly regarded as a straightforward plagiarism of his machine by Clarke, with a letter to the *Philosophical Magazine*, exposing Clarke's duplicity:

> A reader unacquainted with the progress which magneto-electricity has made since this new path of science was opened by the beautiful and unexpected discoveries of Faraday, might be misled, from the paper I have alluded to, to believe that the electro-magnetic machine there represented was the invention of the writer, and that the experiments there mentioned were for the first time made by its means. No conclusion, however, would be more erroneous. The machine which Mr. Clarke calls his invention, differs from mine only in a slight variation in the situation of its parts, and is in no respect superior to it. The experiments which he states in such a manner as to insinuate that they are capable of being made only by his machine, have every one been long since performed with my instrument, and Mr. Clarke has had every opportunity of knowing the truth of this statement.[116]

He reminded the *Philosophical Magazine*'s readers that his own apparatus was still on display at the Adelaide Gallery, and that his claims as its inventor had been acknowledged in public by Faraday, Daniell, and Charles Wheatstone.

He gave a detailed description of his machine as it appeared at the Adelaide Gallery, explaining its performance and mode of operation. Saxton emphasized that his magnetoelectric machine, like Clarke's magnetic electric device, now also contained contrivances for displaying the effects of both quantity and intensity. This double armature, "producing at pleasure, either the most brilliant sparks and strongest heating power, or the most violent shocks and effective chemical decompositions," had been added to the instrument in December 1835, following his hitherto unpublished experiments on different lengths of coils.[117]

As he made clear, those experiments had been made as a result of the trial conducted at the Adelaide Gallery between Saxton's own machine and Pixii's in November 1833. On that occasion Saxton's had produced the more brilliant sparks, while Pixii's machine had been superior in inflicting shocks and affecting the electrometer. Pixii's machine had possessed a much longer coil than Saxton's and his experiments had convinced him that this was the cause of the difference in performance. He pointed out that recent experiments by Joseph Henry, Faraday, and others had firmly established this point: "Mr. Clarke has no more claim to the application of the double armature to the magnet than he has to the discovery of the facts which suggested that application."[118]

The status of Saxton's trials with different kinds of coils on his machine was changing during the course of the dispute. In order to defend his claims to inventor's rights over the magnetoelectric machine, he had to show that he had preempted Clarke's alleged improvements and already incorporated them into his own device. What had been *private* speculations and manipulations of his apparatus were translated into a story of *public* discovery and therefore of experiment.[119] His tinkering with the machine in order to improve its performance became an experimental trial that clarified facts about the nature of electricity. In this process, features of the machine such as the difference in performance with different types of coil became characteristics of the natural world. The difference between *intensity* and *quantity* of electricity was established as part of the process of deciding who had invented the machine.

The notions of quantity and intensity of electricity were fundamental for early-nineteenth-century British electricians. By the 1820s their use had solidified around the material practices of battery making, being used to account for the different phenomena associated with different kinds of battery.[120] Simply speaking, a quantity battery was usually composed of a single pair of large plates, while an intensity battery consisted of a number of pairs of small plates linked together in series. A quantity battery was used to produce effects such as sparks and heat, while intensity batteries were used for producing shocks and chemical decompositions. The priority dispute between Clarke and Saxton was part of the process whereby these terms of the battery-maker's art were appropriated to describe the characteristics of new sources of electricity.

Clarke responded to Saxton's attack a few months later in the same journal. This time he was explicit that his claim to originality and invention lay in the superiority of his machine's performance as compared to Saxton's. He reminded Saxton and his readers sarcastically of the great difference that "a slight variation in the situation of its parts" might make to the working of any machine. He drew attention to the praise his version of the machine had been given by popular lecturers such as William Sturgeon and George Bachhoffner. He also pointed to its use by the medical profession, some of whom, he claimed, had requested that he modify the Saxton machines that they owned according to his own principles. There was no further reply from Saxton in the pages of the *Philosophical Magazine*.[121]

This series of disputes concerning the novelty and superiority of Saxton's invention brings to the fore a number of the electrical instrument-making community's often conflicting views and assumptions concerning the attribution of originality and the ways in which claims to individual inventions and discoveries could be defended and sustained in that culture. It also makes clear the role played by institutions such as the Adelaide Gallery in providing that community with material and social resources. In particular it makes clear how claims to philosophical discovery for this community were deeply grounded in, and often inextricably linked with, claims to the invention of machines. The status of men such as Clarke, Saxton, and Sturgeon as natural philosophers was the product of the ways in which their machines performed in public.

In this way inventions and their inventors emerged together. Exhibition halls such as the Adelaide Gallery were crucial for such functions as providing the spaces within which machines could appear as commodities and their makers as entrepreneurs. The characteristics of the good invention and the characteristics of the good inventor were negotiated simultaneously in these spaces.

Such attributions were the outcomes of competitions to attract public attention and sell a product. As such, inventions and their inventors changed during the process of invention. Saxton moved from being a successful performer to being a charlatan as his machine's capacities shifted. Clarke moved from being a dabbler in trivial philosophical toys to being the inventor of the magnetic electric machine and a highly successful businessman. By the mid-1840s he was the owner of an ironworks, and by the 1850s he was the proprietor of the Royal Panopticon of Science.[122]

## CONCLUSION

Despite the highly public culture that they inhabited, however, and their emphasis on the spectacular display of their experimental productions, these men clearly regarded their machines and their experiments as private property. While these devices and their attendant phenomena were very much designed for public consumption, they were also to remain the private property of their originators. Saxton, for example, did not object to Clarke's making the machine he claimed as his own, but objected rather to its appropriation as Clarke's own invention: "the magneto-electric machine which Mr. Clarke has brought forward . . . is a piracy of mine; the piracy consisting not in manufacturing the instrument, for everyone is at full liberty to do so, but in calling it an invention of his own and suppressing all mention of my name as connected with it."[123]

What was at stake was the maintenance in the public eye of the unique relationship between the experimenter and the experiment. The inventor's aim was to retain property rights over a particular machine and the phenomena it produced even while the machine was being reproduced by others. It was important to Saxton, for example, that his magnetoelectric machine should always be labeled as his machine. This is clear from instrument-maker's catalogs of the period. Apparatus would more often than not be identified with their originators, even when they were not the actual makers of the physical item being advertised. E. M. Clarke, for example, would advertise an item of apparatus as being Barlow's wheel, Shillibeer's battery or Professor Henry's voltaic magnets to remind the consumer that the products remained the private property of their inventors.[124] The display of machines at places such as the Adelaide Gallery aided this process of identification. It helped maintain the link in the public eye between the machine and its maker.

Clearly at stake here was the artisan-inventor's property of skill. The skills of the maker embodied in the machine, as Thomas Hodgskin had argued, made that machine the property of its maker.[125] This notion of property was not couched in terms of physical ownership but rather in terms of a right to a recognition of the relationship between the invention and the inventor.[126] It was this relationship, embedded in artisanal values, that defined what it meant to be an inventor. Invention, however, was also defined by the spaces in which its products appeared as public commodities. In places like the Adelaide

Gallery, invention was a matter of performance and showmanship as well. Production and display could not easily be dissociated. Making the machine work and presenting it as a salable commodity were components of the same process.

The electricians who worked and displayed their wares at the galleries of science were a part of this entrepreneurial showman's and artisan's culture. It was in this context that they earned their often precarious living and, more important, forged their own sense of identity. The Adelaide Gallery provided Saxton, for example, with material resources and the opportunity to advertise his skills and his products. His time was spent with the gallery's other workers and promoters. When he tried to patent one of his inventions, it was J. S. Brickwood, one of the Adelaide Gallery's Council, who provided the finances.[127] Electricity in these places was a matter of performance and craft skills. Its practitioners were machine-makers. Those who visited the galleries saw electricity take its place in a world of spectacular engineering feats and machinery, defining the new and progressive factory system.[128] Like other mechanics and artisans, electricians tried to use that world to celebrate and protect their skill and craftmanship. The aim of their practical science was to give them status as producers of machines and of useful knowledge.[129]

Galleries such as the Adelaide and the Polytechnic were crucial resources for electricians trying to find a place for themselves within the networks of London natural philosophy. Through such places they could have access to the materials they needed to ply their trade and to audiences who could appreciate their performances. The galleries provided links to the wider world of London labor in two ways. On the one hand, through contacts forged at the exhibitions, electricians might have access to the workshops and suppliers who provided the galleries with raw materials. On the other, their experiments and instruments were on show alongside the machines and artifacts that emerged from London's manufactories. They were there to be perceived as being part of the world of labor. Electricity, therefore, might be perceived in such places as being a matter of both work and display. Electrical inventions, like the machines with which they appeared, were seen as being embedded in the networks of commodity production and consumption of which the galleries were part.[130]

# A Science of Experiment and Observation: The Rise and Fall of the London Electrical Society

THE FIRST HALF of the nineteenth century witnessed the proliferation of specialist scientific societies in London. Many historians have noted this process as central to the development of autonomous scientific professions in England. These new institutions were largely the preserve of vocationally minded gentlemen, middle-class and professional in their attitudes, who wished to redefine the boundaries of natural philosophy. *Specialization and the formation of specialist bodies were rapidly becoming the marks of scientific discipline.*[1] The Geological Society, founded in 1807, and the Astronomical Society, founded in 1820, exemplified for many the new direction of natural philosophy. For some contemporary commentators, specialization was the offspring of the division of mental labor that would revolutionize the sciences just as its manual counterpart had revolutionized the factory system. The aristocratic dilettantes who had dominated the Royal Society under Sir Joseph Banks's twenty-year presidency were to be swept aside in a tide of reformist zeal.[2]

It was against this background of new disciplinary formation that a group of electricians met in the spring of 1837 to found the London Electrical Society. The first suggestion that such an institution was required was made in conversation during a course of lectures on electricity delivered by William Sturgeon at Edward Clarke's "Laboratory of Science" on the Lowther Arcade. The aim was to formalize the informal links that had already developed among various experimenters through precisely such gatherings as Sturgeon's lectures. A meeting of various "gentlemen" was held on 16 May at Clarke's establishment to launch the new society.

The humble surroundings in which these "gentlemen" met serves to underline the gulf that separated them from the founders of the other specialist organizations. None of the Electrical Society's founders were fellows of the Royal Society. Unlike their counterparts in the Geological or Astronomical Societies, they had little or no contact with the elite of scientific London. Sturgeon, who chaired the opening meeting, was a lecturer at the East India Company's Military College at Addiscombe, William Leithead was an operative chemist, and John Peter Gassiot was a wealthy wine merchant with a passion for electricity. Their links were with the Adelaide Gallery, where Sturgeon also lectured, and with London's instrument-makers, rather than with the Royal Institution and the gentlemanly specialists.

In terms of institutional precedents, the London Electrical Society had far more in common with the short-lived London Chemical Society of the mid-1820s than with the prestigious gentlemanly societies. The Chemical Society was founded in 1824 following enthusiastic correspondence in the *Chemist* concerning the need for such a body.[3] The enthusiasts were artisans and mechanics anxious to improve themselves through education in chemistry. They aimed at forming a small and self-contained association wherein they could instruct one another in the science. It was to be "a society of young chemists, who might, at their common expense, purchase chemical tests, instruments, &c., as they want them, and that without bringing down ruin on any of them."[4] Others proposed that a monthly subscription of a shilling should be devoted to purchasing books and apparatus selected through a ballot of all the members.[5] In many ways, the projected organization was to be a typical artisan gathering.[6]

At the first meeting of the London Electrical Society, it was "proposed by the chairman, seconded by Mr. Gassiot, and resolved that the Society be denominated 'The Electrical Society of London' its objects to be the experimental investigation of Electrical Science in all its various branches and its advancement not only by pursuing original paths of investigation, but also by testing the experiments of other enquirers."[7] Gassiot was elected as honorary treasurer and Thomas Patrick as honorary secretary. A provisional committee was set up to prepare a list of temporary regulations for the administration of the society until a formal structure could be set up.[8]

Over the next few months meetings of the society were held at regular intervals at E. M. Clarke's "Laboratory of Science" in the Lowther Arcade. Meetings consisted of the reading and discussion of papers by members and their guests. The provisional committee reported on its deliberations and produced a list of rules for the regulation of the society, which were published in the *Annals of Electricity*, edited by Sturgeon. The rules established the format of meetings, announced the membership fees for the different categories of members, and set forth the conditions to be fulfilled before the society was to reform its administrative structure on a permanent basis. The society's meetings were to consist of the reading of papers submitted to the secretary. Two categories of members were established: resident (being defined as living within twenty miles of London) and nonresident. The membership fee was set at two guineas for resident and one guinea for nonresident members. Finally, the rules stipulated that elections for a permanent council and officers would not be held until the society had acquired fifty members.[9]

The infant society's first annual general meeting was held on 7 October. It was an optimistic occasion. Over the summer the increase in attendance at the weekly meetings had encouraged the management committee to search for more commodious accommodation. They announced that in future meetings would be held at the Lecture Room of the Adelaide Gallery. The Council of the Society for the Illustration and Encouragement of Practical Science (the gallery's proprietors) were warmly thanked for their liberality and their offer to make the gallery's electrical apparatus freely available for the use of the society's members.[10]

One other announcement was made at the meeting. The committee suggested "that it would materially assist the objects for which the society was originally instituted, if the number of resident members was increased to 100, previous to the election of officers."[11] Sturgeon spelled out the logic behind this decision in his chairman's address. The aim was to provide ample opportunity for the metropolis's professorial and gentlemanly elite to join the society and take an active role in its administration.

As Sturgeon pointed out, "hitherto, the management of the different philosophical societies of London has generally fallen on the same individuals. That the prominent and most efficient members of the Royal Society are also those of the Astronomical, Geological, and other Societies, and that it is impossible these gentlemen can devote any more of their valuable time to the services of any other, however important its object, or however much it may be required." The delay was designed to allow the "distinguished individuals as now adorn the councils of those societies"[12] an opportunity to involve themselves with the Electrical Society's activities.

Sturgeon remained sanguine, however, that the Electrical Society's future was rosy in any event. If the elite maintained their distance, the society was still assured the support of those "individuals previously unknown in the annals of science, who have, within the last few years, devoted their time as well as pecuniary means to the cultivation of electricity." The society would become "a parent to foster and cherish their investigations; a grand storehouse in which they may repose the rich productions of their labours, and a temple for their kindred spirits' resort."[13]

Sturgeon ridiculed the opinion that progress in electricity could be made only by the highly trained elite, "deeply skilled in experimental investigation," as being "as groundless as it is detrimental to the progress of any particular branch [of science]." Everyone was in a position to contribute to the advance of knowledge, even those whose interest in electricity was limited to fact gathering and the observation of novel and spectacular phenomena. "Let no one imagine that he cannot render science a service. I have already stated that electricity is a science of experiment and observation; and therefore, this Society cannot receive greater assistance than by the communication of such facts as, from time to time, come under the notice of its members."[14]

The tenor of Sturgeon's speech shows clearly the ambivalence of his relationship toward the scientific elite, those men who ran London's prestigious scientific institutions. While he was quite explicitly inviting their support, several features of his address were hardly calculated to gain it. The main thrust of his appeals for support was aimed at those who were outside elite metropolitan culture, and his barbed remarks concerning the pretensions of those "deeply skilled in experimental investigation" constituted a clear attack on the gentlemanly specialists. While they asserted that only their brand of specialized and disciplined expertise could generate scientific progress, Sturgeon claimed that the activities of lay observers were just as valid. Elite specialists, typified maybe in this instance by John Herschel, were firm adherents of empiricism but were emphatic that such observation could have value only if conducted under the direction of a trained philosopher.[15]

Again, the position of the London Electrical Society's leadership bears comparison with the pronouncements a decade earlier of the *Chemist* and the Chemical Society's activists. They, too, were committed to the notion that experiment should be an egalitarian enterprise open to all. Thomas Hodgskin in the *Chemist* was particularly virulent in his attacks on elite practitioners who pandered to the aristocracy.[16] Others among the first promoters of the Chemical Society insisted that there was nothing impractical about the notion that the members could teach themselves with no need for elite guidance.[17]

There were organizational similarities too. The London Electrical Society never appointed a president, and the chair was elected by ballot at each meeting.[18] The Chemical Society's founders also considered such a structure. This is not to suggest that their organizations were identical or that the issue was one of a straightforward distinction between artisan and bourgeois cultures. The London Electrical Society, while in practice never appointing a president, clearly articulated an intention of so doing. The Chemical Society, on the other hand, did eventually appoint a president and a council.[19] Tensions concerning the proper way of organizing such institutions obviously existed within both societies.[20]

Sturgeon presented an interesting genealogy for his claims and for the fledgling society. The inspiration for the society's foundation had come from no less controversial a source than Joseph Priestley, "the first electrician of the age." In the preface to his *History of Electricity*, Priestley had argued that electricity, "the youngest daughter of the sciences," should lead the way in the division of scientific labor:

> Let philosophers now begin to subdivide themselves, and enter into smaller combinations. Let the several companies make smaller funds, and appoint a director of experiments. Let every member have a right to appoint the trial of experiments in some proportion to the sum he subscribes, and let a periodic account be published of the result of them all, successful or unsuccessful. In this manner, the powers of all the members would be united, and increased.[21]

Sturgeon presented the London Electrical Society as the culmination of Priestley's dream. The science of electricity was a collective enterprise that could best be pursued by the combination and "united labours" of like-minded men.

Without such a combination, Sturgeon suggested, electricity would be "left to the caprice of individuals" who, "having fettered themselves to certain hypotheses, appear more intent on advocating particular theories than in furthering the cause of science." The only useful role to be played by such "hypothetical incongruities" was to "show how cautious we ought to be in receiving as axioms the opinions of individuals, however eminent they may be esteemed for their scientific attainments." Without combination and collaboration, electricians would be "left to flutter on the wings of vain hope, and ponder away their valuable time in whimsical hypotheses, which have no reality in nature."[22]

The choice of Priestley as father figure was potentially dangerous for the London Electrical Society. His reputation remained tarnished by heterodox religious opinions and his outspoken support for the French Revolution. In view of his own statements it was very difficult to dissociate his natural philosophy from such dubious associations.[23] Writing during the 1840s, the prominent Whig politician Lord Brougham had to work hard to make such a distinction:

> In turning, however, to recount the events of [Priestley's] life, we make a somewhat painful transition from contemplating in its perfection the philosophic character, to follow the course of one who united in his own person the part of the experimental inquirer after physical truth with that of the angry polemic and fiery politician, leading sometimes the life of a sage, though never free from rooted and perverted prejudice—sometimes that of a zealot against received creeds and established institutions, and in consequence of his intemperance, alternately the exciter and the victim of persecution.[24]

Following Brougham's cautious dissection of his career, only Priestley's claim to the discovery of oxygen remained of his philosophical reputation. His *History of Electricity* was dismissed as "a careless and superficial work, hastily written . . . and the original experiments afforded no new information of any value."[25]

The centenary of Priestley's birth had indeed been commemorated a few years before the founding of the London Electrical Society. A gathering of some of the "big guns" of London science had assembled at the Freemason's Hall on 25 March 1833 to toast the "founder of pneumatic chemistry." Like Brougham a decade later, the distinguished gathering were unanimous that Priestley's claim to scientific immortality lay almost entirely in his discovery of oxygen. Many of those present solemnly congratulated themselves on their ability to rise above mere party politics and praise Priestley without being tainted by his doctrines.[26]

Priestley was therefore a dangerous, or at least ambivalent, father figure for the London Electrical Society. This was doubly the case when what was being cited was his view concerning the need for association and combination for the progress of science. By the 1830s the dominant ethos had shifted away from Enlightenment ideals. Scientific discovery and progress were now held to emerge from the workings of isolated genius rather than from dubious cabals such as the Lunar Society with which Priestley had been associated.[27] It would have been very easy for Sturgeon's audience to read his remarks as an attack on the Royal Institution, where the cult of individual genius was even then being fostered by Michael Faraday, who was following in the footsteps of his mentor Humphry Davy. They would certainly have been aware that Sturgeon was even then embroiled in dispute with Faraday in the pages of his *Annals of Electricity*.[28]

In any case, Sturgeon's appeals for collaboration and the participation of untrained observers were a far cry from the pronouncements of a scientific gentleman such as John Herschel. In works such as the *Preliminary Discourse* the latter emphasized the role of observation in the sciences, but the main thrust of his argument was that such observation was useful only if conducted in a disciplined and methodical manner.[29] There was no room for the haphazard accumulation of fact in his schema. Neither was there any room for the kind of apprenticeship into the mysteries of electricity that members of the London Electrical Society held to be necessary. He inveighed against the "whole tendency of empirical art" to "bury itself in technicalities, and to place its pride in particular short cuts and mysteries known only to adepts."[30]

For Sturgeon, one of the main arguments in favour of association was that only thus could the skills and practices of experienced electricians be transferred to others. The management of electrical instruments and apparatus was regarded as a craft that could not be adequately conveyed without direct contact. Others in the society shared Sturgeon's opinion. The popular science writer Henry Minchin Noad proudly advertised his membership in the society on the title page of his *Lectures on Electricity*.[31] He held that such a forum was essential for both the production and the dissemination of knowledge:

> At the present time, peculiar advantages are laid open to the tyro. An electrical society has been recently established in London, and if the future may be prognosticated from the present, its success will be triumphant; rarely a month passes, without some new and important fact being announced, or some new apparatus being exhibited; conversation is unrestricted; and here the beginner may, from the experience of the more advanced enquirer, get his difficulties removed . . . the information derived from books, though it may do well for the closet, will generally be far from satisfactory in the laboratory; here more detailed instruction, particularly in manipulation is required, and it is by actually witnessing the various operations performed that the necessary information can be acquired; hence the great advantage of a society, in which there is a community of taste and feeling, and in which knowledge is unrestrictedly communicated.[32]

The aim of the Electrical Society was to promote the science of electricity by initiating new members into the art of experimental manipulation.

From this perspective, the "technicalities" and "particular short cuts" condemned by Herschel were an integral part of electrical experimentation. Electricity could be made open to budding new practitioners only by the kind of association that the London Electrical Society hoped to achieve. In the terminology of social constructivism, this might be interpreted as a recognition by these practitioners of the crucial role of enculturation in the transmission of skill. Such skills were inarticulable and could be passed along only by direct contact between participants.[33] In contemporary terms, Noad recognized that

4.1 Frontispiece of Henry M. Noad, *Lectures on
Electricity, Comprising Galvanism, Magnetism, and
Electro-magnetism* (London, 1844), showing an
idealized electrician's laboratory. At center stage
is an Armstrong hydro-electric machine. There is a
Cooke and Wheatstone five-needle telegraph in
the background.

experiment, too, required something like the artisanal property of skill celebrated by radical commentators. On this reading, what the London Electrical
Society offered to its members was something very much like an apprenticeship in the craft of electrical experimentation.

## SPREADING THE WORD

The main activity of the London Electrical Society was, of course, the organization of regular meetings where members and others would deliver papers or
perform experimental demonstrations. Meetings were held on Saturday evenings in the Adelaide Gallery's lecture room. Typically two or three papers

would be delivered during the course of a session, followed by discussion. New members would also be introduced to the society at these meetings. The society published its transactions; initially, only selected papers were published, but it was soon decided instead that abstracts of all communications to the society would be published. Reports of the society's activities appeared on a regular basis both in natural philosophical journals such as the *Philosophical Magazine* and in more general publications such as the weekly *Literary Gazette*. William Sturgeon's *Annals of Electricity*, in particular, paid much attention and devoted much space to the new institution and its activities.

The papers presented at the meetings, as listed in the published *Proceedings*, varied enormously in range, although the pool of authors presenting them was initially quite small. A large number of the papers presented concerned themselves with the practical matter of improving apparatus and instrumentation. Such a concern was typical of the electricians' interest in making the technology of electricity easily manipulable and available to all. The first contribution recorded in the *Proceedings*, at the start of the General Meeting of 7 October 1837, was an exhibition by Thomas Bradley, superintendent of the Adelaide Gallery, of a thermoelectric battery.

Bradley's exhibition was a typical example of the society's activities with regard to instrumentation. The apparatus he presented was a battery of fifty-six pairs of elements embedded in plaster of paris. His aim, on the one hand, was to use this equipment to elucidate a disputed point in electrical theory. Was the electricity produced in such a battery solely the result of thermoelectricity, or did voltaic (i.e., chemical) action also play some role? The plaster of paris was intended to insulate the elements from each other so as to make voltaic action impossible. At the same time, of course, another function of the apparatus was to provide a spectacular display of thermoelectricity.[34]

Other contributions that fit easily into this category included E. M. Clarke's display of a new magnetoelectric machine; descriptions of a new version of the voltaic battery by William Leithead; accounts of more convenient laboratory apparatus such as Golding Bird's instrument for breaking contact between the primary wire of an electromagnetic coil and an electrometer "without the interposition of manual labour"; or George Bachhoffner's contribution on the different arrangements of the electromagnetic coil and the use of a spiral conductor in increasing the power of a voltaic circuit.[35]

Clarke had been developing new versions of the magnetoelectric machine since 1835. The model he demonstrated to the society on 4 September 1838 and the experiments he performed with it were presumably similar to those he had previously published. With this machine, which consisted of two coils of wire that could be rotated in front of a powerful permanent magnet, Clarke could demonstrate a spectacular array of experiments. When he employed his intensity armature, which consisted of fifteen hundred yards of fine copper wire, the "effect . . . on the nervous and muscular system is such, that no person out of the hundreds who have tried it could possibly endure the intense agony

*To render soft iron magnetic.*

Fig. 8. A, a piece of iron bent as in the figure. B, a soft iron keeper, which adheres to the iron on the connexion being made as represented, so long as the machine is in action.

Fig. 8.

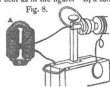

*To obtain sparks of various colours by the use of different metals.*

Remove the break, and substitute the brass piece B. Into the small hole insert a piece of wire C, of any metal, for instance gold. Let the extremity of the spring Q be also of gold. On rotating the machine, sparks of a purple colour will be obtained.

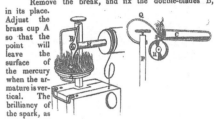

*To exemplify the disadvantages attending the mercury flood.*

Remove the break, and fix the double-blades B, in its place. Adjust the brass cup A so that the point will leave the surface of the mercury when the armature is vertical. The brilliancy of the spark, as

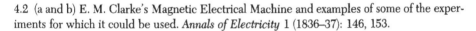

4.2 (a and b) E. M. Clarke's Magnetic Electrical Machine and examples of some of the experiments for which it could be used. *Annals of Electricity* 1 (1836–37): 146, 153.

it is capable of producing."[36] His quantity armature, a coil of forty yards of thick copper bell-wire, gave "large and brilliant sparks, sufficiently so that a person can read small print by the light it produces."[37]

Contributions such as Clarke's show clearly the ways in which members of the London Electrical Society regarded experiment as performance. The aim of such performances was to produce spectacular display as much as to demonstrate new phenomena. They were also geared toward utility. One of the potential markets that Clarke envisaged for his machine was medical practitioners, and some of its features were specifically designed with that market in mind. Electricity could cure the sick as well as dazzle audiences. Golding Bird, himself a young physician at Guy's Hospital, also aimed his device at a medical market. William Sturgeon's exhibition of a "contrivance for multiplying the discharges of an Electro-magnetic Coil, so as to cause more frequent sparks and shocks," was also designed for the production of spectacular displays in the lecture theater as much for as the production of new experimental knowledge.[38]

The interest of members of the London Electrical Society in instrumentation and its uses was not restricted to the improvement of laboratory apparatus but embraced the wider role of electricity in society. The utility of science for industry was a common rhetorical trope. Such arguments were of rather dubious status among the scientific elite, who were far more concerned to privilege the status of science as an abstract and disinterested body of knowledge.[39] Such distinctions were of little concern to members of the London Electrical Society. Martyn Roberts's "On the Application of Galvanism to the Blasting of Rocks" or Coombs's "Account of Messrs. Davenport and Cook's Electromotive Machine" were perfectly acceptable and clearly regarded as having the same status as a contribution such as Sturgeon's "Experimental and Theoretical Researches in Electricity &c." During the period 1837–39 many of these "utilitarian" contributions concerned issues such as the design of lightning conductors (especially for use on ships). They were aimed at a particular audience within the naval establishment. By late 1839–40 this focus was replaced by a more manufacturing-oriented interest in the use of electricity in such processes as engraving and plating.[40] Demonstrations of the utility of electrical science made effective displays, as many of the Adelaide Gallery's exhibits illustrated.[41]

These kinds of demonstrations, however, were not entirely unproblematic. Coombs's account of Davenport's and Cook's machine, for example, caused considerable controversy. At this meeting, on 17 July 1838, Coombs first of all demonstrated a working model of the "electro-magnetic locomotive machine." The locomotive, pulling a total weight, including the batteries, of seventy-three pounds, traveled around a forty-three-foot circular track at the speed of about two miles an hour. Coombs assured his audience that the performance would be much improved had not the Adelaide Gallery's supply of chemicals for charging the batteries run out.[42]

The controversy arose when Coombs proceeded to give a historical account of the development of electromagnetic machines, claiming for Davenport the distinction of having been the first to produce rotary motion by means of electromagnetism. Davenport's machine had been invented in 1834, but both William Sturgeon and William Ritchie had published details of electromagnetic rotary machines as early as 1832.[43] As the *Literary Gazette* correspondent put it, "the reading of the paper gave rise to a warm discussion." An unnamed member of the society's managing committee (quite possibly Sturgeon himself) described the paper as "unphilosophical" and objected strongly to its having been read before the society. The *Literary Gazette* condescendingly advised the "young and important Society" to have a care to avoid such controversies in future.[44] Here, as in the case of the bitter dispute between Clarke and Saxton, priority in invention and the demarcation and differentiation of candidate inventions were delicate and hotly contested issues.[45]

Another interest that may be reconstructed from the society's *Proceedings* was in cross-disciplinary applications—in other words, extensions of electrical science and its authority into discourses that had already been appropriated by other disciplines. The leadership of the London Electrical Society were more than happy to describe their science of electricity as the "queen of the sciences" and to promote incursions across disciplinary boundaries. The closing remarks made by H. M. Noad in his *A Course of Eight Lectures on Electricity, Galvanism, Magnetism, and Electro-magnetism* are typical:

> when we reflect on the prospects, which are held out by the attentive study of electricity, of eventually giving us a closer insight into the operations of nature as connected with the animal, vegetable, and mineral kingdoms—when we consider the almost certainty of its usurping, at no very distant period, the place of steam, as a mechanical agent, and thus being made, in the most extensive manner, subservient to the uses, and under the control of man, little more inducement, will, I imagine, be wanting to increase the body of its cultivators, and to assign it, its proper place, among the most important of the physical sciences.[46]

Such superpositions of natural and useful knowledge were commonplace in the rhetoric of popular science and place members of the London Electrical Society such as Noad firmly in that context.

The status claimed for electrical studies by members of the London Electrical Society may be compared to the far more humble role accorded to such investigations by William Whewell. The polymathic reformer of the Cambridge mathematics tripos, coiner of the term *scientist*, and future master of Trinity College aimed to classify and tame the sciences.[47] The *History of the Inductive Sciences* (1837) relegated electricity, magnetism, and galvanism to the category of mechanicochemical sciences. They appeared third on a scale that accorded first place to the mechanical sciences. Whewell relegated electrochemistry to an even more lowly position, reserving his discussion of its history to the section on chemical sciences.[48]

Whewell admitted that the electrical sciences presented great difficulties to the taxonomist. He suggested that the problems surrounding galvanism's place in his scheme were a consequence of the lack of clarity in its central ideas, particularly the idea of polarity in galvanism.[49] Whewell regarded the electrical sciences in terms of the contents and clarity of their ideas. As a result he perceived electricity to be problematic. Noad and others like him regarded electricity in terms of its instrumental, technological content and its ability to subsume other disciplines. As a result they perceived their discipline to be thriving and expanding.

One prominent contributor to this effect in the society's proceedings in its early days was the operative chemist Thomas Pollock.[50] Many of his papers were aimed at articulating and defending a vibratory theory of electricity, using that theory to posit links between electricity and other modes of force such as heat or chemical action. According to Pollock, electricity as manifested in voltaic batteries consisted in the vibration of an electric fluid that also resulted in the phenomena of heat and chemical action which attended the battery's action. Applying this theory, in a paper read before the society on 18 November 1837, he proceeded to explain the differences between the actions of intensity and quantity batteries.

According to Pollock, quantity batteries, which could heat, oxidize, and even burn metals while being unable to produce chemical action or cause shocks, did so because as a result of their intensity of vibration they gave more fluid to surrounding bodies than they abstracted from them. Intensity batteries worked in the reverse manner. Similarly, small batteries were more efficient in producing electromagnetism since their vibrations produced less of the heat that would otherwise interfere with the electrical action.[51] Theories such as these made electricity the ontological cornerstone of any natural philosophical inquiry, referring a diverse range of phenomena back to a fundamental underlying electrical action.

Another good example may be found in the papers delivered by William Leithead on the relationship between electricity and the prevalence of various diseases. His claims were based on an electrical model of nervous action that linked changes in atmospheric electricity to changes in the electrical constitution of the body. The failure of medical practitioners to appreciate this connection was attributed to their neglect of electricity. Leithead waxed lyrical concerning "the superiority of the satisfaction derived from administering to the comforts, or alleviating the sufferings of our fellow-creatures, over the wide-spread glory of a warrior," which his electrophysiological theory would bring about.[52]

In a subsequent paper, Leithead noted the necessity of oxygen to life and proceeded to suggest that the presence of oxygen was also essential for electrical action. He raised the question of whether oxygen could be regarded as a compound of light and electricity combined with a base, and hydrogen as a combination of heat and electricity with a base. The *Literary Gazette* applauded this production as tending to "strengthen the bonds of union of the sciences, chemistry, electricity, and physiology."[53] Again electricity was being promoted as the fundamental cause of natural phenomena and the missing link required to forge the sciences together.

Probably the most notorious example of the London Electrical Society's cross-disciplinary incursions was the episode of Andrew Crosse's electrical production of insects—the *Acarus crossii*—in late 1836.[54] Crosse, an enthusiastic amateur electrician who had only recently gained nationwide prominence following his lionization at the Bristol meeting of the British Association for the Advancement of Science in 1836, had been carrying out a series of experiments to determine the effects of small, constant currents of electricity over long periods of time on mineral samples. Insects appeared inside the apparatus, apparently emerging from the mineral samples themselves. These results, when they were reported in the press, caused some controversy, since they were widely perceived as bearing on the highly contentious question of the origins of life and the legitimacy of natural, as opposed to theological, answers to that question.

The first detailed account by Crosse himself, who had been a member of the London Electrical Society almost from its inception, of the experimental procedures that had produced the *Acarus crossii*, was in a paper delivered to the society.[55] It is not clear whether Crosse's results convinced members (in fact some of the medical contingent were very hostile toward Crosse in the medical press), but they were prepared to count such work as a legitimate activity. This was in stark contrast to the attitude of Michael Faraday, who made every effort to disassociate himself both publicly and privately from the episode.[56]

Faraday was perfectly well aware of the perils of transgressing disciplinary boundaries, particularly on dangerous and contentious issues such as the production of life. Such worries were not shared by members of the London Electrical Society who took work such as Crosse's to be a legitimate part of their discourse. When W. H. Weekes in 1842 successfully replicated Crosse's experiments, C. V. Walker, then honorary secretary, routinely included his paper in the list of the year's important contributions, along with "The Voltaic Process for Etching Daguerreotype Plates, by Prof. Grove . . . [and] The description of Voltaic Apparatus, by Mr. Joule."[57]

One interesting feature of much of this cross-disciplinary work, which casts light on the electricians' tendency to regard their apparatus as *constituting* nature, is the extent to which experiments were quite explicitly conceived of as reproductions of nature. Both Andrew Crosse and Robert Were Fox, in their experiments on the formation of mineral structures by means of electricity, were straightforwardly trying to experimentally reproduce the natural environment in which such structures might have been formed.[58] In his work on electrobiology, Alfred Smee attempted to electrically replicate the sensory phenomena of sight, smell, taste, hearing, and touch by means of electrical apparatus.[59]

According to the ever enthusiastic *Literary Gazette* correspondent, Crosse, in his ongoing experiments on crystallization, "works on a grand scale, and wisely endeavours to imitate nature as closely as possible."[60] His experiments were certainly gargantuan in scale. The water battery for the experiments about which the *Literary Gazette* enthused consisted of three thousand pairs. They were designed to investigate the effects of passing very low intensities of electricity through different minerals immersed in various salt solutions. The aim was to determine whether this would result in the artificial production of crystals through electricity over a period of many months.

The results of these experiments were spectacular. They produced not only the insects mentioned previously but also beautiful conglomerations of crystals. In his first communication to the Electrical Society, Crosse described the ways in which he had produced "perfect rhomboidal crystals of selenite . . . which exactly resemble nature," and "covered a piece of quartz with crystalline sulphate of lead."[61] A year later he proudly displayed actual specimens of his productions to the society. Crystals of selenite, sulfurets of zinc, and acetates of copper could all be produced by electrical action. Crosse held that "every substance found in the earth, gems or what not, can be formed in the union of pressure, heat, electricity, and absence of light." The *Literary Gazette* drily commented that they preferred "Mr. Crosse, as a crystal-maker, to Mr. Crosse the insect-maker, as a *nomme de guerre*," but nevertheless applauded his manifestation of "the constant characteristics of a true philosopher."[62]

Experiments such as these illustrate how the Electrical Society's technological and cross-disciplinary concerns meshed together. These men regarded electricity as the universal fluid that underlay all physical action. As Crosse expressed it, electricity was "the secondary cause of every change in the animal, mineral, vegetable, and gaseous systems."[63] In other words, nature operated in much the same ways as did the electrical machines and instruments that formed the stock-in-trade of members of the Electrical Society. Understanding the technological details of electrical instruments was therefore a good way of finding out how the world worked. Spectacular displays of the actions of electrical machines were simultaneously displays of nature.

WITNESSING THE WORLD

Holding regular meetings at which papers and experiments were presented and discussed was not the only activity in which the London Electrical Society and its members were engaged. Particularly under the auspices of their wealthy wine-merchant treasurer, John Peter Gassiot, informal "Electrical Soirées" were occasionally held. One such event was held at the parochial schoolroom, at Clapham Common, at the conclusion of a series of lectures held there with Gassiot's support. The evening was devoted to the spectacular display of electrical phenomena. A series of a hundred Grove cells were used to demonstrate the heating powers of electricity by brilliantly igniting nine feet of iron wire; samples of electrotypes, electrotints, electrochromes, and electroplating were on show; some of Crosse's crystals, including carbonate of lime "like a group of white fir trees," could also be seen. In short, "every thing connected with this interesting science" was on display.[64]

A similar event was hosted by Gassiot a year later to celebrate a visit to London by the eminent Genevan electrician Auguste de La Rive. The highlight on this occasion was the use of the hundred pairs of Grove cells to produce a spark between carbon points:

> To look at it direct was painful. Its effect, however, we fully appreciated, by observing the brilliancy it imparted to the natural colours of foreign moths and butterflies in a case suspended against the wall. Had they been in fluttering existence, winging their way through tropical sunlight, they could not have looked more bright or beautiful. Another pleasing proof of the power of the electrical light was the distance, through the window, it penetrated the outer darkness, shooting over the lawn; but now softened into the sweetest moonlight, and yet clothing the shrubs and turf with intense green.[65]

Again, the full panoply of electrical effects was on display, including a Wheatstone electromagnetic machine, electrotypes, and specimens of the *Acarus crossii*.

These extravagant displays were not simply frivolous. They had a role to play in defining the electrical world and its possibilities. In the same way that the

massive exhibitions of the second half of the century made sense of technological progress, shows such as Gassiot's celebrated and circumscribed the achieve-
ments of electricity.[66] Showmanship was an integral feature of early Victorian
technological progress. It was through display that the possibilities of electrical
machines and products could be made visible and fetishized. Gassiot's
"soirées" showed how nature was electrical and could therefore be harnessed
and commodified.

Gassiot had access to resources that few others could match. His large private income allowed him to purchase and construct apparatus on a massive
scale. It was this that made it possible for him to organize spectacular extrava-
ganzas such as his "electrical soirées."[67] His resources could also be mobilized
for other purposes. On more than one occasion his house and laboratory at
Clapham Common became the forum for collaborative ventures by members
of the Electrical Society to carry out experiments with huge and expensive
apparatus far beyond the means of most electricians.

One such collaborative experimental project commenced at Clapham Common on Saturday, 16 September 1838. Gassiot, Thomas Mason, William Sturgeon, and Charles V. Walker gathered together to conduct a number of experiments using an array of 160 Daniell cells as improved by modifications
suggested by Mason. According to Walker, as a result of Mason's modifications
the cells could be constructed at a cost of no more than a shilling each.[68] The
individual cells consisted of half-pint earthenware vessels in which cylinders of
copper and zinc were placed, the zinc being on the inside. The zinc was
wrapped in brown paper so as to separate the salt solution in which it was
immersed from the copper sulfate solution surrounding the copper cylinder.
The batteries were placed on slips of glass covered in shellac to ensure their
insulation and were arrayed in groups of twenty cells in series.

Walker, in his account of the experiments presented to the Electrical Society
on 16 October, went into considerable detail concerning the battery's construction. As he reminded his audience, "often, in tracing the labours of our
predecessors, we are prevented [from] repeating their experiments, through
want of such precise description; or if we venture to repeat them, how often do
we arrive at conclusions different from theirs, not through careless manipulation on either hand, but simply because we see only in part their mode of
proceeding, and hence omit some necessary element in the apparatus."[69]
Walker was keen to emphasize to his audience the *economy*, the *power*, and the
*convenience* of this form of the apparatus. These were the attributes most
highly prized by practical electricians.

An episode such as this collaborative project provides us with a way of looking at the aims of experimentation from the point of view of members of the
Electrical Society. It shows what kinds of skills were required and how the
practical division of labor could be accomplished. Given the controversy
that soon surrounded at least some aspects of this experimental collaboration,
it also highlights some of the fragilities of experimental culture. Walker's detailed reportage to the society's meetings of the aims and conduct of these

experiments makes clear the practical, philosophical, and social problems of experimentation.

From the perspective of a culture for whom a supply of electricity is usually only the flick of a switch away, it can be hard to imagine how difficult it might be to secure such a resource. Walker's first priority in his communication, as already noted, was to provide detailed accounts of the supply of power. Many commentators on electrical experiments during this period noted that one of the major difficulties facing the hopeful experimenter was the production of a steady, constant supply of electricity. George Bachhoffner lamented to Sturgeon the difficulties of making a battery that was both constant and economical.[70] Economy was crucial. John Shillibeer, advocating his own combination, emphasized that "it requires but little food, and with that little will perform a good honest day's work."[71] As Walker himself pointed out, only the development of the Constant Battery by Daniell during the mid-1830s provided an adequate resource with which to counter this problem.[72] Following Daniell's invention of his battery, Gassiot even suggested to Faraday that Daniell should be canonized for his achievement.[73]

But Daniell's battery was by no means readily available. Daniell's own account of the battery's construction and its virtues had been published in the *Philosophical Transactions* of the Royal Society.[74] While this was without question the most prestigious possible locus of publication, it was also the most exclusive. The *Philosophical Transactions* were not widely disseminated beyond the confines of the Royal Society's fellowship, and few electricians would have had easy access to them. As a result the details of Daniell's battery were not at well known among many of his contemporaries.[75] This was one of the motives for Walker's own detailed account of the batteries used on this occasion.

The first set of experiments to be described took place as the different sets of twenty cells were being individually charged with copper sulfate and salt solution. In this sequence he tested the decomposing, deflecting, and heating powers of each set of twenty cells to display the batteries' capacities and to check the integrity of their performance. As Walker noted, the initial results of the decomposition experiments suggested major variations among the batteries.

In some cases the experimenter could avoid discrepancies by checking the connections between the individual cells. In others it was found that time was needed for the solutions to saturate the brown paper separating the zinc from the copper before constant results could be achieved. In some recalcitrant cases it was eventually noticed that copper sulfate was leaking through the brown paper, through holes made in the process of preparation, and affecting the results. These holes were plugged with sealing wax before further experimentation took place. In short, preparing the batteries for experimentation was an extensive and labor-intensive process that required detailed practical knowledge of the probable result and of the obstacles that might prevent the batteries' proper operation.

Another obstacle was the absence of standardized apparatus and supplies. Walker warned his audience that chemicals obtained commercially were rarely pure, and that as a result the unwary experimenter would frequently find that such supplies, purchased at great expense, were inadequate for proper experimentation. This warning also applied to commercially acquired instruments. Walker noted of the measuring apparatus used in this sequence of experiments "that the graduated inch of no two Voltameters was the same; so that it was necessary to assume one as a standard, and calculate the gases evolved in the others by this standard."[76] In presenting results that could at least potentially be replicated, therefore, the experimenter had to provide minute and detailed descriptions of the equipment used. Without such detail, in the absence of agreed standards of apparatus, meaningful comparison of results from different sources was impossible.[77] Attempts to produce standard apparatus were common during the 1830s. Without such standards experimental knowledge could not easily be transmitted outside the closed boundaries of small groups such as the Electrical Society.[78]

The main aim of this afternoon of experimentation, once experiment and modification had assured the experimenters that the batteries were performing adequately, was to compare the relative decomposing, deflecting, and heating powers of increasingly large arrays of cells. To this end, they tested the batteries in multiples of twenty from one to eight, using voltameter, galvanometer, and lengths of wire. In this way it was shown that decomposing power decreased with the size of the battery, deflecting power stayed comparatively constant, and heating power increased. Heating power was measured as the length of platinum wire that could be made visibly red-hot when it was connected to the poles of the battery. The experiment seemed to indicate that heating power was directly proportional to the number of batteries.

The final sequence of experiments did not take place until after half past ten at night. In these experiments the entire battery of 160 cells was put in operation so that its effects could be shown off. Mercury was brought to combustion; the spark from charcoal points was used to fuse glass and steel, and the "physiological effects were exceedingly powerful: it required the strongest nerves to volunteer the experiment."[79] One final result was of "peculiar interest." Walker described it as follows:

> When the ends of the main wires were placed across each other, (at about one or two inches from their extremities,) not touching, but with an intervening stratum of air, the striking distance, through which the electricity passed, producing a brilliant light, that wire connected with the *positive* end of the battery became red-hot, from the point of crossing to the extremity. The corresponding portion of the other wire remained comparatively cold.[80]

This novel and unexpected result, which as Walker noted, *"perhaps* was then observed for the first time," was to give rise to some controversy, focusing on the authorship and proper conduct of experimentation.

This new phenomenon, which had been presented by Walker as an incidental, though interesting, by-product of the main sequence of experiments, rapidly changed its status in others' accounts of the collaborative experimental project. In a letter to Edward Brayley, one of the *Philosophical Magazine's* assistant editors, John Peter Gassiot presented the observation as the climax of the day's experimentation. The note was duly published, with a footnote by Brayley certifying that he had personally witnessed the "remarkable and hitherto . . . unnoticed difference in the temperature of the positive and negative electrodes" in the company of Faraday, Daniell, and Jonathan Pereira. In this account the discovery of the new phenomenon was firmly attributed to Gassiot himself.[81]

William Sturgeon had also communicated news of the day's experimentation in a letter to Benjamin Silliman, editor and founder of the *American Journal of Science*.[82] His account focused on the spectacular displays of fire and sparks with which the day's work had culminated. He noted the "exceedingly curious and interesting" heating phenomenon but showed no sign of attributing any special significance to it. It was only one among the many spectacular effects that could be obtained from a large and powerful voltaic battery.[83]

His next communication to Silliman, however, placed a very different gloss on the day's proceedings. In this letter, dated 6 August 1839, he gave a very detailed account of the way in which the day's work had been organized and the labor divided, and he attributed the discovery of the new phenomenon firmly to himself. His stated motive for communicating these new details was the reported failure of various Continental philosophers such as La Rive to replicate the experiment and his apprehension that this was due to the lack of detail in Walker's account. His new revelations would, he hoped, be sufficient to ensure that American philosophers could carry out the experiment successfully.

According to Sturgeon, it had taken at least six hours to prepare the large battery for experimentation. Walker had commenced work on preparing the apparatus between eight and nine in the morning. Mason and Gassiot had set to work at about eleven, and he himself had arrived about half past one in the afternoon. The experiments did not commence until after three o'clock. While the majority of the decomposition experiments were proceeding, Sturgeon and Mason alternated as experimenters, while Gassiot marked the time taken for decomposition with a stopwatch and Walker recorded the results.

Sturgeon's account of the crucial experiment was heavily loaded. He had already noted that he had been requested beforehand to prepare a list of experiments, but that few of these had been possible owing to the time taken to complete the exhaustive decomposition tests.

> With regard to the experiment *in which I discovered* the great difference produced in the two polar wires, it was undertaken from the views which I had long entertained concerning the non-identity of the *electric* and *calorific* matter. . . . It was late in the evening before I had any opportunity of making the experiment. The rest of the party were engaged in something else at the time, and the battery was in series of one hundred and sixty pairs.[84]

He then detailed the procedures required to demonstrate the phenomenon successfully, emphasizing that he had drawn the attention of his coworkers to the experiment only after he had completed it himself.

Two features of Sturgeon's account are worth emphasizing. First, he claimed sole credit and responsibility for the actual conduct of the experiment. It had been performed by him alone with no assistance from his fellow experimenters, who were aware of the experiment only after Sturgeon had brought it to a successful conclusion. Second, the experiment was represented as having been premeditated. The discovery was not the result of accidental observation but rather the consequence of a deliberate test of theories already put forward by Sturgeon himself.

The experiment was represented as a decisive proof of the singularity of the electric fluid and its nonidentity with caloric. The heating of the positive pole was the result of the electric fluid's driving the calorific matter before it as it flowed down the wire. The caloric became trapped in the extreme end of the positive wire where it could move no further, thus causing the unusual heating effect. "Nothing can be more simple to explain; nor do I know of an experiment that tends more to support the doctrine of *one species* of electric matter only; and that it moves through the voltaic conducting wires, *from* the positive *to* the negative pole."[85] In this way the novel effect became Sturgeon's property. It was the result of his own experimental dexterity and the consequence of theories that he had already formulated.

Sturgeon's interpretation of events did not remain unchallenged. Charles V. Walker replied with an angry letter to Silliman, which strenuously challenged Sturgeon's account in almost every particular.[86] Walker denied outright that Sturgeon had any right to claim individual credit for the discovery of this phenomenon. His main complaint, as he expressed it, was "the frequent recurrence of the pronoun *I*" in Sturgeon's account of the proceedings. He suggested that for a variety of reasons, Sturgeon's claim to sole credit for the discovery was unacceptable.

His main assertion was that the collaborative nature of the experiments made any individual claim illegitimate:

> The sole object was to advance the interests of science, through the medium of the London Electrical Society, and not to found individual claims to individual experiments, when each by agreement was contributing his own share to the common stock; you may judge, therefore, of the surprise with which I saw the experiment in question, not only claimed by Mr. Sturgeon as his, but also as being undertaken from certain views which he had long entertained.[87]

Walker's implication was that the fruits of their joint labor could not be divided up that easily. Regardless of which particular individual had performed which task, credit for any discovery was to be shared equally. In any case, since the experiments had been carried out under the aegis of the Electrical Society, that institution was also entitled to any credit that might ensue.

Walker professed himself to be dubious of Sturgeon's assertion that the experiment was deliberately planned to test a particular theoretical claim. Had this been so, he suggested, Sturgeon would surely have revealed his plans to his fellow laborers and urged the company to carry out the experiment sooner.

If he had entertained these views, he had a marvelous manner of concealing the experiments he had based on them; we, in our innocence of what good things were in store, were plodding on through that extended series of experiments on decomposition, with such a battery as had *never* been excited before, and yet our chief man (for he was the only scientific man by profession among us) is unable to avail himself of the first opportunity that *ever* occurred to him of bringing his views to the test.[88]

Viewed from this perspective, Sturgeon was guilty of improper conduct on one of two counts. Either he had lied to Silliman in claiming that the experiment was preconceived, or he had deliberately withheld information from his fellow experimenters.

Walker's account of the way in which the phenomenon had actually come about also differed substantially from Sturgeon's. While Sturgeon claimed that all his coworkers were otherwise occupied when he conducted the experiment, Walker asserted that at least one other had been present and that the experiment was not preplanned. "With regard to the experiment in question, it appears to have resulted, like many others in all the sciences, from merely fortuitous circumstances. He and Mr. Mason were amusing themselves with the wires, and observing the length of the arc of flame, and the phenomenon of the heated electrode presented itself; but neither knew which electrode it was until they had examined."[89] Walker suggested that it was only after La Rive had failed in his attempt to reproduce the phenomenon that Sturgeon had recognized its significance and then decided to assert his own individual right to credit for the discovery.

Competing accounts such as these reveal some of the tensions of an experimental culture. Different participants even in the same experiment could have quite different notions of what was going on, and what the proper interpretation of events might be. The status of participants in the experiment was evidently negotiable. While Sturgeon regarded his coworkers as witnesses and assistants to his own discovery, Walker regarded himself and the others as equally engaged in the process of experiment and therefore as equally deserving of credit for any discovery. The witnessing of an experiment was a crucial feature of its validation, as evidenced by Gassiot's display of the new phenomenon to Faraday, Daniell, Pereira, and Brayley. To convince, the display of a new phenomenon had to be accompanied by the authoritative presence of those whose testimony was trustworthy. The dispute between Sturgeon and Walker, however, reveals some of the dangers inherent in the presence of such witnesses. The line between participant and observer could always be differently drawn.[90]

The dispute also highlights the issue of resources, in terms of time, space, and money. All accounts of the experiments agree on the point that their performance required hours of intensive labor by a number of participants. Disagreement arose over the question of the status and reward of the participants. The same point may be made concerning the space in which the experiment was performed and the apparatus used for the performance. The laboratory and the apparatus were the property of Gassiot rather than Sturgeon, who as a result did not have full control over their deployment. The result was ambiguity concerning the authorship of the experiment.

Such a difficulty would have been unthinkable in the case of Michael Faraday once he had established himself.[91] He had full control over the resources of the Royal Institution. There could be no question of his status with regard to his assistants. It was fully understood, for example, that Sergeant Anderson would be rewarded with cash for his services while Faraday would get the kudos.[92] Similarly, no one doubted that the laboratory of the Royal Institution and its apparatus were at Faraday's command. A successful claim to the authorship of an experiment required control over the resources used to perform the experiment.[93] The ambiguity of Sturgeon's relationship with his fellow workers, the laboratory, and the apparatus could be used to undermine his claims to authorship.

Electricians such as Sturgeon and his associates at the London Electrical Society regarded experiment as being primarily a matter of public display. Sturgeon and others who made their living through lecturing depended for their livelihood on their skills at public performing. This again made control over space and resources an important matter. Without access to these resources an electrician could not make a new display part of his repertoire. It is worth noting that Gassiot, who controlled these resources, was the one who first demonstrated the new phenomenon to the metropolis's elite and convinced them of its reality. Sturgeon was left trying to defend his claims in the obscure pages of the *American Journal of Science*, far removed from the centers of authority.

One general point about this experiment, however, brings out the sense in which it was typical of the electricians' culture of the London Electrical Society. Simply put, the experiment was performed *on* the apparatus as much as performed *with* the apparatus. The battery and the measuring equipment were objects of research. The aim of experiment was to gain a practical understanding of the ways in which these machines operated, and how their effects might be maximized. Experimenting with electricity was, for these men, a demonstrative art. It could be done successfully only by those who were intimately acquainted with their apparatus and its idiosyncrasies. Its aim was to display the powers of their machines and hence the powers of nature.

## A Failed Experiment

For the London Electrical Society, electricity was a populist, utilitarian, and interdisciplinary science. This was the evidential context that rendered their activities meaningful.[94] As such it was calculated to appeal to a wide variety of interests. Its members included wealthy businessmen such as J. P. Gassiot, medical practitioners such as Golding Bird, country gentlemen such as Andrew Crosse, and instrument-makers such as E. M. Clarke. By 1840 the society was apparently well-established. Regular meetings were held and reported upon in the popular press, and the *Proceedings* were published on a regular basis.

By 1843, however, the project had ended in utter and abject failure. The first sign in the records that something was amiss is the unexplained gap in the Minute Book for the period between 4 July 1839 and 10 March 1841. The Minute Book itself contains no explanation for the absence of any record for this twenty-month gap. The published *Proceedings* of the society continued until 19 May 1840, but they then stopped with equal abruptness and were not resumed until 20 April 1841.

By the end of the 1830s the society's membership, although apparently active, remained small. In 1839 the membership was no greater than eighty, of whom only about half were resident members. Reminiscing almost half a century later, Walker recalled that membership had peaked at seventy-six. As a result no attempt had been made to place the society's administration on a permanent footing. The attempts to recruit members of the metropolis's professorial and gentlemanly elite into the society's ranks had not been successful.

A crucial blow must have been the loss of William Sturgeon, who in 1840 left London for Manchester where he had been appointed superintendent of the Royal Victoria Gallery for Practical Science: an enterprise modeled on the Adelaide Gallery.[95] Sturgeon had been a frequent contributor to the meetings and was probably a frequent chairman of meetings. After his removal to Manchester, Sturgeon made no further contribution to the society's *Proceedings*. Absence from London did not preclude such contributions; some of the society's more prolific contributors such as Martyn Roberts, Robert Were Fox, or W. H. Weekes were nonresident members. It is clear that there was some connection between Sturgeon's departure and his quarrel with Walker.[96] Failure to come to agreement concerning the proper allocation of credit in experiment therefore had disastrous consequences for the London Electrical Society.

The first few entries of the resumed Minute Book show that for the approximately twelve months preceding the meeting of 3 March 1841, no meetings of the London Electrical Society had been held at all. The first meeting was itself a Special General Meeting, with J. P. Gassiot in the chair, held to consider the precarious financial and administrative position of the society. At this meeting and the next, Gassiot outlined the society's difficulties, estimating that by the end of the next year the society would be in debt to the tune of seventy-three pounds. An interesting indicator of the extent to which the society's sights had been lowered by circumstances was that this estimated debt was based on the assumption that at the end of the year the society would have only twenty resident members and a further twenty nonresident members.[97] This would suggest that the society's membership had more than halved during the hiatus in its existence.

The steps taken by the meeting in response to Gassiot's suggestions to restore the fortunes of the London Electrical Society were extraordinary. The provisional Management Committee was disbanded, and Gassiot himself resigned as honorary treasurer. Charles Vincent Walker, who at some stage during the past year had replaced Leithead as honorary secretary, was asked to continue in this post and take over in addition the duties of the treasurer and take any steps he saw fit to continue the society. The whole administration of the London Electrical Society was placed under his control. As the honorary secretary he was already responsible for editing the society's *Proceedings*. Finally, the seriousness of the situation was underlined by a call from the leadership for members willing to underwrite the London Electrical Society's accounts to the extent of up to ten pounds each in case of serious debt in the near future. Gassiot took the lead in putting his name forward.[98]

Charles V. Walker undertook his daunting task of restoration with considerable enthusiasm. He immediately perceived that if the revitalized society were to prosper, it required a far higher public profile. He undertook to make the quarterly issues of the *Proceedings* available on sale to the public at a price of 2s. 6d. Such a step would increase the society's revenue and bring its very existence to the notice of a wider audience.[99] He authorized the printing of 250 extra copies of volume 1 of the *Proceedings* "in order to promote the interests of the Society, by presenting copies to all the Scientific Institutions in the Kingdom, as well as to some of those on the Continent." Furthermore, he drew up a list of thirty-one bodies who should be sent copies of everything published by the society. The list included the Royal Society, the Royal Institution, the London Institution, and the Athenaeum Club.

At first this new strategy appeared to be successful. The entries in the Minute Book show that a constant influx of new members were being introduced at the society's meetings; letters of thanks were received from other institutions. These included correspondence from Faraday at the Royal Institution and Grove at the London Institution, and many from foreign institutions. The *Proceedings* show that a far higher proportion of contributions were from foreign philosophers. Walker's reports in the Minute Book certainly seemed to indicate that the society's fortunes were improving. The *Proceedings* were selling so well that a further print run would be required to satisfy the demand. At the next Annual General Meeting he informed the members that the London Electrical Society's debt was only £15.3.6: far less than had been anticipated twelve months earlier.[100]

A few months later Walker called another General Meeting of the society to report that he had taken the important new step of acquiring apartments for the society's own use, rented from the proprietors of the Royal Polytechnic Institution. He explained that to an extent the decision had been forced upon him owing to the fact that the Adelaide Gallery now remained open to the public during the evening and was thus no longer a convenient venue. He emphasized, however, that the decision represented a dramatic stride forward. It would now be possible to set up the society's library and its collection of apparatus on a permanent basis. Members would have free access to the Royal Polytechnic Institution and its facilities, and Walker was sufficiently optimistic concerning the society's future to offer free use of his own apparatus to the members. Rent for the new apartments would add twenty-five pounds per annum to the society's expenses, but Walker was confident that the increased membership would more than cover the extra outlay. A subscription fund was set up to finance the refurbishment of the apartments at no extra cost to the society.[101]

However, before the next Annual General Meeting, due on 8 April 1843, disaster struck. The accounts now showed a deficit of £159.0.9. Walker's strategy had ended in failure while costing the society far more than its annual income allowed. Most of the debt was owed to the society's printers, for although membership remained at twenty-six resident and twenty-four nonresident members, Walker had been authorizing print runs of about five hundred on publications to be sold to the public and distributed free to the influential. The constant influx of new members had only disguised the fact that old members were failing to renew their subscriptions. The society was now unable to pay the rent on its new apartments. Walker had no choice but to bare his breast before the membership. His final report was bitter:

> You placed me in almost despotic power over your interests: you gave me no Council to guide, and no Committee to direct; you entrusted the management of your affairs to my unexercised judgement, and the expansion of the Society to my limited influence. Two years' experience has taught me that I am not sufficient to bear this double burden. You have had my best exertions—would they had been better!—you have had my best wishes,—would they had been consummated!—and the result of the protracted experiment appears in the present condition of the Society. I had hoped for better things, I had encouraged a feeling of pride at the early promise of the Society, but this has been entirely dampened by the powerful assurance that there is a limit, beyond which my individual power to advance is of no avail.[102]

Walker complained angrily about lack of attendance at meetings and lack of support from the membership. He implied that one of the major factors underlying the society's present state was the membership's failure to introduce new members. There was no real option other than dissolving the society and setting up a Provisional Committee to deal with the outstanding debt.

It is difficult to determine the immediate cause of the society's final failure. Financial mismanagement by Walker obviously played some part: his predictions of the society's expansion proved far too optimistic. The efforts to recruit members from the elite ended in failure. Philosophers such as Daniell, Faraday, Wheatstone, Grove, or Herschel never joined. Without their experience in running the affairs of societies, the London Electrical Society was forced to compete with established institutions without a knowledge of how such an enterprise should be regulated. Grove did indeed present a paper at the London Electrical Society, read out by Gassiot, but there is no record of his ever having been a member.[103] Gassiot's resignation as treasurer and subsequent withdrawal from a large part of the society's affairs coincided with his own election as a fellow of the Royal Society. Gassiot, entering the elite community, was presumably anxious to distance himself from his associations with the London Electrical Society. The society's leadership had also obviously overestimated

popular interest in electrical science. The Adelaide Gallery itself was in financial difficulties—a a situation that forced the London Electrical Society into
additional expenditure on its own apartments. The Royal Victoria Gallery in
Manchester was to close within a year. The society's appeal to a popular and
egalitarian inductivism had alienated the elite while failing to attract a new
constituency.

The extent of the alienation is apparent in a letter from William Snow Harris
to Faraday: "Daniell & Wheatstone have blamed me for meddling with Mr
Walkers paper. But I do not know when so great a public question is concerned
that I have any pretention to hold my head so high as to think others beneath
notice—My Talents do not place me in the position which yours do or
Daniells."[104] This was a reference to Snow Harris's "Observations by W. Snow
Harris, Esq., F.R.S., &c., on a Paper entitled 'On the Action of Lightning Conductors' by Charles V. Walker, Esq., Hon. Sec. London Electrical Society,"
which was communicated by Gassiot to a meeting of the society on 20 September 1842 in response to Walker's paper of the previous month.[105] Daniell,
Wheatstone, and (by implication) Faraday were evidently explicitly elitist in
their choice of forum and protagonists. Participation in a debate at the London
Electrical Society was to be characterized as "meddling" rather than as proper
philosophical behavior.

Walker, however, was convinced that the constituency was there to be cultivated. Three months after the society's dissolution he applied the experience
he had gained as editor of their *Proceedings* to the launching of his own *Electrical Magazine*, which he explicitly claimed in an "Address to the Reader" as the
legitimate successor to the *Proceedings* of the London Electrical Society. The
attempt was short-lived. The *Electrical Magazine* ceased to appear in 1846.[106]

CONCLUSION

The London Electrical Society stood for a particular vision of the role of the
experimenter in society. Its practitioners privileged the variety, rather than the
singularity, of electrical effects and held that the performance of electrical experiments was in principle open to all. In this they stood in stark contrast to
Faraday, who notoriously maintained his laboratory as a very private space into
which only he and his staff of assistants would normally enter.[107] As the acrimonious exchanges between Sturgeon and Walker illustrate, however, an egalitarian perception of experimental practice and discovery had its problems. It
produced a context where the distinction between experimenter and audience
could become blurred, and where the allocation of credit was no longer clear-
cut. Different participants in the same experiment could have very different
views as to what was going on.

This chapter has provided an outline of the society's origins, the aspirations
of its founders, and the activities that characterized its proceedings. In its
origins, insofar as it was founded during a period when specialist scientific

societies devoted to the promotion of a particular discipline were proliferating, it might appear to have been part of that trend. A consideration of its early membership and the stated aims and aspirations of the founders prove that appearance misleading, however. Its founders and members did not on the whole come from the same constituency as provided the membership of new disciplinary institutions such as the Astronomical or the Geological Society. Their position was rather more marginal in the world of metropolitan science. To a degree they had more in common with the mechanics' institutes and the even more short-lived, radical Chemical Society. While the founders of more elite disciplinary bodies advocated a vision of science as a disciplined, expert vocation, the London Electrical Society's leaders adhered to a more egalitarian and craft-based vision of the scientific enterprise.

Their proceedings blurred many of the distinctions between science and art that members of the gentlemanly elite were anxious to foster. Electricity at the London Electrical Society's meetings or at the informal electrical soirees of their treasurer John Peter Gassiot was a matter of spectacular display as much as precise and accurate measurement. Their concern with the technology of electricity, the apparatus and instruments through which the electric fluid was made visible, was embedded in this culture of showmanship. The collaborative experiment described at length in this chapter, for example, was aimed at assessing and improving the capacity of a battery to make electricity visible to spectacular effect. That episode highlights, however, the fragilities of this egalitarian electricians' culture. The disputes that arose concerning the ownership of those experimental results demonstrate the difficulties of distinguishing between individual and collective claims to discovery and invention. It seems likely that the tensions arising from this experiment, leading as they did to William Sturgeon's resignation from the society, were instrumental in bringing about the London Electrical Society's downfall. Its membership proved unable to sustain it as a space for the practical realization of their particular vision of the business of electrical experimentation.

# CHAPTER 5

# The Right Arm of God: Electricity and the Experimental Production of Life

ONE HIGHLY significant feature of electricians' work during the 1830s and beyond was the attention devoted to the universality of electricity. In particular, electricians devoted much attention to the connections of electricity with life. Discussions and debates concerning these issues proliferated in a variety of settings, drawing on a wide range of resources. To bring the first part of the book to a close, this chapter discusses some of these debates to show how different accounts of electricity in its connections with nature and culture could be put together in different social settings. Such issues were often strongly contested and serve to highlight some of the fragilities of the electricians' culture and some of the larger society's hopes and fears concerning the future of electricity.

In 1843, William Sturgeon had published the lectures on galvanism that he had for the past few years been delivering in his new capacity as lecturer to the Manchester Institute of Natural and Experimental Science.[1] These lectures provided a genealogy for the experimental study and understanding of current electricity that contrasts strongly with the usual historical account. Sturgeon staunchly defended the primacy of Luigi Galvani's claims to the discovery of a specific animal electricity and dismissed Alessandro Volta's assertions to have refuted Galvani's discoveries as inadequate and unfair.[2] At best, Sturgeon argued, Volta's experiments showed only that electricity *could* be produced by the contact of two dissimilar metals, *as well* as being a product of the animal body.

Sturgeon utilized a variety of resources to construct a particular vision of the past and present state of galvanism and its cultural connections. It is not clear to what extent Sturgeon's published lectures reflected the contents of his public performances, or where those performances took place. If we accept the account of the lectures' history that he presented in the book's preface, they had been delivered regularly for a number of years. They were, he asserted, "arranged in nearly the same order as that in which I am in the habit of delivering them." They were delivered, presumably, to audiences at the Royal Victoria Gallery in Manchester, where Sturgeon was superintendent.[3] They had also possibly been delivered before London audiences prior to Sturgeon's move to the provinces.

It is worth noting the distinction drawn by Sturgeon between electricity and galvanism. He had previously published a course of lectures on electricity in which he had limited himself to a discussion of "ordinary" electricity as produced by the electrical machine and stored in the Leyden jar.[4] Galvanism, however, consisted of the electricity derived, on the one hand, from the animal body and, on the other, from the voltaic battery. Sturgeon also alluded to his lectures on "Magnetic Electricity," which formed yet another separate division. Clearly, the specificity of the different subjects was constituted by the different *machines* used to produce the electrical fluid rather than in terms of differences in the fluid's action. The galvanic and the electric fluid were not regarded as being different in themselves but were treated separately because they were *produced* through the use of different items of apparatus. This division of the subject was not unique to Sturgeon. Instrument-makers' catalogs of the period routinely divided their apparatus in much the same way into electrical, galvanic (or voltaic), and electromagnetic.[5] Electrical apparatus included machines, jars, and demonstration devices such as the thunder house; galvanic apparatus included batteries and decomposition instruments; electromagnetic apparatus ranged from devices to demonstrate electromagnetic rotations, such as Barlow's Wheel, to electromagnets, induction coils, and electromagnetic machines.

Galvanism, as represented by Sturgeon at least, was intimately linked to the processes of life. It arose directly, if serendipitously, from Galvani's interest in the relationship between muscular action and electrical agency. Sturgeon related the accidental observation, by Madame Galvani, that frogs' legs prepared for cooking twitched when sparks were drawn from the prime conductor of a nearby electrical machine. Being already interested in the links of electricity with muscular action, Galvani repeated and refined the observations, conducting a number of experiments which ultimately convinced him that electricity was produced between the muscles and the nerves of the frogs' legs. He came to regard "the whole animal frame as a natural electrical machine, in a continual state of excitement; and which, like the rubber and glass parts of the ordinary electric machine, has its muscular and nervous systems in different electrical conditions."[6]

Sturgeon then outlined Volta's objections to Galvani's experiments and the investigations that led him to the invention of the voltaic pile. He insisted, however, than none of Volta's experiments conclusively disproved Galvani's hypothesis of animal electricity. An "electric logician," he suggested with some irony, would have difficulty following Volta's reasoning. What Volta had done was discover a different source of electricity: "By Galvani's method the muscles and the nerves were the source of electric action; whilst by Volta's 'artifice,' (a method of experimenting first employed by Galvani,) the dissimilar metals constituted the electric source."[7] Sturgeon did give Volta some credit, however. Despite the fact that he "perverted" the application of his discovery by using it to attack Galvani, his work "put into the hands of philosophers the most valuable electrical apparatus they ever yet possessed."[8] This was, of course, the voltaic pile.

In subsequent lectures Sturgeon proceeded to give a variety of experimental illustrations of galvanism, drawing heavily on Galvani's own experiments, and also on the work of Galvani's nephew and main partisan, Giovanni Aldini. Galvani's experiments had been conducted almost entirely with frogs. Aldini, however, had been more flamboyant in his experiments and had experimented on a larger scale, utilizing the heads of oxen and on several occasions the bodies of executed criminals. Sturgeon gave detailed descriptions of Aldini's demonstrations while admitting that he had never had the opportunity of performing them himself.

Aldini had been an assiduous defender of his uncle's name and reputation.[9] Between 1800 and 1805 he traveled extensively throughout Europe, repeating his spectacular performances and spreading the creed of animal electricity. In 1802 he appeared before the Société Galvanique in Paris, shortly after Volta had visited the city to defend and propagate his own version of the new fluid.[10] Later that year he visited London and performed before the Royal Society.[11] He utilized the heads of freshly slaughtered oxen and dissected frogs to produce the galvanic fluid.

Aldini was capable of producing the signs of electricity from animal tissues without the use of any metals: a crucial requirement if Volta's claim that such signs were the products of metallic contact was to be defeated. His audience at the Royal Society were certainly impressed: "Here then we have the most decided substitution of the organized animal system in the place of the metallic pile: *it is an animal pile; and the production of the galvanic fluid, or electricity, by the direct or independent energy of life in animals, can no longer be doubted.*"[12] The enthusiastic commentator in Nicholson's *Journal of Natural Philosophy* was convinced that "Galvanism is by these facts shewn to be animal electricity, not merely passive, but most probably performing the most important functions in the animal economy."[13]

Aldini's "courage" in performing on the bodies of executed criminals was applauded. On 17 January 1803, he had his opportunity to experiment on a human subject. In the presence of Thomas Keate, president of the Royal College of Surgeons, the body of Forster, executed at Newgate for murder, was taken down and made available for experimentation.[14] The poles of a large galvanic battery, consisting of 120 plates of copper and zinc, were connected with various parts of the dead man's anatomy. Connections were made between his ears, between his mouth and one of his ears, and between one of his ears and his anus. The result was a startling exhibition of contractions and convulsions:

> On the first application of the process to the face, the jaw of the deceased criminal began to quiver, the adjoining muscles were horribly contorted, and one eye was actually opened. In the subsequent part of the process, the right hand was raised and clenched, and the legs and thighs were set in motion. It appeared to the uninformed part of the by-standers as if the wretched man was on the eve of being restored to life.[15]

As the newspaper reporter drily remarked, however, that would have been an unlikely eventuality: "several of his friends, who were under the scaffold, had violently pulled his legs, in order to bring a more speedy termination to his sufferings."[16]

Aldini's experiments were at the time a minor sensation. They were reported in the press, and his public displays were patronized by the fashionable, including the Prince of Wales. Contemporary commentators seemed agreed that Aldini had indeed proved the truth of his uncle's theories and that animal electricity did exist. The anonymous commentator in Nicholson's *Journal* asserted that while it had been "indistinctly apprehended or conjectured in the way of theory, that the galvanic or electric matter was excited, collected or generated in the bodies of animals, where it was considered as the great cause or instrument of muscular motion, sensation, and other effects highly interesting," Aldini's experiments had been successful in "having placed this proposition in the rank of established truths."[17] The experiments were also seen as having established the possibility of restoring life in cases of drowning and suffocation. They were touted as providing a possible cure in cases of insanity and apoplexy.

There was nothing radical, or even heterodox, about Aldini's performances. They were certainly flamboyant and impressive, but the spaces where they took place and the function they served sanitized them of dangerous political connotations. On the contrary, in performing on the corpse of an executed criminal, Aldini was fitting in neatly with the medicojuridical role of Regency surgeons. Their license to take down and experiment upon the bodies of the condemned was explicitly regarded as a continuation of the punitive sentence handed down by the State.[18] His experiments were supported by the Royal College of Surgeons and applauded by the Royal Humane Society for the possibility they offered of restoring life.

Sturgeon drew extensively on these resources in his lectures to display the intimate connection of electricity and vitality. Some of these experiments Sturgeon made clear he had not performed himself, but asserted that the arrangements described by Aldini were "so judiciously chosen" as to be guaranteed success. In his preface he explained that while he had been unable to perform Aldini's experiments on the human body, he had been given the opportunity of galvanizing the bodies of four young men who had drowned after falling through the ice at Woolwich. Although this attempt at resuscitation had been unsuccessful, Sturgeon was confident that this was simply the result of the length of time elapsed before the experiment took place. There could be no doubt that in the proper circumstances galvanism was capable of restoring life.

*Sturgeon drew attention to other more recent experiments to support his* contention. Twenty-five years previously, in 1818, the Glasgow natural philosopher and chemist Andrew Ure had conducted a celebrated series of experiments on the murderer Clydesdale, who had been executed by hanging in Glasgow on 4 November 1818.[19] Clydesdale was brought to the anatomical theater within ten minutes of being cut down from the gallows where he had been hanging for nearly an hour. A variety of electrical experiments were then performed, involving the connection of the poles of a large, 270-pair battery across various parts of the man's body.

When one pole was connected to the spinal marrow and the other to the sciatic nerve, for example, "every muscle in the body was immediately agitated

with convulsive movements, resembling a violent shuddering from cold."
When the phrenic nerve was exposed and the battery connected between it
and the diaphragm, after various adjustments to maximize the power, "full, nay,
laborious breathing, instantly commenced. The chest heaved, and fell; the belly
was protruded, and again collapsed, with the relaxing and retiring diaphragm."
Ure recorded the conviction of the "many scientific gentlemen who witnessed
the scene" that this "respiratory experiment was perhaps the most striking ever
made with a philosophical apparatus."

The most spectacular and disturbing demonstration was yet to come. When
the supraorbital nerve on the forehead was exposed and electric discharges
passed between it and the heel, extraordinary effects were produced:

> every muscle in his countenance was simultaneously thrown into fearful action; rage,
> horror, despair, anguish, and ghastly smiles, united their hideous expression in the
> murderer's face, surpassing far the wildest representations of a Fuseli or a Kean. At
> this period several of the spectators were forced to leave the apartment from terror or
> sickness, and one gentleman fainted.[20]

Some of the possibilities of galvanism were evidently too much even for the
stomachs of hardened Glaswegian medical men.

For Ure, as for Sturgeon a quarter century later, the experiment pointed
toward a rosy future for galvanism, and the prospect of "raising this wonderful
agent to its expected rank, among the ministers of health and life to man." No
skills were required beyond those possessed by any competent medical practitioner. Any danger to the patient was outweighed by the possibility of saving
life: "it is surely criminal to spare any pains which may contribute, in the slightest degree, to recal the fleeting breath of man to its cherished mansion."[21] Such
experiments were also, of course, proof positive of galvanism's connections
with vitality.

Sturgeon's lectures constructed a strikingly novel genealogy for galvanism.
Rather than tracing the development of experiments and ideas from Volta's pile
through Humphry Davy's heroic experiments, he pointed to a different history.[22] His heroes were Galvani and Aldini, Ritter and Lichtenberg, Cruickshank and the modern battery-makers such as Grove and Smee. The history of
galvanism, on the one hand, was the history of the discovery and demonstration
of electricity's connections with and role in the animal economy. On the other
hand, it was also the history of galvanism's instrumentation and the devices
used to make the subtle fluid more visible and therefore more public.

That instrumentation was not simply the voltaic pile and its descendants. It
also included the frogs, rabbits, dogs, and oxen manipulated to produce the
galvanic current. Also to be listed among galvanism's instruments were the
bodies of Forster and Clydesdale, which had been deployed to render the fluid
spectacularly visible in their facial and bodily gesticulations. Ure's descriptions
of Clydesdale's contortions in particular were deliberately couched in the language of artifice. His body was represented as an automaton reproducing the
theatrical posturing of the Regency stage: its movements were like the "wildest

representations of a Fuseli or a Kean."[23] In this way the presentation of galvanism as simultaneously the product of animal bodies and of machines served as a way of representing natural economies as mechanical, galvanic systems.

In Ure's experiments as represented by Sturgeon, electricity was assuming the role of regulation. Aldini's experiments were also being reread in this way. In its interaction with the body, electricity was the force that regulated and controlled bodily functions and actions. In Aldini's and Ure's experiments those functions and actions had been reproduced through electricity. By introducing this notion of electricity as a force that regulated in the same way that a machine was regulated by its moving power, electricity in its relationship to the body could be seen as a matter for a theory of labor rather than of vitality. The galvanized body was an object of political economy. Ure in particular had strong views concerning the self-regulation of machines and their capacity to regulate their human attendants. Galvanizing the body could be a way of subjecting it to the same discipline.[24]

REPRODUCING NATURE

Human and animal bodies were not the only elements in the natural economy to be galvanized. Everything from geological structures to vegetables was available for electrification. Discussion of electricity's role in powering the natural economy could take place in a variety of different contexts and in styles ranging from the wildly and speculatively romantic to the soberly empirical. The flamboyant Colonel Francis Maceroni, for example, steam locomotive entrepreneur and one of the *Mechanics' Magazine*'s more garrulous and prolific correspondents, chose the occasion of a trip to the Mediterranean to wax lyrical on electricity in nature.

Maceroni had a colourful past that provided him with ample opportunity to study spectacular phenomena. He spent much of his early life in Rome and in Naples, where he was aide-de-camp to Murat, king of Naples. Following Murat's defeat and the end of the Napoleonic Wars, he spent time as a mercenary soldier in Spain and the Americas. During the 1820s and 1830s he tried to make his fortune through invention. The steam coach that he patented in partnership with a Paddington Green factory owner was widely and approvingly cited in the popular press. He was a physical force radical. A few months before the passing of the Reform Bill, he had published a pamphlet describing the construction of weapons for street fighting.[25] This was reprinted as a special issue of the *Poor Man's Guardian*. He was an enthusiastic dabbler in science, having studied medicine and anatomy during his years in Italy.[26]

The colonel had been impressed by the violence of the thunderstorms off the coast of Italy and by the coincidence between flashes of lightning and volcanic eruptions, such as he had witnessed on Vesuvius in 1810. He speculated concerning what he described as "the chemistry, or rather *physiology*, of the matter," seeing "the electrical decomposing and secreting operation . . . as inherent and necessary to the development, growth, constitution, and vital career of the identity we call 'our' globe." He castigated previous commentators, including "Sir William Hamilton, and several other respectable old ladies amongst the F.R.S." for failing to recognize the significance of electricity's central role in the production of such phenomena.[27]

Maceroni proceeded to suggest a number of zoomorphic analogies interconnecting the system of the earth, animal structures, and electrical apparatus. The alternating layers of air, clouds, water, and earth were like the alternating layers of a voltaic battery. They were also analogous to the structures of "organized" beings, "with their skins, cellular membranes, muscles covered by aponeuroses, continually rubbing against each other; fat, internal membranes, with intervening fluids always in motion."[28] The analogy extended to the nervous system:

> Are not brains, nervous ganglions, and nerves, which are evidently the seat of vital action, in the identities we call animal, real electrical machines; similar in principle, as they are similar in substance and in structure, to the electrical discharging apparatus of the gymnotus and torpedo, which consist of large brain-like ganglions connected with the spinal cord?[29]

In short, everything from "vital" phenomena to the eruptions of volcanoes could be seen as the result of galvanic action.

Accounts as all-embracing and radical as Maceroni's were rare, but by the mid-1830s discussions of electricity as a crucial agent in the natural economy were quite common in the pages of the *Mechanics' Magazine*. Thomas Pine of Maidstone in Kent, for example, made numerous contributions concerning the connections of electricity, particularly atmospheric electricity, with vegetation. His writings reveal the extent to which informal networks of practitioners collaborated in the production of galvanic knowledge. These men held in common a view of electricity as an all-pervasive element holding together the natural economy. The *Mechanics' Magazine* itself clearly had an important role to play in the production and maintenance of this electrical world. It provided a forum within which such informal networks could be sustained and extended.

Pine first appeared in the pages of the *Mechanics' Magazine* in a communication by his friend and fellow electrician W. H. Weekes. Weekes was writing to inform the magazine's readers of a new method of constructing a mercurial trough.[30] He mentioned in passing that the trough had originally been developed by Pine "for the purpose of verifying, by a series of actual experiments on the respiration of plants, an ingenious theory, entitled . . . 'Electro-Vegetation.'"[31] Having received publicity in this fashion, Pine himself hastened to communicate a detailed account of the theory to the *Mechanics' Magazine*'s readership.

Pine's main interest in his experiments was to "show the agency of electricity through the various stages of vegetation."[32] To this end he had carried out a number of experiments to test the conductibility of the pointed extremities of plants when they were held up to the prime conductor of an electrical machine or connected to a charged Leyden jar. He established that "they are the most potent of conductors, not excepting the most acute metal points."[33] In this communication Pine also drew attention to a series of experiments carried out by W. H. Weekes in which that electrician had used electrometers armed with vegetable points as highly sensitive detectors of atmospheric electricity.[34]

Pine saw these experiments as unmistakable evidence for the universal role of electricity. The sun was to be regarded as the source of electricity, as well as of heat and light. In fact, "light, heat, and electricity, are no other than so many effects or results from the operations of one common and inconceivably subtle fluid, of which the sun is the source or centre."[35] This fluid operated everywhere, "whether accumulated in the natural artillery of the heavens—in the volcanic bowels of the earth—or concentrated in the artificial arrangements of a voltaic battery."[36] The structure of plants in particular was designed, on the one hand, to exploit electricity to promote their growth and, on the other, to prevent "dangerous accumulations of the electric matter in the atmosphere" by transferring it to the ground. This was their role in maintaining the "equilibrium on which the general harmony and tranquility of nature depends."[37]

In a number of subsequent contributions Pine expanded on his notion of the role of electricity in vegetable life. In so doing, he also attempted to tie in his work with the research and interests of other electricians, in particular William Sturgeon. Sturgeon had his own interests in atmospheric electricity and had worked hard to establish its role in the machinery of nature. Pine's efforts provided a resource that he was happy to exploit. At the same time Pine was delighted at the opportunity of using Sturgeon's experimental researches to corroborate his own. In his second communication to the *Mechanics' Magazine* Pine quoted at length a letter from Sturgeon, defending his own view—against Humphry Davy among others—that the atmosphere was in general electrically positive with respect to the ground.[38]

Another enthusiastic investigator of atmospheric electricity was Andrew Crosse, with whom Weekes, Sturgeon, and probably Pine had close links, first informally and through correspondence and later under the auspices of the London Electrical Society, of which they were all members. In his isolated Somerset mansion, Fyne Court, Crosse was notorious for his electrical experiments. One-third of a mile of wire was strung from trees and poles on his estate to draw lightning from the heavens.[39] The accumulated electricity was stored in a large battery of fifty Leyden jars. The current could sometimes be sufficient "to charge and discharge the great battery 20 times in a minute, with reports as loud as a cannon, which being continuous were so terrible to strangers that they always fled."[40]

Crosse combined his interest in atmospheric electricity with an enthusiasm for experimenting on the role of electricity in the formation of geological structures. It was in this capacity that he had been introduced at the Bristol meeting of the British Association in 1836.[41] His experiments, which required large and expensive arrays of voltaic batteries working for long periods of time, were aimed in particular at the production of artificial crystals. At the Bristol meeting, for example, he claimed to have produced samples of "quartz, arragonite, malachite &c."[42] Similar samples of crystals produced by Crosse were later to be displayed at meetings of the London Electrical Society and by John Peter Gassiot at his electrical soirees.[43]

Crosse's enthusiasm for the production of artificial crystals was not isolated. The Cornish Quaker Robert Were Fox engaged in similar experiments as part of his project to map the role of electricity in the formation of mineral veins in the earth.[44] Fox had been conducting experiments in Cornish mines for a number of years. During the 1820s, for example, he had carried out extensive research on the increase with depth of the temperature in mines.[45] About 1830 he commenced a series of researches to establish the existence of electricity in metalliferous veins and speculated concerning its role in their formation.

Fox was impressed both by the analogy between mineral veins and voltaic combinations and by the fact that their direction typically lay at right angles to the magnetic meridian. He speculated that such veins probably originated in the form of fissures in the surrounding rock that became filled with water containing the various minerals in solution. Under the influence of the earth's magnetism, electric currents would be produced in these mineral solutions that would in turn lead to the formation of mineral-bearing veins in those fissures running at right angles to the meridian.

These currents in the earth were related to similar currents in the atmosphere that produced the aurora borealis, which was also at right angles to the magnetic meridian:

> The aurora may, therefore, I think, be considered an exhibition of electric currents at a great height, which are connected with others nearly parallel to them, in the interior of the earth. Whether, however, we regard terrestrial magnetism as the effect or cause of the direction of electric currents, it cannot be doubted that these phenomena are in harmony with each other. . . .[46]

It was to substantiate claims such as these that Fox, like Crosse, carried out experiments on the formation of artificial crystals by means of electricity.

Fox was impressed by the way in which electricity not only acted to maintain the natural economy but sustained political economy as well:

> We observe that many of the most useful metals are the most abundant:—and the fact that they are generally confined to certain veins, and to certain portions of them only, is perhaps, of greater import, than we might suppose:—for had they been disseminated in the strata, or even dispersed throughout all mineral veins, the labour required to obtain them, would have rendered them practically useless:—or had they, on the other hand, been much more concentrated, their rapid exhaustion might entail incalculable injury on future generations.[47]

Electricity therefore had a role to play throughout nature and culture. As Fox noted, life could not exist without it: "should the hand that produced it, suspend its operation but for one moment, animal and vegetable life would be universally extinguished."[48] Like a well-designed engine, there was in nature a "harmony of parts, and a consistency of operation," which guaranteed proper performance.[49]

Images of the world or of the universe as an electrical machine permeated many of these discussions of electricity's role in governing the natural economy. In fact such images were central to the ways in which experimental practice was conducted. Electricity's role in the natural economy was established through the building of machines that reproduced these processes. Crosse and Fox established the formation of metalliferous veins in the earth through voltaic action by building apparatus in their laboratories that artificially reproduced the natural process. The production of crystals through the use of galvanic batteries meant that the earth, too, was such an item of apparatus. Maceroni saw the different layers of earth, water, and air as being the same as the different layers of a voltaic pile.

This was particularly clear in William Sturgeon's account of the origins of terrestrial magnetism. Drawing on Ampère's electrical theory of magnetism, his own early researches on electromagnetic rotations, and Fox's experiments in the Cornish mines, he declared the evidence for the existence, role, and operation of terrestrial electricity to be irresistible.

> That electrical currents are continually flowing in the earth must appear obvious to every one conversant with voltaic electricity. The materials which form our batteries, and display electric streams at our pleasure, have all been brought from this exhaustless source. Nature's laboratory is well stored with apparatus of this kind, aptly fitted for incessant action, and the production of immense electrical tides; and the insignificancy of our puny contrivances to mimic nature's operations, must be amply apparent when compared with the magnificent apparatus of the earth.[50]

Sturgeon's concern with forging links between cosmological phenomena and electrical machines stretched back to some of his earliest experiments.[51] His experiments on thermoelectric rotations were aimed largely at showing that this was the mechanism whereby the earth rotated. The way to show that the world was an electrical machine was to build electrical machines that could mimic the "vast apparatus of nature."[52]

MODELING THE UNIVERSE

Accounts of the earth or of the universe as a vast electrical machine or system could clearly serve a variety of interests. Such accounts were as conducive to the purposes of the politically radical and romantic Francis Maceroni as they were to those of the sober Quaker Robert Were Fox. The particular uses these different individuals made of electricity were, of course, contingent. Maceroni used an electrical account of the world to promote a pantheistic vision.[53] Were Fox, on the other hand, used the results of his electrical experiments to celebrate the benevolence of the "Almighty Author."[54] Electricity and galvanism did not carry their own cultural connotations with them fully formed. Their cultural uses had to be fashioned and could as easily be conservative as they could be radical. The uses to which electricity could be put had much to do with where it was made and by whom.

Many electrical accounts of the universe, however, carried implicit, and sometimes explicit, criticism of Newtonian hegemony. Such critiques, carrying with them assaults on established tradition and authority, could very easily be put to subversive use. A case in point is that of Thomas Simmons Mackintosh, a radical and outspoken Owenite lecturer, who during the 1830s produced a long series of articles in the *Mechanics' Magazine*, articulating and staunchly defending an "electrical theory of the universe" that explicitly aimed to demolish long taken for granted Newtonian shibboleths.

Mackintosh took advantage of the reappearance of Halley's comet in 1835 to inform the *Mechanics' Magazine*'s readers of his electrical theory. This theory, as first presented, consisted of three propositions:

1st. That comets are immense volumes of aeriform matter discharged from the sun by the agency of electricity.

2d. That comets are gradually condensed, and eventually become planets.

3d. That planets move in spiral orbits, and ultimately fall back into the body of the sun.[55]

The sun, according to Mackintosh, was to be regarded as a vast spherical conductor, charged with electricity, that extended its influence to the limits of the solar system. Only electrical agency could possibly be the cause of the sun's intense internal energy. The discharge of a comet from the sun could be regarded "merely as a spark drawn from the prime conductor of the solar system."[56]

In subsequent contributions to the magazine, Mackintosh developed his electrical account and spelled out its ramifications. He was unequivocal that all "motion throughout the solar system is effected by the agency of electricity."[57] This claim extended not only to the movements of planetary bodies, which were maintained at their respective distances from the sun by a balancing of positive and negative electricity, but to the "minuter processes of vegetation, oxidation, and vitrification." Different modes of motion were caused by "specifically distinct modifications of electricity."[58]

One consequence of this theory was that it posited a definite and limited lifespan for the solar system. As Mackintosh made clear, by "carrying on the vegetative and other processes of nature," the natural electricity of the earth and other planets was gradually dissipated. As a result the repulsive power between the sun and the planets was gradually weakened such that all the planets would eventually fall into the sun. The planets were in effect "propelled down an inclined plane by the power of electricity."[59] Mackintosh pointed to large bodies of evidence suggesting the gradual decrease of the earth's electricity; he referred to Pine's experiments on vegetation, Fox's experiments on electricity in the earth, and work on chemical action to show how the earth's electricity would decrease until that globe "ultimately falls back into the sun a worn out and exhausted planet."[60]

In a final contribution, Mackintosh drew on a wealth of geological evidence to support his theory by showing that in the past the earth had possessed at least five moons which had produced various geological strata as they fell to earth with their electricity expended.[61] Concluding his account, he expressed confidence that his claims would not be found at variance with "the geometrical demonstrations of Newtonian philosophy."[62] His account differed from the Newtonian orthodoxy simply in ascribing effects to electricity, "a power of whose existence we are assured by the evidence of the senses," rather than to gravitation, "a conventional term adopted for the purpose of explaining a certain effect resulting from no known cause."[63] His theory constituted an appeal to the common man's everyday senses and a stand against obscurantism and deference to old authority. He was certain that it would "at some future date be received as the true system of the universe."[64]

Mackintosh's arguments produced a lively and often acrimonious debate in the pages of the *Mechanics' Magazine*. Correspondents lined up to defend Newtonian orthodoxy, on the one side, and the new electrical heresy, on the other. "Ursa Major," one of the more prolific contributors in defense of Newtonianism, thundered that "when the great truths of philosophy are in question, I cannot afford to be polite."[65] He admitted that the electrical theory seemed plausible and ingenious enough, but denied that it could be reconciled with Newtonian philosophy, and that therefore it must be false. Similarly, "Kinclaven" asserted that as he found "so many parts of the electrical theory opposed to that of universal gravitation. . . . I would have neglected my duty had I allowed the subject to pass unnoticed." Mackintosh's galvanic circles were no better than "the whirlpools and vortices of Cleanthes and Descartes, and the celestial machinery of Eudoxes and Callipus."[66]

"Zeta" was just as damning, describing Mackintosh's theory as "a revelry carried to inebriation, and corresponding collapse." He ridiculed the model's complex electrical machinery:

> The solar system, an immense electrifying machine! The sun its prime conductor! and the planets, (our unhappy earth among them) pieces of elder pith skipping between it and the nether regions, at short intervals of millions of years! What an idea for an inheritor of mortality! and that Newton, too, should have "failed in not tracing the cause of motion to the all-pervading power of electricity!" But where is the battery—where the discharging rod—where the musical bells?[67]

Such attempts to trespass beyond the bounds of "legitimate reasoning" were "senseless imaginings" and signs of mental disorder: "it is the faculties playing at *leap-frog* with reasoning undermost."[68]

Mackintosh's response was to reaffirm his principles and defend the legitimacy of his researches. He argued that the publication of speculations such as his in the pages of the *Mechanics' Magazine* was a central part of that journal's mission:

By observing this connexion between the theories of science and the practices of art, the general reader will be enabled to perceive that this subject is not of a character merely speculative from which nothing of a practical and profitable nature may be expected to result; on the contrary, it may be anticipated, when the laws which regulate the evolution of galvanic electricity shall have been more thoroughly and carefully investigated, that a *new motive power* will be placed at our disposal, which may perhaps eventually supercede the steam-engine.[69]

Understanding the electrical nature of the universe was essential for the artisan who wished to use his inventive skill to exploit new sources of power.

Such invention could also raise the social and intellectual status of the artisan. By exploiting electricity for the purpose of producing economical (and economic) power, he would be displaying his ultimate mastery over nature.

And if it shall eventually be made manifest that, by the agency of electricity, variously combined and modified, the whole material world is kept in motion, how sublime, how ennobling the reflection, that man, by the innate strength of his intellectual faculties, has raised and exalted his species from the condition of a shivering and helpless savage to the commanding eminence of science and civilization—and has placed at his disposal, and obedient to his will, not only the subordinate elements of nature, but even the *primum mobile* itself.

Bringing the galvanic engine to perfection would be the *"ne plus ultra* of mechanical invention."[70] Those who wished to attack, rather than explore and improve upon, his theory were, Mackintosh implied, standing in the path of the material progress of mankind and the artisan's opportunities for self-improvement.

Other correspondents also came to Mackintosh's defense. Thomas Pine argued that his own work on electrovegetation supported the electrical hypothesis.[71] Francis Maceroni took advantage of the debate to introduce again his own views on the universal role of electricity. He attacked Mackintosh's opponents for their deference to authority: "Objections and special pleadings against facts, or alleged facts, merely because the admission of them would militate against the received theories of Newton or Kepler, or any other name, however illustrious, forms a species of controversy more calculated to check inquiry and delay the discovery of truth than to promote it."[72] His own system he admitted to be purely speculative but asserted that such schemes should be taken seriously as part of the search for truth.

The debate over Mackintosh's electrical speculations continued in the *Mechanics' Magazine* for a number of years. Specific issues igniting controversy included the question of the nature of the galvanic fluid, the possibility of the moon's falling to earth, the speculation that former satellites colliding with the earth had produced the various geological strata, and the viability of perpetual motion. The magazine was not Mackintosh's only forum. Articles appearing in the Owenite *New Moral World* also provided detailed accounts of his electrical theory.[73] He delivered at least one series of lectures on his theory at the Hall of

Science on London's City Road in 1837.[74] It was part of the repertoire of lectures he delivered at halls of science throughout the Midlands and the north of England. It played its role in Owenite campaigns to refound science on rational, socialist principles. By all accounts his lectures were popular. A correspondent in Liverpool enthused, "The attendance at the institution here is on the increase, in consequence of Mr. Simmons Mackintosh, of London, having commenced a course of six lectures on the 'Electrical Theory of the Universe.'"[75] At Ashton a month or so later it was reported that one of Mackintosh's lectures "was better attended than any previous lecture in this place."[76] At Salford his lecture "on the attractive and repulsive moral forces . . . illustrated by several very humerous drawings" was said to have conveyed "much instruction and gratification to the audience."[77]

During the late 1830s he also published a book articulating and defending his position, challenging the Royal Society to respond to his theories.[78] In the preface Mackintosh was explicit that this was "a work which professes to show that the Newtonian Philosophy is based upon false principles."[79] He cited Crosse, Fox, Pine, Ure, and Weekes to establish that "electrical action is incessantly going on in the animal, vegetable, and mineral kingdoms, that animals, vegetables, and minerals are produced and maintained each in its own appropriate condition by its agency."[80] This physical basis of life had moral consequences:

> It is better to view man as an organized machine, and to search for the seat of those impulses in the functions of his physical nature, where assuredly they are to be found, than to trace them to sources beyond our knowledge and above our control. If the roots of moral action be in the physical organization, in proportion to our knowledge of that organization we can stimulate or retard that action, and turn it into courses which would lead to a larger amount of happiness both to the individual and to the society of which he forms a member.[81]

Understanding electricity meant understanding the human machine and with it the human soul.

Understanding the universe as a machine had far-reaching consequences. While it held out the possibility of utopia, it also limited salvation to the terrestrial sphere. No engine could continue to operate forever.

> The river flows because it is running down; the clock moves because it is running down; the planetary system moves because it is running down; every system, every motion, every process, is progressing towards a point in which it will terminate; and life is a process which only exists by a continual approach towards death. Eternal life and perpetual motion are almost, or altogether, synonymous.[82]

Replacing the Newtonian philosophy with the electrical theory of the universe meant replacing the whole social, political, and religious order that underpinned early-nineteenth-century life. If man and the universe were electrical machines, then the Kingdom of Heaven could be founded only on Earth.

Mackintosh was by no means the only promoter of an electrical universe, though few others drew such extreme and politically subversive conclusions from their claims. William Leithead, one of the London Electrical Society's secretaries, promulgated a theory that similarly established electricity as the central universal agent.[83] His main aim was to demonstrate the role of electricity in the production of disease. Fevers, for example, could be caused by the accumulation or diminution of the galvanic fluid in the brain. The electrical condition of the atmosphere could also affect the health of the human body.[84] Leithead was not alone in his claims. In the wake of cholera outbreaks during the 1840s, the eminent Irish surgeon Sir James Murray traced the cause of the disease to electricity, suggesting that the illness could be prevented by the electrical insulation of houses.[85]

Modeling the universe as a vast electrical apparatus could therefore address a variety of interests. For Mackintosh and Maceroni an electrical universe was an element in a radical political philosophy. For Leithead and Murray it provided a new understanding of disease. Numerous resources could be put to work in the production of such models. Sturgeon's electromagnetic account of terrestrial rotation, Pine's electrical vegetables, Crosse's and Fox's electrically generated crystals could be and were all utilized in the construction of an electrical world. New developments and new inventions such as electromagnetic engines and telegraph apparatus could also take their place in this galvanic ontology, buttressing it and deriving significance from it at the same time.

THE NEW FRANKENSTEIN

In late 1836, only a few months after his rapturous reception at the Bristol meeting of the British Association, a startling new addition to the electrical universe emerged in Andrew Crosse's Somerset laboratory. On returning to Fyne Court, Crosse had continued his experiments on the production of artificial crystals that had so impressed the Geological Section.[86] In this particular experiment a dilute solution of silicate of potash, saturated with muriatic acid, was dripped slowly over "a piece of somewhat porous red oxide of iron from Vesuvius." Two platinum wires were connected to this piece of volcanic rock, which were in turn connected to the poles of a voltaic battery. By this means the stone was kept constantly electrified as the fluid was dripped over it. The aim of the experiment was to produce artificial crystals of silica.

As was usual in Crosse's experiments on mineral crystallization, this apparatus was kept working for a long period. Crosse recorded the results as follows:

> At the end of fourteen days, two or three very minute white specks or nipples were visible on the surface of the stone, between the two wires, by means of a lens. On the eighteenth day these nipples elongated and were covered with fine filaments. On the twenty-second day their size and elongation increased, and on the twenty-sixth day

each figure assumed the form of a perfect insect, standing on a few bristles which formed its tail. On the twenty-eighth day these insects moved their legs, and in the course of a few days more, detached themselves from the stone, and moved over its surface at pleasure, although in general they appeared averse to motion, more particularly when first born.[87]

Over the next few weeks about a hundred of these insects appeared. Repeating the experiment brought about the same results.

News of these experiments was spread rapidly by journalists in the popular press, who regarded Crosse as hot property following his triumphal appearance at the Bristol meeting only a few months previously.[88] Reactions were predictably mixed. For some, Crosse had conclusively established the material basis of life. Others condemned him for his blasphemous meddlings with God's laws.[89] Crosse himself was carefully circumspect about the way in which the insects might have been produced, and certainly avoided any suggestion that they were an original creation. Several commentators, notably the geologist William Buckland, suggested that electricity had revivified fossil insect eggs embedded in the volcanic rock used in the experiment.[90] Such a reading had the virtue of preserving Crosse's reputation as an experimenter while avoiding dangerously materialist interpretations.

For Crosse, electricity was central to the universe's operations. It was "a property belonging to all matter, perhaps ranging through all space, from sun to sun, from planet to planet." Establishing this claim and making electricity useful was the task of the London Electrical Society.[91] Those who stood in his way stood in the way of progress:

> I have met with so much virulence and abuse, so much calumny and misrepresentation, in consequence of the experiments I am about to detail, and which it seems in this *nineteenth century* a crime to have made, that I must state . . . that I am neither an "Atheist," nor a "Materialist," nor a "self-imagined creator," but a humble and lowly reverencer of that Great Being, whose laws my accusers seem wholly to have lost sight of. More than this, it is my conviction that science is only valuable as a means to a greater end.[92]

Others, however, could give very different accounts of the moral consequences of electrical experiments and Crosse's endeavors.

A satirical short story in the scurrilous and Tory *Fraser's Magazine* soon made explicit some of the ways in which experiments such as Crosse's could be regarded as subversive. The hero of this tale, "The New Frankenstein," was a German student obsessed with galvanism.[93] He had made crystals by galvanism and produced insects from volcanic rock. This was an unambiguous reference to Crosse and his recent experiments. This student encountered Frankenstein's monster, which had on him "the effect produced at Guy's Hospital on the medical students, when the corpse of a criminal, under the effect of a powerful galvanic battery, opened his eyes, made one step from the table against which he was placed, erect, and stiff, and fell among them."[94]

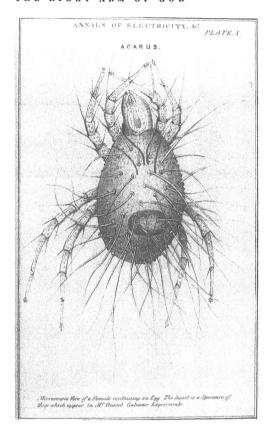

ANNALS OF ELECTRICITY, &c.
PLATE X.

ACARUS.

*Microscopic View of a Female containing an Egg. The Insect is a Specimen of those which appear in Mr Crosse's Galvanic Experiments.*

5.1 Andrew Crosse's electrical insect. *Annals of Electricity* 2 (1838): pl. 10.

He discovered that the monster, though capable of movement by electricity, lacked a mind, and proceeded therefore to ransack the brains of Europe's greatest philosophers, from Coleridge to Goethe, stealing their thoughts to instill into the monster's brain. For this he used an apparatus of his own invention, similar to the infamous Perkins's Tractors. The result, of course, was that the poor monster went insane. Discovering now that his monster lacked a soul, the impious student sought to provide one by looting an Egyptian tomb. As he did so, the ground opened beneath his feet as Satan rose to claim him and his creation: an outcome that was prevented only by his awakening and discovering that it had all been a dream!

The satire's message was clear. Producing life by electricity was nothing more than a parody of nature. Thomas Carlyle used the term *galvanic* in much the same way to signify the life-in-death of machine culture.[95] Even if the galvanic power was sufficient to simulate the physical properties of life, it could not hope to reproduce the moral and the spiritual. An attempt to do so would in any case be blasphemous. Others, however, responded enthusiastically to the new discovery. Francis Maceroni in the *Mechanics' Magazine* welcomed Crosse's experiment as a vindication of his own views. It was a "clear experimental proof of . . . the identity of all *life*, with the electric, galvanic, magnetic, or solar substance."[96] The *Analyst* applauded his efforts, repeating his strictures against his opponents.[97] The *Lancet* too, was enthusiastic. Crosse was described as "the Galvani-Volta of England" and held up as an example to the "[q]uacks of the scientific world, medical as well as physical."[98] Another correspondent was as enthusiastic, calling for more attention to electrical matters on the part of medical practitioners.[99]

Electricians on the whole welcomed the announcement of Crosse's experiment, albeit with varying degrees of caution. The *Magazine of Popular Science*, house journal of the Adelaide Gallery, praised Crosse's experimental acumen and condemned the "scoffers" and "scorners" who had attacked his work without waiting to hear its details.[100] Henry Noad included details of Crosse's work in the second edition of his *Lectures on Electricity*, discussing also W. H. Weekes's successful replication of Crosse's experiments in 1841.[101] Charles Vincent Walker, then secretary of the London Electrical Society, hailed Weekes's replication as one of the major electrical events of the year.[102]

While electricians regarded Crosse's work as legitimate, if controversial, more elite practitioners found his contribution considerably more problematic. Following a display of Crosse's insects at a Friday Evening Discourse at the Royal Institution by William Clift, reports appeared in the press that Faraday himself had succeeded in replicating the experiments. Faraday was obliged to respond, denying that he had conducted such experiments, and indeed denying that he recognized such an engagement as appropriate. "I am reported to have said, that by experiment I have fully confirmed the extraordinary results of Mr. Crosse. . . . What I said was almost the reverse of this, for I merely stated, upon the occasion of the insects being shown by Mr. Clift and Mr. Owen at the Royal Institution, that we wished it to be directly understood we had no opinion to give respecting their mode of production."[103] As James Secord points out, Faraday was particularly sensitive to the problems of transgressing the disciplinary boundaries then being drawn by the gentlemen of science, particularly on dangerous and contentious issues such as the production of life.[104] Many electricians were, however, unconcerned by such issues.

Faraday was obliged to position himself carefully on the matter so that he could appear authoritative without linking himself too closely to potentially heretical doctrines. He took advantage of the arrival of a *Gymnotus electricus* at the Adelaide Gallery to conduct his own experiments on electricity and life. It was an unusual step for Faraday. While he was a frequent visitor at the Adelaide's exhibitions, he did not normally carry out his own experiments in such places. He had little choice, however, since the Adelaide owned the only living electrical eel in London. His requests to the Colonial Office for specimens of his own had proven unsuccessful.

Faraday used these experiments, which were duly presented to the Royal Society and published in the *Philosophical Transactions*, to carefully demarcate the proper role of experiment in relationship to life.[105] Faraday argued that his experiments on the electrical eel were "upon the threshold of what we may, without presumption, believe man is permitted to know of this matter." Experimentation on the relationship of electricity with the nervous system and with life was of "surpassing interest" but was at the same time to be approached with due humility and a recognition of the limits of man's capacities.[106] The electrical eel was a good site for experiments on such matters since it was a living creature that appeared to produce electricity by its own volition. As such, it could be used to clarify the nature of the relationship between electricity and nervous force.

Most of these experiments were designed to establish the identity of the eel's electricity with that from other sources, such as a battery or electrical machine. It was already known that the eel could administer a shock. Faraday proceeded to demonstrate that the other effects of voltaic and common electricity could also be obtained. Sparks were produced, a galvanometer needle was deflected, a steel needle was magnetized, and a solution iodide of potassium was decomposed. Other experiments established the direction of the current and its distribution in the water surrounding the eel. Faraday suggested that it should be possible to map the lines of force surrounding the eel, and that they should "resemble generally, in disposition, the magnetic curves of a magnet, having the same curved or straight shape as the animal."[107] He was bringing the eel inside his own world of electrical lines of force in space.

Faraday speculated that since the production of electricity evidently exhausted the eel's nervous power, it might be possible to restore nervous power by passing electricity through the eel. He worked by the analogy of electricity's relationship to heat and magnetism. He emphasized, however, that this was only an assumption made to guide experiment and not a claim that electricity and nervous power were analogous. In any case, "with respect to the nature of nervous power, that exertion of it which is conveyed along the nerves to the various organs which they excite into action, is not the direct principle of *life*; and therefore I see no natural *reason* why we should not be allowed in certain cases to *determine* as well as observe its course."[108] Electricity could be rendered safe for use in experimentation with organic creatures if the nervous power were brought out of the realm of the organic and treated as "an inorganic force."

Faraday's attempts to divorce himself from Crosse's experiments were not wholly successful. More than a decade later, a Mr. Wild, lecturing on Paddington Green against the Old Testament account of creation, referred to no less an authority than Faraday as having "demonstrated that life was but electricity, by producing through its agency animalcules, maggots, &c." He was alleged to have remarked, "Gentlemen, there is life, and, for aught I can tell, man was so created."[109] Outside the confines of the Royal Institution and its polite circles, it was difficult for Faraday to control the uses of his experiments. He tried hard, however, to divorce electrical experimentation on life from its populist and materialist contexts, making it a safe science for gentlemen.

MEDICAL GALVANISM

There were an increasing number of places in early Victorian England where the body could be galvanized. Medical electricity had been an occasional feature of the therapeutic repertoire of practitioners since the eighteenth century. During the 1830s concerted efforts were made to integrate electricity into hospital regimes,[110] while at the same time the proliferation of new galvanic technologies, partly driven by the needs of a medical market, led to a more widespread availability of the galvanic fluid. Electricians hawking their latest inventions would often target a medical audience.[111] The instrument-maker Edward Clarke, for example, announcing his controversial Magnetic Electrical Machine during the mid-30s, was quite emphatic that medical gentlemen would find his invention of great utility.[112] Instrument-makers' catalogs of the period frequently included electrical apparatus adapted for medical purposes in their range of advertised products.[113]

Application of electricity for therapeutic purposes certainly seems to have been a routine activity on the part of electricians. William Sturgeon mentions that he had been called upon to administer electricity to some young men who had fallen through the ice in the river near Woolwich. His efforts were unsuccessful owing, he recorded, to "two hours having been occupied, by the usual routine of medical treatment, before the batteries were employed."[114] The notorious Andrew Crosse was also reputed to have offered his services as a medical electrician to his less affluent neighbors.[115] The administration of shocks for the purpose of entertainment was also a routine part of electrical performances at popular venues such as the Adelaide Gallery.

A number of professional "medical galvanists" also offered their services to the public. One of the more successful of these practitioners was William Hooper Halse, who settled in London during the early 1840s, having previously pursued his career as a medical galvanist in Devonshire. He was associated with the radical publisher Richard Carlile, who had his own interests in electrical cosmologies and medical galvanism.[116] He had links as well with at least some members of London's electrical community, having published on several occasions in William Sturgeon's *Annals of Electricity*. In one contribution to Sturgeon's *Annals* he detailed a number of experiments in which newborn puppies were immersed in water until drowned before being returned to life by means of electricity.[117] He recommended his experiments to the medical profession as proof positive of the "astonishing powers of galvanism" in supplying the nervous fluid. A summary of his experiments was also printed in the Owenite *New Moral World*.[118]

After arriving in the metropolis in 1843, Halse spared no effort in publicizing his activities. His pamphlets were packed with grateful testimonials and stories of miraculous cures from his days in Devon.[119] He also apparently dabbled in mesmerism, describing himself in his broadsheet as a "medical galvanist, proprietor of the galvanic family pill."[120] Potential patients wishing to be galvanized had the choice of either attending at Halse's residence, at the cost of one guinea a week for half an hour's daily treatment, or purchasing one of his machines for their own use at the price of ten guineas. This, incidentally, was almost twice the price of an electromagnetic machine or induction coil purchased from an ordinary instrument-maker.

The most striking feature of Halse's account was the way in which the sick body was represented. Simply speaking, the sick body was out of control. It was subject to convulsions, spasms, and twitches in much the same way that the corpses electrified by Aldini and Ure had been. Or rather, in this case, the sick body did to itself what could also be done with electricity, and galvanism was the means of *regaining* control. The body was represented as an automaton divorced from the mind's domination.[121] Halse specified a wide range of complaints for which galvanism was the cure. He listed "all kinds of nervous disorders, asthma, rheumatism, sciatica, tic doloureux, paralysis, spinal complaints, long-standing headaches, deficiency of nervous energy, deafness, dulness of sight, liver complaints, general debility, indigestion, stiff joints, epilepsy, and recent cases of consumption."[122]

It was of particular use in cases of nervous disorder, since in such circumstances his galvanic apparatus was "actually supplying the nerves with their proper stimulus during the whole of the operation." In almost all these cases, the patient's body was in some way out of control. Mrs. S. of Torquay, for example, consulted Halse concerning her daughter,

> whose face on one side was dreadfully distorted, the right side of the mouth and right eye being so contracted that they almost touched. She could not close her eye either by day or night, and, as may be supposed, it had a very awful appearance.[123]

Halse galvanized her face every day for a month, and her face "resumed its natural appearance." Electricity had been a means of regaining control over her body.

Another such example was an Oxford clergyman suffering from indigestion, the result, Halse suggested, of too much studying:

> He was highly nervous, and totally unfit to enjoy society; his eye was dull in the extreme; he was subject to spasms, costiveness, headache, extreme languor, lowness of spirits, very excitable in sleep, feet and legs like ice, and not the least appetite. He frequently saw the vision of a white cat running about the house, and he declared to me that it appeared just as real to him as if it had actually been a cat.[124]

The clergyman had completely lost the ability to regulate his own body and mind. Again, it took a month of daily galvanism before the symptoms, including the white cat, eventually disappeared.

Pamphlets such as Halse's were explicit in their condemnation of the medical profession, who were represented as being both mercenary and unscientific. Halse's patients had almost invariably suffered at the hands of unscrupulous physicians before being saved by his galvanic ministrations.[125] They were unscientific because they failed to realize the therapeutic efficacy of galvanism and the scientific doctrines upon which its application was based. Even galvanism's disrepute as a therapeutic agent could be laid at the door of some of the profession's botched efforts to use it. A large number of them were no more than "a set of bunglers who have recourse to this agent, and who know no more how to manage it than a donkey knows how to manage a musket."[126]

More orthodox practitioners using electricity were, of course, just as scornful of the pretensions of irregular medical galvanists. Golding Bird, assistant physician at Guy's Hospital and supervisor there of their "electrical room," fulminated against his "quack" competitors:

> These medical galvanists lay heavy contributions on the public, and one of them in the west end of the metropolis, regularly pockets some thousands per year by his practice. There is scarcely a part of London where these irregulars do not exist—the small suburb of Pentonville alone rejoices in two, if not three.[127]

For Bird, the supposed electrical competence of men such as Halse was no compensation for their ignorance of medicine. Galvanism was viable as a therapeutic agent only under medical direction.

Bird and his colleagues at Guy's did, however, share with the medical galvanists a common perception of the disorders that could be treated galvanically, and of the kinds of bodies that required treatment. Electricity was a therapy for nervous disorders ranging from hysteria to paralysis. Bodies that were subjected to galvanism were out of control, and the role of the galvanic fluid was to restore regulation. Electricity could ensure not only that patients regained control of their own bodies. It also guaranteed that the doctor could control the patient's body as well.

This regulatory function of electrical treatment for Golding Bird was quite explicit in some of his remarks on hysterical patients, particularly in cases where imposture was suggested: "In more than one case of want of power in emptying the bladder in hysterical girls, I have succeeded in curing this annoying symptom by passing a pretty strong current from the sacrum to the pubes. My own impression has been, however, that the pain of the current and dread of its repetition have constituted the real elements of success in these cases."[128] For Bird it was ultimately irrelevant whether the loss of control was real or feigned. Electricity would cure it:

> If the patient simulates paralysis . . . she can seldom resist the pain and surprise of the shock, and the previously rigid limb will generally instantly move. On the other hand, in hysterical paralysis, where the affection, however excited at first, is now uninfluenced by the will, there are few curative remedies so important as the electro-magnetic current.[129]

Electricity's great advantage was that it could restore the regulatory power over the body of both the patient's will and the doctor's authority.

Despite this seeming consensus as to the regulatory power of electricity over the body, there was no such agreement concerning the way in which that power was exerted. Of particular interest here is the conflicting attitude toward pain of the irregular medical galvanists and their professional opponents. Galvanists such as Halse, while they made every effort to dramatize the extent of their patients' sufferings, were adamant that electricity properly applied was completely painless. Orthodox practitioners such as Bird, however, took the opposite view, regarding the pain inflicted by the therapy as integral to its action. This was a reflection of differing notions of what the relationship between healer and patient might be.

## THE BODY ELECTRIC

One of the most far-reaching and comprehensive attempts to establish electrical regulation over the body was that of the young surgeon Alfred Smee during the 1840s. Smee, the son of the accountant-general at the Bank of England, had received his medical education at King's College, London, before entering St. Bartholomew's Hospital where he was dresser to the infamous William Lawrence during the late 1830s.[130] Lawrence had been accused of blasphemy and materialism in the course of disputes with John Abernethy two decades previously.[131] Interestingly, as Marilyn Butler has recently argued, these disputes constituted one possible source of material for Mary Shelley's *Frankenstein*.[132] The relationship did not damage Smee's career, however. He became a member of the Royal College of Surgeons in 1840, and a year later, following his invention of a cheap and durable voltaic battery, he was elected a fellow of the Royal Society.

Smee's writings on electrical matters appeared throughout the following decade, culminating in 1849 with the publication of his *Elements of Electrobiology*.[133] Smee was no friend of reform. He vigorously opposed the reforms of the Royal Society during the mid- and late 1840s. In later life he stood unsuccessfully several times as Tory M.P. for Rochester. In many respects his science of electrobiology was offered as an explanation of and antidote to revolutionary enthusiasms.[134] He argued that a proper electrical understanding of the structure of the human brain as a series of connected batteries made plain the causes of such cases of "perverted reasoning."

Smee provided a systematic account of the sources of electricity within human and animal bodies and of the way in which body and brain were organized into a number of different batteries. He provided a new nomenclature to classify the different functions of body and mind, attempting to show how all those functions could be produced by means of electrical action. Following his teacher, William Lawrence, he insisted that life was a function of organization, but held that the organization should be understood as a highly complex voltaic circuit. All human and animal functions, from nutrition to reason, were solely dependent on physical laws.

A number of experiments were introduced to illustrate the ways in which electricity could reproduce the sensations of seeing, hearing, smelling, and so forth. Smee made the human body electrical by showing the ways whereby an

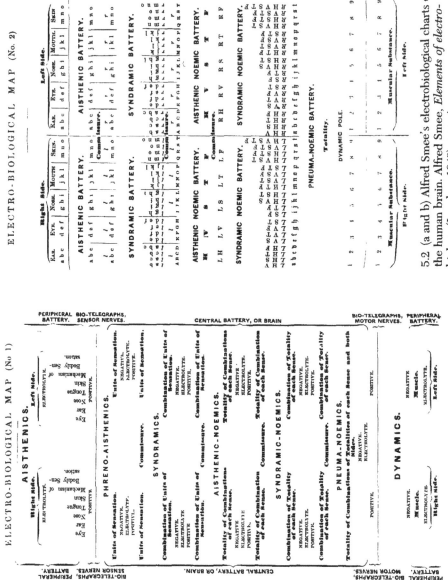

5.2 (a and b) Alfred Smee's electrobiological charts of the human brain. Alfred Smee, *Elements of electro-biology* (London, 1849), app. A.

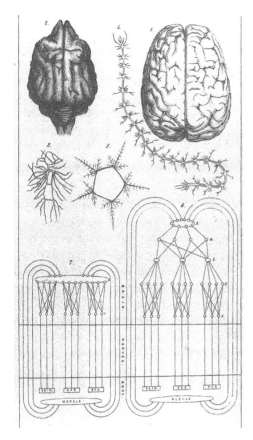

5.3 Schematic diagrams of the
electrical circuitry in the human
(on the right) and animal brains.
Alfred Smee, *Instinct and Reason*
(London, 1850), facing p. 210.

electrical experiment could reproduce its functions. This was typical of the kind
of argument deployed by Smee's fellow electricians. William Sturgeon, for example, had demonstrated that the earth rotated by means of electricity by constructing an artificial globe that did just that. Similarly, Andrew Crosse had
shown that crystals in the earth were the result of galvanic action by producing
such crystals artificially with his electrical apparatus. Such arguments worked
both ways. The success of the experiments buttressed the electrical ontology,
while the ontology suggested that it was the electricians and their experiments
who had the necessary skill and authority to investigate nature.[135]

Smee was trying to deploy the same strategy to validate his claims concerning the electrical nature of life. Having demonstrated with his ingenious experimental apparatus that electricity could indeed simulate the senses, he could
then plausibly argue for electrical structures in the brain. According to Smee's
model, the peripheral battery of the body was linked by "bio-telegraphs" (or
nerves) to the central battery of the brain, which acted as a sorting and combining system for the vast array of sense impressions that it received. The crucial
feature of the human brain, which differentiated it from that of an animal, was
the existence of an additional structure, the "pneuma-noemic battery," that was

responsible for combining the "totality of combinations of totalities of each sense and both sides."[136]

According to Smee's model, *all* sense impressions ever received were permanently stored in the central battery of the brain. These impressions then determined any individual's future behavior. The implications of this fact were spelled out in Smee's next publication, *Instinct and Reason: Deduced from Electro-biology.*[137] Here he showed how all animal and human behavior, from the simplest to the most complex, stemmed from electrobiology. Aberrant human behavior, ranging from Chartist agitation through religious enthusiasm to female hysteria, was the result of a combination of bad impressions in previous life and faulty electrical circuitry in the brain.

> A great stimulus always acts upon some persons more or less as a charm, but under its excitement the faculty of judgement is much interfered with. We find that under such circumstances large communities, or even whole nations, have their cool unbiased judgement deranged. I need hardly remind my readers of the Puseyite epidemic, under which so many gave up their faith and all their property to the insatiable jaws of the Roman Catholic church. I need hardly remind my readers of the revolutionary epidemic of 1847 [*sic*], which spread from country to country, so that even the sober-minded English, on the memorable 10th of April, were so affected, that London was placed under military defence, and the public buildings were converted into fortifications, armed and provisioned to stand a siege.[138]

Diseases such as hysteria could be cured by electrotherapy. More serious disorders such as Chartism required the imposition of new impressions to counteract those previously received. Electrobiology was a recipe for the strict regulation of everyday life. In all such cases of "perverted reason" there was "an irregular action of the brain, analogous to the irregular action of machinery."[139] As an orthodox Anglican, Smee was keen to emphasize that the science of electrobiology offered positive proof of Christian doctrine.[140]

Smee's claims were not universally well received. The *Literary Gazette* simply dismissed his work as "fatuous."[141] The *London Journal of Medicine* was more scathing. Smee was a "man of one idea" who held that "man (who has been called the 'prime work of God') is neither more nor less than an animated voltaic battery." They remarked that "in our judgement, the language is not only singularly incorrect, but the reasoning is very bad, and most of the reasonings are crude, unphilosophical, and unsupported by facts."[142] They complained, too, of the "fashion very prevalent at the present day, of disseminating doctrines nearly allied to materialism," accusing Smee of bad faith in denying the materialist implications of his own work.

The fiercest attack came from the *British and Foreign Medico-Chirurgical Review.*[143] The reviewer sarcastically suggested "that the author is either a wonderful genius, or deserves a precisely opposite character."[144] The remainder of the review made clear which of these two characterizations was accurate. Discussing Smee's experiments on electrical vision, the writer suggested that he should persevere until able to construct genuine *seeing* eyes, capable of being supplied at a low price to the unfortunates who stand in need of them. We do not see why, by the proper application of electric circuits, every body who likes it should not have an additional pair of eyes fixed to the back of his head; the great utility of which, as endowing the possessor with a marvellous power of *circumspection*, need not be enlarged upon.[145]

The review poured scorn in particular on the suggestion that morality was a product of electricity. It concluded by suggesting that Smee was "either a second Newton, or a shallow pretender," leaving the readers to draw the obvious conclusion themselves.

The *British and Foreign* returned to the attack a year later, reviewing *Instinct and Reason*.[146] Poking fun at Smee's attempts to build "a living, moving, feeling, thinking, moral, and religious man, by a combination of voltaic circuits," the review dismissed the work as an exercise by Smee in "the free indulgence of his propensity to vanity and egoism."[147] Smee was no more than a self-serving charlatan: "One thing he takes every opportunity of impressing upon the minds of his readers,—namely that Electro-Biology is to effect the great *moral* revolution by which the whole world is to be renovated; when, of course, Mr. Alfred Smee will take rank as a sort of second Messiah, whose work is to complete that which was left imperfect by the first."[148] Smee's prescription for electrically monitoring and regulating society was too outlandish for the mild, reformist constituency of the *British and Foreign*.

Conflicts such as these were, of course, about authority. Different constituencies were urging their own claims to expertise over the body. The matter of who might count as being the radical in such circumstances is not entirely straightforward. Smee responded furiously in the *Lancet* to the *British and Foreign*'s attack, naming the reviewer as William Benjamin Carpenter and accusing him in turn of arrant quackery in supporting homeopathy.[149] For Smee, Carpenter's attack was evidence that he too was subject to that "perverted reason" which electrobiology both explained and promised to cure.

### Conclusion

It would be a mistake to suppose that the apparent heterodoxy of figures such as Smee was the cause of their marginalization. It is equally arguable that their increasing marginalization resulted in increased heterodoxy as they sought different audiences for their work. In any case attributions such as these were as negotiable then as they are now. Electricity and its role in regulating the human body were valuable resources for a variety of interests. Even Carpenter, despite his disdain for Smee's electrobiology, was happy to invoke human electricity on occasion. Others, such as Golding Bird, labored hard to dissociate their work from that of the irregulars to make electricity safe for hospital consumption. In fact nothing about the electrical body was uniquely heterodox. What mattered was the ways in which it was used and by whom.[150]

This is not to suggest that the electrical body was completely open to interpretation. Smee's dilemma illustrates the fact that some positions were very difficult to maintain. Smee's position might be characterized as that of a "reactionary heretic" in that he was seeking to utilize heterodox medical and galvanic resources to produce a reactionary recipe for social action.[151] He was, however, branded as a radical. It is worth noting that Carpenter, in the passages just quoted attacking Smee, satirized him as a mercenary tradesman and as an enthusiast. This was an easy task for Carpenter since the typical places where electricity was seen in London were sites either of trade or of enthusiastic fervor.[152] Electricity's links with radicalism were forged through social geography. Nothing Smee could say in terms of ideology proved sufficient to break that link.

Electricity's place in early Victorian society was becoming determined by the places where it was seen.[153] Outside the safe setting of the Royal Institution, electricity took its place, on the one hand, in the commercial, entrepreneurial spaces of the Adelaide Gallery or the Royal Polytechnic Institution. On the other hand, its theatricality in those spaces linked it to enthusiasm. These were also, increasingly, the spaces of political and religious radicalism. Using the resources tainted by those places was therefore a dangerous business. A great deal of work was required to make electricity a safe way of regulating the Victorian world. The preceding chapters have tried to show how electrical experiments were bound to particular places. It mattered where and by whom an experiment was made. The cultural contexts of electricity varied enormously, from the Royal Institution, to the Adelaide Gallery, to the consulting room of a medical galvanist, to the lecture hall of a political radical.

This chapter has shown how electricity could, however, be used to breach barriers, between new disciplinary formations such as astronomy and geology, for example, and in particular to breach barriers between nature and society. Electricity seemed in many ways to simulate the properties of vitality. The galvanic fluid could be used to bring back at least the semblance of life to the dead. If some experimenters such as Crosse or Halse were to be believed, electricity could indeed be an agent of resurrection and could produce life from dead matter. Maintaining a strict boundary between the natural and the social mattered for many during this period, however. This was the source of Michael Faraday's efforts to carefully recircumscribe the limits of electricity's capacity to illuminate the mysteries of life. Radical agitators such as Mackintosh (or the mysterious Mr. Wild) were clearly aware of this as they put electricity to work in breaking such barriers. Such efforts were strongly resisted. The universality of electricity had to be kept in check. It turned out that there were very few places where that universality could be maintained.

# PART TWO

# Managing Machine Culture

# INTRODUCTION

## From Performance to Process

ONE OF THE key claims of the first half of this book was that electrical experimentation during the first few decades of the nineteenth century in Britain was quite frequently an irreducibly local performance. It mattered a great deal to Michael Faraday, to name the most famous example, that his experiments took place at the Royal Institution and that his public performances occurred in that illustrious establishment's lecture theater.[1] Thus the first half of this book devoted particular attention to the examination of electrical experimentation's places. Where an experiment took place, as much as who performed it, was a matter of consequence. The way an experiment might look on the stage of the Royal Institution was not at all the same as its appearance at one of John Peter Gassiot's electrical soirees or at the Adelaide Gallery of Practical Science.

Just as places mattered, individual performances mattered too. As many commentators have noted, notions of the self in relation to society were in the process of being reinvented during the first half of the nineteenth century as new groups (and new types of individual) emerged and aimed to forge identities for themselves.[2] The natural sciences were no exception to this trend. The life of an experimental natural philosopher during the first half of the nineteenth century was not straightforwardly mapped out. There was no blueprint of a career profile to follow from classroom through research to academic success. Careful self-fashioning through individual performance was required to make the early-nineteenth-century experimenter.[3] Exhibitionism, in some form or another, was a crucial strategy in establishing both the experimenter and his experiments. His performances at the Royal Institution were a key feature of Faraday's status and of the status of his experiments. In the same way, though aimed at very different audiences, performances at the Adelaide Gallery or the London Electrical Society made experiments as well. Such places provided electricity with a context and therefore with a cultural sense. The significance of an experiment could not be dissociated from its space. Exhibition forged the links interconnecting the experiment, the experimenter, and the nature that was being displayed.

This second half of the book moves toward a different perspective, although many of the themes already emerging will continue. Early Victorian Britain was increasingly a machine culture. Fascination with, and speculation about, machinery and its futures was rapidly becoming a staple part of early Victorian life. Commentators from a wide variety of social backgrounds and cultural perspectives engaged with the machine, seeking to understand the role it would play in transforming their way of life. Machines could be celebrated as the sources of the new wealth that was already in the process of transforming the nation into the nineteenth century's greatest imperial and industrial power.[4] Equally they could be condemned as the source of the poverty visible in the metropolis's and many provincial cities' growing slums. They could be metaphors for progress as much as for spiritual decay. Electricity had an important role to play in this machine culture. The following chapters seek to examine aspects of that role.

A note on terminology may be appropriate here. Much of what will be canvassed in the following chapters would in modern terminology be described as *technology*. Accordingly, that term will be routinely used for purposes of discussion. It should be emphasized, however, that such a term was not at all common in contemporary usage. While modern commentators draw distinctions between science and technology, the nearest equivalent demarcation in the early nineteenth century was that between science and *art*. Art defined those practices that required craft skill in their production, ranging from the activities of the painter and sculptor to those of the machine-maker. This was conventionally contrasted with the abstract system building of the natural philosopher. Institutionally, the practices were represented by the Royal Society and the Royal Society of Arts. The boundaries were not, however, static. Distinctions might be drawn between the fine and the mechanical arts, with the latter being more closely associated with the natural sciences. At the same time, magazines seeking to purvey culture to the middle and higher orders of society frequently invoked both art and science (as well as literature) in their titles.[5]

Machines certainly occupied a contested space that crossed the boundaries of science and art. They were represented in different places and by different commentators as products of both ways of going on. The worlds of natural philosophy and machinery intersected in many ways. Debates raged as to what the role of science in industrial progress and the production of technology might be and what it should be. While some commentators such as William Whewell argued fiercely that natural philosophy could not and should not be tainted by commerce, others such as Charles Babbage were just as adamant that science was central to any such progress and production.[6] His own projected calculating engine was a product of the factory as much as of Cambridge mathematics, and the tables of numbers it was designed to produce were aimed at factory managers as much as at astronomers.[7] On a practical level, the vast range of natural philosophical activities conducted in London simply could not take place without the networks of manufactories and workshops scattered throughout the city. The practice of experiment required a range of human and material resources that few other cities could match. Babbage's engine, for example, would have remained even more of an elusive goal without the expertise and ingenuity of the machine-makers who crafted the parts and designed the tools to build them.[8]

Many commentators identified the capacity for self-regulation as central to their definitions of machinery. The self-regulation extended as well to a regulation of the workforce attending the machine. Issues of autonomy and control over the pace of work were crucial to the self- and group identity of early industrial workers. The rate of labor was to be determined in accordance with the customary tradition of a particular craft rather than by the requirements of an individual master. Competition between workers was strongly discouraged. As masters sought to introduce an increased division of labor and new machinery into the workplace, this tradition was increasingly under attack. New means of production required new forms of discipline. Machines seemed to provide such a disciplinary technology just as much as they provided increased production. For many masters and many analysts of machinery's role in political economy, one of the machine's main virtues was the capacity it provided for massively increased control in the workplace.[9]

Charles Babbage, in his *Economy of Machinery and Manufactures*, emphasized the self-regulating capacity of machinery as one its central features.[10] Machinery was a physical embodiment of the division of labor, a means of internalizing its drive toward specialized efficiency. This was the essence of a machine: "when each process has been reduced to the use of a simple tool, the union of all these tools, actuated by one moving power, constitutes a machine." Machinery enforced uniformity of work and uniformity of product. "Uniformity and steadiness in the rate at which machinery works, are essential both for its effect and its duration." This capacity of the machine could be imposed on its attendant laborers as well: "One great advantage which we may derive from machinery is from the check which it affords against the inattention, the idleness, or the dishonesty of human agents."[11] Machinery thus made possible a rational system of labor. Babbage's favorite example was, of course, his own Calculating Engine, which exemplified an embodiment not just of the division of labor but of the division of mental labor. In a completed engine the whole range of calculational practices associated with the production of numerical tables could be replaced by the simple business of turning a handle.[12]

Babbage's emphasis was on the machine's capacity for regulation. His essay was full of detailed descriptions of those flywheels and governors that guaranteed the uniformity in action and motion of modern machinery. Peter Barlow, in his *Treatise on the Manufactures and Machinery of Great Britain* concurred. Like Babbage he included technical accounts of the intricate means of governing and regulating machinery's movement. The capacity for self-regulation, the embodiment of management, that these devices provided, was a central feature of economical machinery. Indeed regulation was central to the meaning of *economy*: "expense . . . is not . . . all that enters into the question of what is to be understood by the term economy. We have to look at the convenience of application, the steadiness and uniformity of power and motion."[13] Barlow even celebrated the prison treadmill as the epitome of economy. This was a machine that did no productive work but by its uniform, regular performance disciplined the prisoners inside it.[14]

The image of machinery as the linchpin of a system was clear in Andrew Ure's *Philosophy of Manufactures*. Ure was even more emphatic than Babbage and Barlow that machinery's key characteristic was its self-regulation, and that one of its main virtues was its capacity to extend its control of itself to control the workforce as well. He eulogized a utopian vision of the ideal factory.

> The term *Factory* in technology, designates the combined operation of many orders of work-people, adult and young, in tending with assiduous skill a system of productive machines continually impelled by a central power ... this title, in its strictest sense, involves the idea of a vast automaton, composed of various mechanical and intellectual organs, acting in uninterrupted concert for the production of a common object, all of them being subordinated to a self-regulated moving force.[15]

The precise use of words is worth noting. The machines here are *productive*, while the task of the laborer is to *tend* the machines. Ure was emphatic that one of the primary functions of machinery was to displace skilled labor, substituting unskilled, preferably child or female, and therefore docile labor. The gender connotations of *tend* may well be worth noting. Like Barlow, Ure remarked the irony that a modern manufactory, despite the precise etymology of the word, was the last place to look for any work carried out by hand.

It is clear that none of these texts should be read as accounts of actual factory conditions or as reflections of real machine use. They do, however, represent particular accounts of how machinery could be depicted, and in particular how machines were seen as playing a role within a system. More radical commentators could concur with this view, even as they did not condone its effects. Peter Gaskell, for example, largely accepted Ure's view of the factory system and machinery's role within it. In fact he cited Ure extensively in his own account. Machinery was indeed designed to replace skilled artisanal labor with that of women and children more amenable to control.[16] His views on the implications were, however, different. While agreeing that it was "the great aim of machinery to make skill or strength on the part of the workman valueless, and to reduce him to a mere watcher of, and waiter upon, *automata*," he argued that the artisan's sense of pride in his "skill and ingenuity" would be destroyed as he became "as much a part of the machines around him as the wheels or cranks which communicate motion."[17] Unlike Ure, Gaskell saw this systematization of men and machinery as a potential cause of, rather than a panacea for, social disorder.

Machines, by this account, were parts of a process. Their defining features were their uniformity and regularity, their capacity for limitless reproduction. Babbage in particular drew attention to this aspect of machine work: "Nothing is more remarkable, and yet less unexpected, than the perfect identity of things manufactured by the same tool."[18] Individuality was not a feature to be expected or desired in the productions of a machine, or in its performance. Mechanical uniformity was not universally applauded. As noted, radical commentators took exception to the dehumanizing tendencies of the machine. Conservative commentators, notably Thomas Carlyle, could take a similar view.

Machines and the new system of manufacture within which machinery operated had a soulless quality that threatened to spread and infect the nation's
body politic. For Carlyle, there was something monstrous about the power
attributed to the machine.[19]

Interestingly, Carlyle's metaphor for the life of the machine was galvanic.[20]
Galvanic life was artificial, soulless, and contrived. It was the result of the Enlightenment's obsession with experimentation whereby the fabric of society
was imperiled in the search for novelty. Machinery destroyed the social order:

> The huge demon of Mechanism smokes and thunders, panting at his great task, in all
> sections of English land; changing his *shape* like a very Proteus; and infallibly at every
> change of shape, *oversetting* whole multitudes of workmen, as if with the waving of
> his shadow from afar, hurling them asunder, this way and that, in their crowded
> march and curse of work and traffic.[21]

The demon's life was galvanic, by which Carlyle implied its illegitimacy. The
machine age's materialism was in danger of destroying the Anglican religion
that remained the "inmost Pericardial and Nervous Tissue, which ministers
Life and warm Circulation to the whole" of the body politic. Without it, that
body would be "animated only by a Galvanic vitality . . . and Society itself a
dead carcass."[22]

This is much what the authors of *Fraser's Magazine*'s "New Frankenstein"
must have had in mind.[23] Machinery's animation, the life of the factory system,
was a fraud. Frankenstein's monster could indeed be restored to life by galvanic
agency. He could even be provided with a rational mind. The galvanic life was,
however, only an illusion, a simulacrum of reality. Frankenstein's monster had
no soul and could not, therefore, be truly living. To its conservative critics this
seemed to be the problem with the factory system and the machines that animated it. The result of this mechanization was a betrayal of the nation's spirit
and the reduction of its people to nothing more than "galvanized corpses."[24]
The identification of galvanism with the soulless automatism of machinery and
the factory system is significant. During the late 1830s and the 1840s, as such
sentiments were spreading, the electric fluid itself was indeed slowly spreading
into the factory system as well.

This second part of the book deals with the industrialization of electricity
during the 1840s. As the first part of the book suggests, electricity was certainly
already a part of consumer culture and of the world of work as well. Electrical
apparatus and experiments took their place in the Adelaide Gallery or the Polytechnic Institution alongside steam engines, power looms, and mechanical curiosities. The production of electrical effects depended as much on London's
networks of instrument-makers' shops and workshops as it did on laboratories
and exhibition halls. During the 1840s, however, electricity became more
straightforwardly a matter of commercial, financial speculation. Electrical apparatus, experiments, and machines became subjects for patents. They became
commodities in the most unambiguous of senses. Electricity during the 1840s
was brought and sold for economic gain. The focus in many ways, therefore, is

on the material culture of electricity. Both Asa Briggs and Carolyn Marvin have recently drawn attention to the proliferation of things during the Victorian age.[25] This book shares their concern with trying to understand the impact of material goods on nineteenth-century life, as Victorians struggled to make sense of the new technologies and technological products that increasingly surrounded them.

Most histories of electrical technologies pay very little attention to the period covered here. Attention is typically focused on the development of large-scale electrical industries during the second half of the nineteenth century. Earlier work is more often than not briefly covered in an introductory chapter on "origins."[26] Such histories are as a rule internalist, limiting their attention to the linear development and refinement of instruments and technologies through the century. Little if any attention is devoted to the cultural place of electrical technology. While such histories frequently pay lip service to the innovative, spectacular character of developments in electrical technology, for example, no effort is made to draw out the significance of that character for contemporary observers. The first half of this book has demonstrated how much the technology of display mattered for early-nineteenth-century electricians. That theme remains central to the second half of the book as well. Electrical technologies in traditional histories are also seen as being straightforward applications of new developments in electrical science. The model is a simple one of scientific discovery and technological application.

In this book, distinctions between science and technology, between discovery and invention, are taken to be emergent properties. That is to say, they are the products of historically contingent negotiations rather than self-evident properties of specific practices or bodies of knowledge. In this again, the analysis offered here has a Latourian flavor. The focus is on the negotiation that resulted in the attribution of a certain status to particular items of work and particular individuals. The decision as to whether a particular instrument or process should be recognized as an invention, or an individual as an inventor, was a matter of perspective. Moral as much as technical (not to mention financial) matters were at stake in such attributions. None of this is to suggest, however, that such decisions were inconsequential. The boundary between discovery and invention, or for that matter between inventor and charlatan, might be permeable, but it mattered a great deal on which side of the barrier an individual eventually found himself. Such divisions were (and are) perfectly real precisely because of their constructed, social nature.

Chapter 6 outlines some of the issues surrounding electricity's entrance into this commercial, entrepreneurial world and some of the strategies that hopeful inventors adopted in their efforts to straddle the worlds of science and commerce. This task was by no means straightforward. Many gentlemen of science, even those in general anxious to promote the economic utility of their scientific products, were wary of being seen to practice experiment for their own financial gain. Such behavior was regarded as incompatible with the proper pursuit of specialized knowledge as a personal vocation.[27] Those such as William Henry Fox Talbot who wished to be both gentlemen and patent holders found themselves in an ambiguous position. Patents in any case were by no means secure. They were also expensive. Protecting an invention by patent was a potentially risky strategy and did little for those who also aimed to enhance their personal status through the business of invention.

The first section of the chapter outlines the emergence of electrometallurgy as an industrial process from its origins in attempts to improve the performance of electric batteries during the 1830s. By the end of the 1840s, what had once been the side effect of producing constant currents in a battery had become the source of a whole range of luxury consumer goods. Electroplating became a new industry, replacing old craft skills with electrical technology. The status of those aiming to claim the new processes as their own invention was, however, ambiguous. The chapter's second section charts the efforts of Thomas Spencer to be recognized as the inventor of electrometallurgy, the strategies he adopted to advance his claims, and the efforts of others to undermine his reputation and hence his claims to invention. It will become clear that a self-proclaimed inventor had to do rather more than point to a patent, or even to a prior publication, to warrant his claim.

The chapter concludes with a look at some other efforts to produce new electrical technologies during the second quarter of the century. Even as steam was barely established as a locomotive power, hopeful electricians were touting their apparatus as the potential source of a more economical and efficient motive power. To some it seemed that the power available to be harnessed from the electric battery was for all practical purposes indefinite. James Joule's experiments during the late 1830s and early 1840s were efforts to test the capabilities of electromagnetism as an economical competitor of steam.[28] His eventual pessimism did little to prevent others from enthusiastically exhibiting and patenting their own engines and locomotives. Electricity could be a source of light as well as motion. As a producer of locomotive power, luxury goods, and spectacular illumination, the electric force dominating nature was celebrated as a force that could dominate society as well.

The electrical technology that seemed to capture the imagination of the Victorian age above all else was, however, the electric telegraph. The possibility of, for all practical purposes, instantaneous communication across vast distances in space was unprecedented. Commentators waxed lyrical over the potential of this new invention and what it revealed concerning the power of man to dominate his environment. Chapter 7 follows the early development of the electric telegraph during the decade or so following the first patent, the controversies surrounding its invention, and some of the uncertainties that emerged along with this new commodification of information. The chapter aims to make clear the contingencies of the telegraph. Many commentators have noted, for example, the telegraph's close links with the expanding English railway system.[29] While there can be no doubt concerning the role the railways did play in the telegraph's early history, the link was by no means inevitable. The early projectors in their first plans made little or no reference to the telegraph as a railway signaling technology.

The chapter's first section follows the fortunes of Charles Wheatstone and William Fothergill Cooke as they attempted to transform the late-1830s electrical technology of display into robust and reliable equipment that could survive the exigencies of the world outside their laboratories and workshops. The social development of the telegraph went hand in hand with its technical refinement, if such distinctions are at all legitimate. Its promoters approached the railway companies in search of financial support, and the technical appearance of the telegraph shifted as it conformed to the new social roles that were now envisaged for it as part of railway culture. The contingency of success is highlighted by a contrast of Cooke and Wheatstone's efforts with those of their contemporary Edward Davy. His failure to transform his experimental telegraph into a viable, working technology was not the result of technical inadequacy as much as of a failure to deal with the problematics of Victorian financial entrepreneurism.

As with electrometallurgy, the question of telegraphy's invention was fraught. In the case of Wheatstone and Cooke in particular, whose commercial partnership rapidly dissolved into acrimony, the question of priority in invention could not easily be disentangled from the issue of financial control over the telegraph's deployment and use. Part of the problem was the telegraph's very novelty. It simply did not appear to be straightforwardly obvious what exactly was to *count* as the telegraph's invention. While Wheatstone pointed to his scientific attainments as professor of natural philosophy at King's College, London, as the proof of his priority, Cooke claimed that his role in projecting the telegraph as a complete system and his management of its subsequent deployment were proof that his was the superior part of the inventive process. Issues of status were, of course, at stake in all this. The partners clashed over what kind of activity might legitimately be credited as invention.

By the mid-1840s the electrical telegraph was widely recognized as an integral part of any efficient railway network. It provided the long-distance signaling technology that allowed for the systematization of railway travel. Without the telegraph, the railway's reliability could not be guaranteed. At the same time, particularly following the incorporation of the Electric Telegraph Company in 1845, new uses were found for the new technology. The telegraph ceased to be just the signaling arm of the railways and became a means of mass communication. The deployment of the new technology in this way required new understandings and techniques of communication. Information traveling down the telegraph wires had to be packaged in new ways as economically viable units. Victorians had some trouble adapting to a means of communication that violated their culture's usual norms of privacy and of courtesy. Talking through the telegraph needed a new language that standardized speech.

An increasingly common metaphor referred to the electric telegraph as the nervous system of the Victorian state. The metaphor worked both ways. The

expanding telegraph system was also recognized as an apt analogy for the workings of the body's nervous system itself. Chapter 8, addressing the electrical
treatment of nervous disorders, may initially seem out of place as part of the
book's exploration of the commercialization and mechanization of electricity.
However, just as electrometallurgy or telegraphy systematized the practices of
the electricians' technology of display, so did the new medical discipline of
electrotherapeutics.[30] The chapter follows the activities of the London physician Golding Bird, placing his efforts to discipline the practices of medical
electricity in the context of the troubled state of the medical profession itself
and its institutions at this time.[31] Bird aimed to dissociate electricity as therapy
from its unsavory connotations by bringing it inside the hospital where it could
take its place as part of a rationalized regimen of medical care.

As Bird campaigned for electricity's role as part of the standard armory of the
accredited medical practitioner, the technology of electrotherapeutics during
the 1840s also flourished. New apparatus made the new therapy easy to apply
without the craft knowledge of the experienced electrician. Instrument-makers
aimed their wares at a market of medical practitioners who would utilize electricity as part of the ordinary regimen. With this discipline came a focusing of
electricity's role. It was no longer the universal panacea but a therapy appropriate for the cure of a specific range of nervous disorders, particularly those of
women. Bird and his cohorts did not see the medical application of electricity
as part of flamboyant performance but as a carefully regulated process designed
to allow the practitioner to discipline and control the otherwise uncontrolled
performances of recalcitrant female bodies. The need for such control was articulated in terms of the need to maintain and preserve the stability of the family
and therefore the state's social life.

This second half of the book describes, in various contexts, the mechanization and systematization of electrical practices and technologies. Electricity
during the 1840s was disciplined. It could be regulated to generate and regen-
erate mass-produced consumer goods. Electricity was the key to disciplining
railway travel, reducing the danger and uncertainty of high-speed travel on the
rails. It disciplined the workings of Victorian commerce, making carefully packaged and standardized information available to all who were willing to pay the
price. Electricity also disciplined the excesses of the human body, bringing it
back under control and maintaining the stability of Victorian domestic life. In
many ways the two-edged metaphor of nerves and telegraph wires was particularly apt. Electricity provided an especially powerful tool for regulating the
Victorian body as much as the Victorian state. It was an appropriate addition to
Babbage's and Ure's fantasy of the factory system.

# CHAPTER 6

# They Have No Right to Look for Fame:
# The Patenting of Electricity

THIS CHAPTER takes as a starting point the problematic status of invention in early Victorian England. Invention itself was a highly negotiable activity and the status of inventors difficult to manage and negotiate throughout this period. Making distinctions between discovery and invention, for example, was not an easy process.[1] These were cultural, social problems. In many ways they amounted to the issue of what kind of person an inventor (or, for that matter, a discoverer) should be. In other words, they were questions concerning the appropriate social group and place for the inventor. The chapter title reflects one response to the question. To be an inventor, and in particular to aim at personal profit from invention, was to exclude oneself from the company of philosophers. If the inventor aimed for cash, he should not expect to receive kudos as well. Natural philosophy was held by some to be a self-consciously disinterested activity. Others disagreed, seeing no reason why entrepreneurial activity should be a bar to inclusion in a gentlemanly scientific culture.

Invention for such men was not an anonymous activity but entitled them to a name (and fame) as well. More could be at stake in an individual's taking out a patent for an electrical invention, for example, than protecting a potential source of income. Patenting was for some men a possible strategy for gaining entrance into the philosophical community. It is worth noting that priority disputes of the kind to be discussed in this chapter were at least sometimes without a direct financial stake. What mattered to the participants was the status they believed their position as the inventor of a new machine or process should allow them. The gentlemanly model of the disinterested investigator of nature was not the only one available. It is arguable that the cult of personal disinterestedness, with its rigid distinctions between science and art, fostered by early-nineteenth-century gentlemen of science was as much as anything else a reaction to a prolific culture of entrepreneurship, in science as in all other cultural forums.[2]

Patents proliferated during the first half of the nineteenth century. Hopeful inventors sought to make their fortune by protecting their innovations from piracy and encouraging entrepreneurs to develop and finance their new machines and processes. In 1800, 96 patents were sealed in England and Wales; by 1850, the number had increased to 513.[3] Much ink has been spilled over the question of the relationship among patents, increase in inventive activity during this period, and the industrial revolution itself. The aim of this chapter is not to engage in this debate. Rather the aim is to consider the taking out of patents, protecting an invention from piracy, as one of a range of options available to an electrical inventor as a means of establishing himself as *the* inventor of a new device or process. Quite simply as well, the aim is to outline how electricity during the 1840s was increasingly becoming a matter of commercial, economic speculation.

Taking out a protective patent during this period was a risky business. To all intents and purposes, in law patents were still governed by the 1624 Statute of Monopolies. Other than that, very little statute law governed their administration, with the result that they were largely subject to common law and the whims of successive generations of judges. While by the late eighteenth century the attorney general was responsible for deciding whether or not a patent should be granted, establishing its validity against attempts at infringement was a matter for the courts.[4] By the 1830s, given prevailing laissez-faire sentiment and a common view of patents as monopolies under another name, sustaining a patent against infringement in court was notoriously difficult. A litigating patentee was as likely to have his patent declared void as he was to have his rights protected.[5]

Patents were also very expensive to acquire. Estimates as to cost differ, but particularly in view of the desirability of also taking out separate patents for Scotland and Ireland, an enforceable patent would require several hundred pounds in expenditure. A patentee had to be confident that his patent was both safe and potentially lucrative before committing himself to what was often a considerable financial outlay.[6] By the 1820s, a new profession of patent agents was being established to guide prospective patentees through the process. Joseph Robertson, editor of the *Mechanics' Magazine*, was one of the more prominent of this new breed.

There was a self-consciousness about the category of inventor as well. Not only did the *Mechanics' Magazine* aim itself at such practitioners of practical science, but new journals also set out to exploit what seemed to be a new category of potential readers. Magazines such as the *Inventor's Advocate* or the *Patent Journal* aimed to provide their audience with accounts of new developments in the sciences, descriptions of new machines and processes, and news relevant to their concerns with invention. Societies such as the Inventors' Aid Association or the Inventors' Law Reform League were established to define and protect inventors' interests in ongoing disputes concerning reform of the patent laws.[7] The *Mechanics' Magazine* celebrated the "classlessness" of its contributors, drawn as they were from all sections of society.[8] These were the kinds of men to whom Thomas Hodgskin referred as "men who labour a little, by, or in conjunction with, this machinery, who are at once labourers and capitalists" and would "extinguish both the mere slave-labourer and the mere idle slothful dolts, who live on the rent of land or the interest of money."[9] Electrical inventors joined a group of men who saw themselves as having common ground in the business of making and profiting from inventions.

Taking out a patent was only one in a range of possibilities facing an electrical inventor wishing to protect and profit from his work. Exhibition was a means of acquiring entrepreneurial attention, although an exhibitor had to take care to avoid charges of prior publication if a patent remained the long-term goal. Exhibition, of course, could be financially rewarding in its own right, precluding the need for a patent to ensure profit. Similarly, simple publication was a possibility, particularly if the inventor were less interested in cash than in kudos. By the 1840s an increasing number of journals and magazines existed through which invention might be made public. Indeed, for those with status, particularly philosophical status, in mind, taking out a patent could be counterproductive. The secrecy implied, and the sordid commercial motive imputed, detracted from what gentlemen of science regarded as proper behavior. Exhibition, particularly at explicitly commercial venues such as the Adelaide Gallery or the Polytechnic Institution, could be equally fraught. The problem was one of being public without overstepping the bounds of appropriate behavior. Strategies for making public claims of priority had to be effective as ways of gaining and maintaining status as well.

William Robert Grove made the gentlemanly position clear:

> It would scarcely add to the dignity of philosophy, or to the reverence due to its votaries, to see them running with their various inventions to the patent office. . . . If parties look to money as their reward, they have no right to look for fame; to those who sell the produce of their brains, the public owes no debt.[10]

Grove, like many gentlemen, was trying to put together a firm demarcation between discovery and invention, philosophy and industry. The issue was one of public demeanor. To be seen profiting from philosophy was simply inappropriate. It was not a matter of denying science's utility, either. Very few if any early Victorian practitioners would make such a claim, though they might deny that utility was philosophy's raison d'être. Gentlemanly natural philosophers sought to make their vocation a source of public good rather than private profit. Others, however, including most of the protagonists of this chapter, resisted the imposition of such a demarcation. While William Sturgeon, for example, might be scornful of the scientific pretensions of the patentees of particular inventions, he did not question the philosophical status of patentees in general.[11] Electricians aiming in increasing numbers during the 1840s to patent and thereby profit financially from their inventions did not mean to abrogate their "right to look for fame" by so doing.

## THE ART OF WORKING IN METAL

While the first patents of items of electrical technology taken out in Britain concerned the electric telegraph, during the first half of the 1840s the majority of electrical patents concerned another genre of techniques and instruments. They aimed to protect a range of electrical technologies to be used primarily in the printing, gilding, and metal-plating industries. They represented the first significant commercial and industrial exploitation of electricians' practical apparatus and know-how. Like the telegraph (the subject of chapter 7) these various technologies were adapted from the routinely used laboratory and popular lecturing paraphernalia developed during the previous decade.[12] In particular, they represented the transformation of the voltaic battery into a producer of material commodities. With electrometallurgy, the battery literally became an economic machine, whose efficiency was judged by its capacity to produce maximum output for a minimum of input.

Much effort during the 1830s had been devoted to the problem of providing a constant and reliable source of electricity. The first voltaic cells, derived originally from Alessandro Volta's *couronne des tasses* arrangement, typically consisted of copper and zinc plates immersed in a dilute acid, such as the Wollaston or Cruickshank arrangements of plates in separate cells of long wooden troughs.[13] They suffered from a variety of problems, the most pressing of which from the perspective of many electricians was the short duration of current. The intensity of the flow tended to fall off very rapidly with time. The result was that in many cases effects were both short-lived and hard to detect. This was a particular problem to those whose living depended on the production of spectacular electrical display. Probably the most famous manifestation of this problem was John Herschel's retrospective assertion that only lack of adequate battery power had prevented his beating Michael Faraday to the magneto-optic effect.[14]

Electricians explained their problem in terms of the difficulty of getting sustained and efficient work from the battery. George Bachhoffner pinpointed the battery's complications as follows:

> On taking a retrospective view of the different arrangements of the galvanic apparatus, from the pile of Volta to the battery of Professor Daniells [*sic*], the following points will, I think, be tacitly agreed to by all experimenters:
> 1st—Their complicated arrangement
> 2nd—The limited duration of their action
> 3rd—Their original expense and cost of repair when out of order.[15]

These issues were economic and pragmatic. Their solution was a matter of making the battery a cost-effective machine.[16] Devising a way of producing a constant current out of a battery was a matter of saving on the amount of fuel that the machine needed to work. A particular cosmology was implicit as well. An efficient battery would be the one that most successfully mirrored the natural economy. Philosophical principles were thus implicit in commercial, economic calculations.

Simply speaking, electroplating and its associated technologies, usually lumped under the heading of electrometallurgy, developed out of the solution to the problem of providing a constant electrical current from a voltaic battery. A commodification of electricity was the outcome of pragmatic tinkering. John Frederic Daniell, professor of chemistry at newly founded King's College, London, had overcome the problem by separating the copper and zinc plates with a porous membrane, immersing the copper in a copper sulfate solution and the zinc in sulfuric acid. As a result, the hydrogen produced by the combination of the zinc with the sulfuric acid, which was the main cause of the problem, now combined with the copper sulfate to form sulfuric acid and was thus removed from the solution.[17] In some circumstances the reduced copper tended to coat the copper plate. Many commentators noted that it was sometimes possible to peel away this new layer and that it would come away showing an exact replica in relief of the plate to which it had adhered.

The publication toward which most British contemporaries pointed as making explicit the commercial potential of electrometallurgy was that of Thomas Spencer of Liverpool in 1839. Spencer was a local Liverpool businessman in the carving and gilding trade with links to the city's philosophical instrument-makers. His paper was read out at a meeting of the Liverpool Polytechnic Society on 12 September and subsequently published at the society's expense.[18] From the beginning, Spencer's announcement was surrounded by at least a certain degree of acrimony. According to his own account, having read in the *Athenaeum* of recent experiments by Jacobi in St. Petersburg that appeared to resemble his own, he delivered a note to the Liverpool Polytechnic Society stating his intention of making his own experiments public at that year's meeting of the British Association for the Advancement of Science, to be held in Birmingham.[19] That association, after its foundation by enthusiastic provincials, was by now securely in the control of a metropolitan and Cantabrigian elite who drew a firm demarcation between their own position as specialist experts and that of their audience. The association's annual gatherings were designed to promote their own vision of science's division of labor.[20] It was for them a self-consciously gentlemanly forum, and the association's leadership kept strict control over the papers presented.

According to Spencer, he had contacted John Phillips, the association's assistant secretary, informing him of his intention to deliver his announcement and two other papers at the meeting and inquiring whether this would be possible. Phillips had replied that his communication on electrometallurgy had been placed on the list for the Mechanical Sciences Section while the other two would be communicated to the Chemical Section. Arriving at the Mechanical Section, he was informed by "a Mr. Carpmall [*sic*]" that his paper was to be the first read on Thursday morning.[21] At the appointed time, however, he was informed by Carpmael, that he could not deliver his paper since he "was quite unknown." As a man of no renown—in particular being unknown both to Carpmael and to Dionysius Lardner, the section president—he could not be permitted to waste their time.[22] Spencer noted bitterly that his paper was replaced by one describing a technique for preventing chimneys from smoking.

Carpmael's appointment as the section's secretary had caused a certain amount of grumbling among provincial members who saw it as another in a long series of examples of undue domination by an elite crew of London-based gentlemen of science.[23] The *Mechanics' Magazine* later saw the incident as yet another instance of the usurpation of the rights of the artisan-inventor by aristocratic science.[24] William Sturgeon expressed "extreme surprise" at Spencer's "extraordinary reception," describing his paper as "without exception, the most interesting one on voltaic electricity, that has hitherto been presented to the notice of the British Association *for the promotion* of Science."[25] Robertson, Sturgeon, and Spencer differed from the association's leadership in their view of what might appropriately count as promoting science. The result, however, was that Spencer was obliged to find a less prestigious and public forum for the announcement of his invention. He chose his local institution, the Liverpool Polytechnic Society.

Spencer's paper described, along with various specimens of the new art, a number of methods for raising patterns in relief on a copper plate. His first attempts had involved covering the plate with cement, drawing patterns through the covering, and using the plate as the pole of a battery. The result was that copper was deposited only in the pattern. The main difficulty here was that the copper pattern tended to come off the plate with the cement. In another method, a copper medal used as a pole could be covered by a new surface that could be removed by heating, producing a mold of the medal. Spencer noted that since lead could be used as a pole, a lead mold produced by pressure could be used to replicate any number of copper, or indeed silver or gold, medals by electrical action.[26]

Accounts of Spencer's invention were rapidly disseminated in a variety of journals. Sturgeon's *Annals of Electricity* was one of the first off the mark, reprinting Spencer's original communication along with a number of appendixes detailing new and improved procedures.[27] An account of the process appeared in the *London Journal of Arts and Sciences*, shortly followed by a brief summary of the invention from Spencer himself. In their next volume, the *London Journal*'s editors did even better. They published the first mass-produced print from an electrotype copy of an original etching, announcing proudly that the "process of Electrotype is now found to be fully competent to effect the object of copying and producing an unlimited number of copper plates, all possessing the identical characters and style of an original engraving, however minute and elaborately wrought."[28] New journals such as these, devoted to the promotion of inventive activity, had a crucial role to play in disseminating news of the latest invention. Sturgeon was not long in reproducing electrotypes in his *Annals*. He printed an example of an address card produced by electrotyping, soon followed by a spectacular engraving of Richard I leaving Cyprus, produced by the same new electrical technology at the Royal Victoria Gallery of Practical Science in Manchester.[29]

Despite the interest, no clear sign had yet emerged that these new experimental techniques would leave the laboratory and become a means of large-scale industrial production. Spencer, himself a master carver and gilder, indeed expressed at least some skepticism as to the present large-scale application of some of his experiments, suggesting that this was a matter for "the practical engraver and printer": "The question with them will be,—Is it cheaper and better than the methods in common use? It may now be answered—Give it a fair trial: the way is pointed out—practice will no doubt enable you to improve upon the methods . . . and most probably may realise an extended field of practical utility for the peculiar mode of operation which has been the result."[30] He expressed only the hope that his "simple discovery" was "destined (at no distant date), to imitate for the uses of humanity, all the most wonderful, but apparently complicated, elaborations of nature."[31] At least the potential of electricity as a mass producer of consumer goods had been established.

Spencer was insistent that he be recognized as the *discoverer* of the new process. He was not simply the improver of established experimental techniques. It was as a discoverer that he stood apart from the practical men whose task it would be to improve on his beginnings:

> I do most confidently expect and hope that the processes here given, will be improved by the practical man in the workshop, inasmuch as they are only given as the best that occurred to me; but it is highly probable that, had I been pursuing this branch of the subject more with a view to pecuniary profit than scientific research, it would have gone from me in a more complete form.[32]

Spencer was making an interesting assertion here. He was claiming discovery or invention as a matter of establishing a principle in experimental electricity. The business of making that principle practical and economical was a matter of mere improvement. Spencer's view makes clear the ambiguities surrounding the notion of invention. He aimed to establish his name by collapsing the distinction between discovery and invention. His contribution had been a matter of experimental philosophy, to be distinguished from the activities of the mere "practical man" whose task it should be to perfect the process. Invention, like discovery, was to be regarded as the business of science rather than art.

Indeed, practical men were cast by Spencer as completely opposed to the enthusiastic inventor:

> When the first results were obtained in 1838, they were shown to several persons in Liverpool, as triumphant realizations of what we might ultimately expect from a further application of electrical forces, which they were inclined to look on as the fond dreams of an enthusiast. . . . Nay, in Birmingham, on showing some medals so obtained, to a eminent die-sinker, I was kindly advised by him not to risk my reputation by a further exhibition, as they were less likely to be deceived *there*, than in such a town as Liverpool.[33]

Birmingham was, of course, the center of England's metalworking industries. On Spencer's showing, its "practical men" were a far cry from the visionary electrical inventor.

It was from Birmingham, however, that the first commercially successful British patents in electrometallurgy came. The main innovator in the field was George Elkington, from a gilt-toy and spectacle-making family of the town. Along with his cousin Henry and others, Elkington took out a series of patents during the late '30s and early '40s, laying the foundations of a new industry. According to Alfred Smee, the first "strictly electro-metallurgical patent" was granted to James Shore, a Birmingham merchant, in March 1840. Less than a month later a patent was granted to George and Henry Elkington for improvements in coating, covering, and plating various metals by means of electricity.[34] The patent included what was to become the crucial feature in electroplating: the use of the cyanides of gold and silver as the solutions in which plating took place.[35] Not everyone agreed as to the merits of the invention. The *Literary Gazette* described it as "a distinction without a difference: that is to say from a scientific point of view." They conceded however, that "legally . . . we believe that if the solution of silver or gold (patented, we presume, as applicable to certain practical purposes by certain parties) were employed extensively by other than they, the distinction and difference would soon be made manifest by injunction or otherwise."[36] The issue of the "distinction and difference," if any, between science and commerce was to be a recurring problematic for the rest of the decade and beyond.

Other patents followed, notably in 1842 the patenting by J. S. Woolrich, again of Birmingham, of an electromagnetic machine specifically for electroplating purposes.[37] Even Thomas Spencer, despite his disdain for "pecuniary profit," took out a patent for an electrical process for etching on metals.[38] Hopeful patentees attempted to carve out places for themselves by adding to the lists of usable solutions or the kinds of articles that could be gilded or plated. Thomas Spencer, for example, took out another patent protecting a process for plating picture frames with copper by means of electricity. His improvements consisted both in the preparation of the frames and other ornaments themselves, and in the types of solutions in which they were placed for electrodeposition.[39] Not only plating but typing processes were being patented. These were processes where a disposable mold was coated through electricity, usually by some precious metal, the mold then being disposed of.[40] For example, one might copy the face of a medallion from a wax mold by impressing the original into the wax, painting the impression with plumbago in solution so that it became a conducting surface, and using it as the positive pole.

Electricity was becoming an agent of mass production. Knives and forks, bowls, plates, candlesticks, trays, and any number of other household goods could be coated with a thin film of gold or silver. Middle-class households could easily afford goods that at least on the surface were indistinguishable from the heirlooms of the wealthy:

> At present a person may enter a room by a door having finger plates of the most costly device, made by the agency of the electric fluid. The walls of the room may be covered with engravings, printed from plates originally etched by galvanism, and multiplied by the same fluid. The chimney piece may be covered with ornaments made in a similar manner. At dinner the plates may have devices given by electrotype engravings, and his salt spoons gilt by the galvanic fluid.[41]

The galvanic fluid was becoming a crucial agent in providing luxurious commodities for a middle class eager to take on the material trappings of their economic power.

Simple patenting was not, of course, enough to bring electricity into commerce. Production as well as consumption was involved here. As a near contemporary pointed out:

> That which appears very simple in the hands of an experimentalist almost invariably becomes much more complex when carried into practice in a manufactory, simply because there is then a greater number of conditions to be fulfilled. Electroplating a piece of steel with silver is to a chemist a very simple matter, because it is of no importance to *him* whether the silver adheres firmly, is of a good colour, or is deposited at a certain cost; but with a *manufacturer*, unless *all* these conditions are fulfilled, the process is a failure.[42]

These were exactly the kinds of problems that plagued the early electroplating industry. Thin silver plate tended to blister or even peel off entirely during the final burnishing process when the product was polished. Without great care in the preparation of the cyanide solution the silver was discolored. In some circumstances, rather than forming a smooth surface, the silver would be deposited in lumps and ridges. A whole new range of practical skills and understandings had to develop to overcome such problems.[43]

There were other problems as well. Competing manufacturers and retailers complained that the new product was much inferior in quality to the old. The silver plate was thinner, for example, and tended to wear out more quickly. Cost was a major concern. The raw materials for the new electrical processes were at the outset more expensive than those required for the old techniques. Workers initially demanded one-third more wages in compensation for having to give up their old trade and acquire new skills. It took time for manufacterers to work out ways of dividing labor in the new processes to maximize output. Ways had to be found of integrating the practices that surrounded electricity in the laboratory with those that would prove profitable in a working manufactory.[44] Yet again, mechanization, or at least electrification, was not undertaken as a straightforward means of increasing profit by replacing labor. According to one estimate, it took seven years for the Elkington manufactory to become profitable.[45] By about 1850, the Elkingtons' firm alone employed more than five hundred workmen. By contemporary standards this was a very large number.[46]

With typical zeal, middle-class manufacturers and their celebrators could in the end point to electrometallurgy as being more than just a cheap way of mass-producing consumer commodities. It was also a means of self-improvement for their workers.[47] The effect of training in electrometallurgy was "of great advantage to the moral standing of the workman" because of the artistic quality of the product. Not only was the labor healthy, the worker being removed from proximity to the noxious fumes and fluids that attended the old techniques, but "the impression produced on the mind and heart, by the study of fine forms of decoration carried out in the highest character of workmanship,

6.1 (a and b) Elkington & Co.'s electroplating workshops and showrooms. From the *Illustrated Exhibitor and Magazine of Art* 1 (1851): 296–97.

lead the artist captive, and prevent him from falling into low pursuits, which would not only degrade him as a man, but destroy his ability to produce the requisite effects in his work."[48] The mass production of commodities was being recharacterized not only as art and as dependent on an artistic capacity, but as a means of also mass-producing the artistic spirit itself.

There was a burgeoning literature on the inherent morality of the factory system into which these kinds of claims fitted well. Andrew Ure, for example, made much of the moral force produced by the disciplinary power of machines. Lancashire cotton mills, Ure's epitome of the factory system, were airy light-filled havens where happy, well-fed urchins frolicked among the purring, self-propelled power looms. "They seemed to be always cheerful and alert, taking pleasure in the light play of their muscles. . . . The work of these lively elves seemed to resemble a sport, in which habit gave them a pleasing dexterity."[49] This was to be contrasted with the drudgery of hand-loom weaving in airless, dark rooms and the moral dangers of overcrowding. Even the routine nature of factory work was regarded as providing an opportunity for the worker to exercise and improve the mind. Accounts of travels through the factory districts emphasized the emancipating and uplifting capacity of the machine.[50] Through the celebration of electrometallurgy's aesthetics, electricity was being integrated into the discourse of the factory system, sharing in its morally progressive ethos.

A number of books and pamphlets were indeed published during the 1840s advocating the practice of electrometallurgy as an artistic recreational pastime for the individual. Texts such as George Shaw's *Manual of Electro-Metallurgy*, Smee's *Elements of Electro-metallurgy*, Spencer's *Instructions for the Multiplication of Works of Art in Metal*, or Charles Vincent Walker's *Electrotype Manipulation* were aimed not only at the potential manufacturer but also, and sometimes quite explicitly, at the individual experimentalist or hobbyist.[51] Electrical science, at the same time as it made possible the mass production of art, also allowed all individuals to engage in a personal artistic activity.

Charles Vincent Walker in particular was keen to emphasize that his pamphlet was written not for the artisan or manufacturer but "for him who delights to devote a portion of his hours of relaxation to the study of those mysteries of nature, into which the eye of science has been able in a degree to penetrate."[52] Above all else Walker characterized this as a *domestic* activity and celebrated the peculiar British attitude toward domesticity:

> there is amongst us as a national love of *home*, and of *home*-occupations. . . . Amongst home attractions is ever found a taste for the fine arts. . . . This taste is abundantly gratified by the discovery of Electrotype; it enables each, who desires it, to furnish himself with durable copies of the finest productions of the chisel and the graver. He finds an exhaustless field open before him;—and, if he devote his time to forming collections, he is animated at every step by the novelty and interest attached to each fresh acquisition.[53]

What was for the laborer in the Elkingtons' manufactory a matter of work was for Walker's (presumably) idealized bourgeois a matter of domestic relaxation. Yet it produced the same end.

The final product of the domestic contemplation of the "mysteries of nature" that was expressed through electrotyping was, like the products of the Elkingtons' laborers' work, a "fresh acquisition." Indeed, for Walker, one of the great advantages of domestic electrotyping was that it doubly affirmed the product as property. "[N]ot the least feature of interest allied to a collection thus formed, is the fact that each specimen is stamped with a double significance of 'mine,'— 'mine' it is by *possession*,—but especially it is 'mine' by *production*,—they are all the work of 'my hands.'"[54] Electricity could allow the bourgeoisie to experience and appropriate the artisan's property of skill in a safe, domestic setting. It was one way of making mass-produced commodities become expressions of individual artistic sensibilities. They could be products of the domestic sphere as well as of the dangerous world of work.

This was an aspect of the contemporary fascination with the new science of photography, closely related to electrometallurgy, as well. Photography shared with electrometallurgy the characteristic of being simultaneously a new mass industry and a new domestic recreation.[55] It may be significant to note that Walker's pamphlet on electrotyping appeared as part of a series of three, under the heading *Scientific Manipulations*. The third pamphlet in the series was George Fisher's *Photogenic Manipulations*, describing the techniques of the new art of photography to the amateur enthusiast.[56] William H. Fox Talbot, the British candidate for inventor of photography, also held a patent for an electrometallurgical process.[57] Photography, even more than electrotyping, was hailed as a product of science allowing nature to duplicate art. Michael Faraday, for example, celebrating Fox Talbot's "photogenic drawings," remarked that "[n]o human hand has hitherto traced out such lines as these drawings displayed; and what man may do, *now that Dame Nature has become his drawing mistress*, it is impossible to predict."[58] Again, nature was being used as a cipher for the labor required to maintain the photographic industry.

Photography and electrometallurgy could be brought into a close technological relationship as well. By the end of the 1840s, electroplating was widely used as the method for covering copper plates with silver for making daguerreotypes. Other connections were possible. William Robert Grove, for example, proposed an electrical process for etching daguerreotype plates so that they could be used for printing.[59] This was hailed by the *Literary Gazette* as the latest step in the application of science to art that had begun with "the wonderful and beautiful processes known as the Electrotype, Daguerreotype, Calotype &c."[60] Grove anticipated that before long, "instead of a plate being inscribed, as 'drawn by Landseer, and engraved by Cousins,' it would be 'drawn by Light, and Engraved by Electricity!'"[61] The remark was presumably intended facetiously but nevertheless predicted the displacement of skilled human labor by an industrial process, disguised as nature.

Such remarks were in fact particularly pointed during the 1840s. As many historians have pointed out, new printing technologies developed during the previous decade were by then having an impact on the status of professional engravers. The availability of cheap mechanized printing led to an explosion of print and the proliferation of printed images.[62] Engravers argued that these new technologies undermined the status of their craft. They were increasingly portrayed as mere mechanics rather than as professional men. Their skills were devalued as they were characterized as mere mechanical reproducers of other men's work rather than as valuable producers themselves.[63] Suggestions such as Grove's could undermine their status even further. By harnessing light and electricity, mass-producing printers could dispense entirely with their services. Their skills of careful and accurate reproduction could become acts of nature.[64]

It was, however, not only the manufacturer or the middle-class amateur enthusiast whom nature and electricity could combine to serve. Electroplating and electrotyping very rapidly became tools in the armory of a very different group. By the mid-1840s galvanic batteries were already a standard part of the equipment of the professional coin-forger. Coining false currency, despite its great danger, could be an extremely lucrative business and was therefore relatively common in early Victorian London.[65] The most common form of forgery was the trade in false silver coinage. The weight of gold coins was hard to mimic, and copper coins had too little value to be worth the effort of counterfeiting. Counterfeit coins were cut out of metal plates of pewter or a cheap silver alloy and impressed with a die. Nitric acid could be used to provide a silvery finish. The advent of electroplating made it possible for counterfeiters to complete the process by coating the false coins with a real silver finish.

Despite its typical base in London's rookeries, silver coining using electricity required some skill and some starting capital. First, plaster of paris molds would be taken with genuine coins—itself a difficult task to carry out with the necessary precision. Molten bronze was then poured into the molds, and when the casts were hardened, they were attached to the negative pole of a galvanic battery immersed in a cyanide solution, a piece of silver being used as the positive pole. As in more legitimate uses of electroplating, judging the proper strength of the solutions and ensuring a smooth plating were difficult matters. Apprehending the culprits could be just as difficult. On top of the usual dangers of physical violence, police attempting to arrest the forgers could well expect a faceful of acid from a galvanic battery.[66]

The art of electrical metalworking was a very fluid technology. It could be a mass-industrial process, a domestic hobby, or an underworld moneymaker. Some of its promoters were certainly aware of the illicit possibilities of electrometallurgy. Alfred Smee in particular warned his readers against the foolishness of imagining that such a moral process could be subverted.[67] Electromellaturgy did, however, represent the first big step for electricity out of the world of the laboratory and the lecture theater. The voltaic battery was transformed from a producer of spectacular shocks and sparks into an item of industrial technology: a producer of commodities. As was not uncommon for the period, this new industrial battery was represented as a piece of self-acting machinery.[68] "Nature" rather than the worker performed the labor in the Elkingtons' manufactory.

During the 1840s electricity was certainly a part of the commercial world. It was a matter for patents and entrepreneurial interest. Electrometallurgy during this period destroyed one set of skills and working practices while providing the focus for the emergence of new crafts. It could be hailed as a decisive example of the progress possible through the harnessing of science to industry. It mattered, therefore, what was to count as science. Was science the practical act of invention or the more abstract business of discovering principles? Was the inventor a man of science or a practical applier of others' principles? Was "practical science" what mattered, or the gentlemanly science of the elite? Thus for reasons ranging from cash to kudos, the question of priority in invention was important. It mattered not only what individual had invented electrometallurgy, but what kind of individual he was. Priority in invention was not just a practical matter; issues of morality and politics were also at stake.

IMAGES OF INVENTION

All kinds of resources could be used to define the inventor and his inventions. Through exhibitions, public lectures, and newspaper and journal reports, issues of property and priority in invention were extremely visible. Patents were, of course, among the resources that could be drawn upon to establish rights of invention. Patents alone, however, were by no means sufficient. Neither, for that matter, was priority in patenting necessarily taken to indicate rights as an inventor. Battles between rival claimants to particular inventions could be bitter. A great deal was at stake in the definition of the kind of person an inventor was, and what claims to fame and public recognition he might legitimately make. The title of inventor of electrometallurgy was clearly worth fighting for.

In Britain at least, Thomas Spencer was widely recognized as the inventor of the art of working with metals through electricity. Spencer himself was adamant that he was the discoverer of the new techniques:

> When the term *discovery* is used, it is wished to be understood literally. . . . I trust this piece of apparent egoism will be excused, when it is mentioned, that my friends have felt more annoyance than myself, at statements made by lecturers on this subject, to the effect, that the facts and their applications have been pointed out frequently, and that I have improved a process already known. I may say however, that *such is not the case*.[69]

It was easy to argue that the new process had emerged naturally from commonplace manipulations of electrical apparatus so that there was no inventor or moment of invention to celebrate. Spencer had to demonstrate the difference between his processes and those of others to make his claim stick.

In some quarters at least, his claim to fame had a decidedly political edge. The radical *Mechanics' Magazine* and its editor Joseph Robertson were among Spencer's strongest partisans, regarding him as an artisan-inventor slighted by elite science. Reviewing recent works on electrometallurgy in 1842, Robertson took advantage of the opportunity to lambaste the elitists—and one of the authors, Alfred Smee, in particular—for their snobbery.[70] The magazine was "sorry to see that there still exists in certain scientific circles the same dogged reluctance, on which we before animadverted, to do justice to the humble 'Carver and Gilder of Liverpool'—for no better reason that we can discover, but the sin of being humble."[71] Smee had failed to make any mention of Spencer at all in his account of electrometallurgy's origins.[72]

The *Mechanics' Magazine* compared Smee's book unfavorably to George Shaw's *Manual of Electro-metallurgy*. Contrasting the two books, the magazine was in no doubt concerning the origins of their different approaches.

> In one, which is a Birmingham production, intended, like most Birmingham wares, for the million, and written by a person, who, like the hero of his subject, boasts the license of no learned society to be useful to his fellow men, the art is honestly admitted to have *"proceeded from the hands of Mr. Spencer,"* . . . while, in the other, which is a London production, dedicated to the Consort of her Majesty, and written by a Fellow of the Royal Society, the name of Mr. Spencer is never once mentioned![73]

The implication was clear. Smee was defending the claims to priority of the Royal Society's brand of gentlemanly science over the artisan science of the "humble" Thomas Spencer. At issue, of course, was the source of invention. Was it the property of artisan or of gentleman? Invention, on the *Mechanics' Magazine*'s showing at least, was democratic in two ways. It was the production of humble mechanics, and it was aimed at consumption by the masses. The opposing picture was that of gentlemanly production designed to be consumed by an equally genteel audience. The contrast was that between an industrial, and this case provincial, populism and an aristocratic and metropolitan elitism.

In Smee's account there was no invention. Electrometallurgy was a natural development of Daniell's battery. Once the observation was made that a layer of fresh copper was deposited on the copper plate during the battery's operation, the step to electrometallurgy was trivial. The *Mechanics' Magazine* differed. It was "an accident of that happy sort which happens to but one or two men in an age, and which by the universal consent of mankind entitles him to whose lot it falls, to be looked upon with all the respect, honour, and gratitude due to the chosen instrument of any great revelation by Nature to her children."[74] For the magazine, the key to Spencer's claim to invention was that he had seen and appreciated the *practical* consequences of the observation. Having observed a process, as Daniell had done, was not enough. The crux of invention was practical application. This, as suggested earlier, however, was not Spencer's own view of the characteristics of invention.

The *Mechanics' Magazine*'s robust defense of Spencer's claims to the invention of electrometallurgy had an unexpected effect a few years later. Spencer's claims as inventor were rigorously and vociferously opposed in a manner that made explicit a whole range of issues concerning the technical, political, and moral status of the inventor. The resources deployed in the dispute laid bare both the fragility and the robustness of an inventor's position—particularly that of an inventor who laid serious claim to *scientific* status in some form. Inventors of electrical technology were among the foremost of this breed. In a variety of ways they sought to combine the moral status of the natural philosopher, the practical knowledge of the mechanic, and the commercial acumen of the entrepreneur. An anatomy of the acrimonious battle that erupted over Spencer's claims to invention can illuminate the ways in which discursive strategies effortlessly blended the technical and the social as the characteristics of the inventor were put together or pulled apart.

The attack on Spencer's position first appeared as a communication in the *Mechanics' Magazine* innocuously titled "Contribution towards a History of Electro-metallurgy," by Henry Dircks, popular lecturer on chemistry and electricity and railway engineer, a former resident of Spencer's native city, Liverpool, then living in London.[75] Dircks alleged, on the one hand, that Spencer's publication of his experiments had been preceded by a previous publication and, on the other, that Spencer failed to adequately acknowledge a variety of sources that had materially aided his own researches in electrometallurgy. He poked fun at the *Mechanics' Magazine*'s designation of Spencer as a "humble carver and gilder," pointing out that he was "a master . . . occupying a large house and fine shop, in one of the leading thoroughfares."[76] The previous publication—surprisingly, given the *Mechanics' Magazine*'s defense of Spencer—was in the pages of that very journal, in the form of a letter from a C. J. Jordan, outlining a process very similar to Spencer's.[77]

In his own communication, Dircks republished Jordan's letter, drawing attention to the facts that like Spencer's memoir it referred to Jacobi's experiments in St. Petersburg, as described in the British press, and used the version of Daniell's battery developed by Golding Bird of Guy's Hospital. On this occasion he simply stated baldly that since Jordan's letter was published before Spencer's communication to the Liverpool Polytechnic Society, Jordan clearly held priority in invention. More damaging were his remarks concerning Spencer's experiments, which he alleged had materially benefited from others' labors that Spencer neglected to mention. The most Dircks was willing to accord to Spencer was that he had improved an already established procedure for electrometallurgy.[78]

In particular, Dircks drew attention to the experiments of the Liverpool instrument-maker John Dancer during the years 1837–38. According to Dircks, following discussions at the Chemical Section of the British Association for the Advancement of Science, held in Liverpool in 1837, a number of his acquaintances, Dancer in particular, had been experimenting with the copper precipitate produced in batteries. Dancer had produced copper plates by this method and had shown the results to, among others, Spencer, who had expressed surprise at their mode of production. When Spencer made his announcement to the Polytechnic Society, Dancer alleged that he had made no comment since "it would look like envy, or a wish to detract from the merits of his experiments, and share the honour without having brought it to practical use."[79] The implications of Dircks's communication was, of course, that Spencer's claims to invention were vitiated not only by previous publication but also by his having acted less than honorably in appropriating the work of others without acknowledgment. Secrecy mattered: Dancer had behaved honorably in sharing his work with others, while, as Dancer put it, if Spencer "had . . . produced compact copper at the time when I showed him the piece in question, he was wrong in allowing me to suppose otherwise."[80]

Dircks alleged not only that Spencer had materially gained from Dancer's experiments but that Spencer's subsequent career displayed his own lack of proper experimental understanding of the invention he claimed, and therefore undermined his pretensions:

> It is a remarkable fact that Mr. Spencer has made no useful or profitable application of the electrotype process, of which his first experiments gave promise; neither did he early secure its applications by patent right. . . . It does not say much for Mr. Spencer's possession of originality of genius and philosophical acumen, which a perusal of his paper on voltaic electricity would persuade us he considers belongs to him, so long to have remained a cypher in an art entirely new, and capable of modifications and applications as yet untried.[81]

Indeed, very little was left of Spencer's reputation, if Dircks was to be believed. His publication was not the first in the field. He had behaved dishonorably either by misappropriating others' work or by failing to be candid about his own progress. His subsequent failure to profit by his "invention" betrayed ignorance of its potential. The *Mechanics' Magazine*'s response to the exposure was unequivocal: "we have made a great mistake in advocating so strenuously the claims of Mr. Spencer to the invention of electrography."[82]

Dircks, whatever his motives, had chosen his strategy well.[83] As far as the *Mechanics' Magazine* was concerned, a major incentive for supporting Spencer's claims to invention was that he could easily be portrayed as the victimized mechanic, the "humble carver and gilder of Liverpool" whose rightful title was being ignored by a privileged and metropolitan scientific elite. Dircks, however, sought to redefine Spencer. He was not a humble mechanic but a "master" with a "large house and fine shop." Furthermore, he was not a victim but a victimizer who had grabbed the laurels of invention from the even more humble brows of the printer C. J. Jordan. Jordan's humility was underlined by the fact that he had made no public claim himself to the status of inventor.

Spencer responded promptly to Dircks allegations. Rather than focusing on the technical details of his apparatus, like Dircks, he focused on the moral context of the dispute, seeking to display Dircks as a duplicitous self-publicist. In his communication to the *Mechanics' Magazine* he included an extract from the *Liverpool Albion*:

> An elaborate paper, being a "Contribution towards the History of Electro Metal-
> lurgy," is just published in the *Mechanics' Magazine* of the present week: it is by Mr.
> Henry Dircks, of London, formerly of this town. Mr. Robertson, the respected and
> talented editor of that magazine, was, until the present time, a warm advocate of
> Mr. Spencer's claims; but, so convinced is that gentleman by Mr. Dircks' state-
> ments and arguments, that he falls in entirely with his views. . . . We imagine this
> paper will make some stir among scientific circles, there never having appeared,
> hitherto, much certainty as to whether or not Mr. Spencer was the original inventor,
> or a mere copyist. We really think Mr. Dircks' paper must be taken as confirmatory
> of the latter opinion.[84]

Spencer revealed that Dircks was in fact the author of this piece of self-
advertisment, describing him as having a "peculiar talent, in thus drawing
attention to his own productions." His motives in communicating with the
*Mechanics' Magazine* Spencer described as being "little . . . short of dishonesty
of purpose."

Spencer's aim was to counter the allegation of dishonesty by portraying
Dircks himself as lacking in integrity. He did claim, however, to have in his
possession documents that would prove beyond doubt his claims as an inventor. Spencer took particular issue with Dircks's inference that his failure to
patent indicated that he was not the inventor.

> Mr. Dircks would have it inferred, that because I have not got rich with the electro-
> type, *ergo*, I am not the inventor. This gentleman is a very Nadgett at worming into
> other people's circumstances with respect to pecuniary matters—it is one of his be-
> setting sins. I thought, however, that the rule was that an inventor did not get rich
> with his own invention, and I had comfortably set myself down as not being destined
> to form any exception.[85]

Spencer was pointing toward the contingency of patents as a means of enforc-
ing the inventor's moral (or indeed commercial) rights.

Spencer requested the *Mechanics' Magazine* to insert an abstract of a paper
he was preparing for the *Philosophical Magazine* outlining his recent researches in electrometallurgy. He also promised that once he had time to put
them in order, he would forward the documents proving his claims. The *Mechanics' Magazine* refused to publish Spencer's abstract on the grounds that it
would be "contrary to editorial etiquette" to publish an abstract of a paper
destined for another journal.[86] It agreed to publish the documents but expressed regret "that Mr. Spencer should have thought it right to have forborne,
from any cause whatever, even for a single day, that complete vindication
which the statements of Mr. Dircks seemed to call from his hands."[87] Having
abandoned its erstwhile protégé, the *Mechanics' Magazine* was clearly not prepared to make his return to the fold at all unproblematic.

Dircks was soon on the offensive again, ridiculing Spencer's defense and
extending his charge to one of outright plagiarism of Jordan's paper. Spencer's
response he characterized as "nothing but abuse and misrepresentations." He

also revealed the reasons behind Spencer's delay in responding fully to the allegations. He was busy conducting mesmeric experiments on a "Mrs. Tod." Dircks clearly felt that this revelation put Spencer's integrity and reputation in its proper context.[88] Spencer was being portrayed as a charlatan. He could not properly be regarded as an inventor since he lacked the appropriate moral qualities. The reader was invited to draw one of two equally damning conclusions. Spencer was either a dupe whose capacity to invent was suspect as a result of his gullibility, or he was another mesmeric fraud and therefore in all likelihood a fraud in electrometallurgy as well.[89] Having prepared the ground, he proceeded "to lay bare such an example of plagiarisms as is perhaps without a parallel in the whole history of science, when honour, more than emolument, has been the object of contest."[90]

Dircks now contended that Spencer had made no significant experiments in electrometallurgy before the publication of Jordan's paper in the *Mechanics' Magazine* and, through a detailed comparison of Spencer's communication with Jordan's, aimed to show that it was nothing but plagiarism. He charged Spencer with disguising also his debt to Golding Bird, whose version of the Daniell cell he claimed to be "the origin of the whole superstructure of electrometallurgy, as well in the hands of others, as of Mr. Spencer." Dircks charged Spencer with disguising his debt to Bird so as to maximize his own significance, claiming that the cells used were identical. Morality and technicality merged in this issue as they did over the issue of similarities between Spencer's experiments and those described in Jordan's letter.[91]

Both Jordan and Spencer had indeed used batteries based on Golding Bird's design in their electrometallurgical experiments, although Jordan did not mention the battery's designer in his communication to the *Mechanics' Magazine*. The battery's distinctive feature was that the different solutions were separated by means of a plug of plaster of paris covering one end of the glass cylinder in which the copper plate was placed.[92] Bird's communication concerning his experiments with this battery made no reference to the deposit of copper on the copper plate. His main claim was that while the battery was in action, crystals of copper were deposited in the plaster of paris plug.

Dircks charged that it was no coincidence that both Jordan and Spencer used the same battery. He alleged that Spencer had merely copied the experiments from Jordan's letter, and went through both communications line by line to substantiate his claim that "when the letter of the one and the pamphlet of the other are placed in juxtaposition, we trace that similarity of design and execution which surely must be more than the result of sheer accident."[93] His explanation of Spencer's failure to patent his invention was straightforward: "he knew better than to patent a recorded discovery, an invention not his own, and wisely confined himself to patenting what he afterwards did, perhaps, discover—a simple branch of its application; his right to which claim, however, I am informed is, curiously enough, at present a matter of legal contest against him."[94]

Spencer in response could do little but simply deny Dircks claims. He suggested, however, that as a means of resolving the issue, Dircks should publish his allegations in full in the *Mechanics' Magazine*, to which he would then respond. The matter should thereafter be placed before a "committee of gentlemen of acknowledged competence" to finally decide the matter. Dircks and the *Mechanics' Magazine* rejected the proposal, at which point Spencer pointedly retired from the dispute, accusing editor Robertson, of gross partiality and partisanship.[95] Ironically, in the third edition of Alfred Smee's *Elements of Electro-metallurgy*, the review of which in the *Mechanics' Magazine* had sparked the debate, Smee, presumably well aware of where that journal now stood on the matter, gave full credit to Spencer for the invention of electrometallurgy.[96]

The debate highlights the problematics of invention at this period. It illustrates, too, that patenting was only one of a range of strategies that an individual could adopt to attempt the appropriation of an invention. In Spencer's case his failure to patent could be used against him, but he could also respond by claiming the moral high ground and asserting that true inventors did not aim at financial reward for their activities. The technical and the moral blended in this debate and others like it. The moral status of the inventor hinged on the technical details of his work. Spencer's status as inventor could depend on the difference between the battery he used and that of Golding Bird. This came in the end to an apparently trivial matter of the thickness of a plaster of paris plug regulating the flow of current. Bird in his experiment had obtained crystals of pure copper embedded in his plaster of paris plug. Spencer had experimented with different thicknesses of plaster to obtain a thin layer of copper deposited on the battery's copper pole. His claim to novelty and to the moral ground of invention rested on this distinction. His moral status depended on the weight given to technical detail. The two could not easily be disentangled.

Establishing rights to an invention was more than simply a matter of financial reward: in the vitriolic dispute between Dircks and Spencer there seems to have been no money at stake at all. Being an inventor was a matter of status as much as anything else. This was why the link between the "practical" matter of invention and natural philosophy was significant, particularly to those such as the editor of the *Mechanics' Magazine* who wished to define the inventor as the artisan. To this end, the magazine could go as far as comparing Spencer, as inventor of electrometallurgy, to Newton, discoverer of gravity.[97] This made patenting an awkward option. The man of science was not meant to invent for personal monetary profit. The possession of a patent was insufficient as a defense of one's status as an inventor in any case. No one, for example, ever described the Elkingtons as the inventors of electrometallurgy. Indeed, as noted previously, it was possible to be distinctly dismissive of anything other than their merely commercial claims to property.[98] While patenting remained only one of several strategies for the defense of an inventor's rights, it was one that more and more electricians adopted during the 1840s.

Usurping the Place of Steam

Electricity was increasingly patentable as the 1840s progressed. The majority of patents were for electrometallurgical processes and increasingly from mid-decade for telegraphy. These were by no means the only categories, however. Systems of electric light and schemes for electric locomotion were being put forward in significant numbers. Electricity was widely regarded as holding out the hope for universal power. William Robert Grove extolled its virtues at the London Institution.[99] The institution itself had, of course, been founded precisely to foster the intermingling of science and commerce that electricity seemed set to make a reality.[100] Alfred Smee felt that even the successes of electrometallurgy were as nothing compared with the future's possibilities:

> we must still look forward to the most important properties of the electric current derived from the galvanic battery; for although great and glorious are the triumphs of science detailed in this work, yet the prospect of obtaining a power which shall supersede steam, exceeds in value all these applications. For to cross the seas, to traverse the roads, and to work machinery by galvanism, or rather electro-magnetism, will certainly, if executed, be the most noble achievement ever performed by man.[101]

Smee was being unusually cautious in saying *if* rather than *when*.

Events like John Peter Gassiot's electrical soirees, marveled at in the pages of the *Literary Gazette*, served to underline the link between electricity's flamboyant technology of display and commercial progress. By putting electrotypes and electromagnetic machines on show alongside spectacular shocks and sparks or crystals made by electricity, the exhibitions were at once naturalizing the ones while commodifying the others. Electrical machines were as much a part of nature as a stroke of lightning was part of the progressive forces of Victorian commercial culture.[102] Electricity was easily seen as the universal power of both nature and industry. It was a progressive force that dominated the natural and could be harnessed to dominate the social world as well. Electricity held out the prospect of "a closer insight into the operations of nature as connected with the animal, vegetable and mineral kingdoms" and at the same time the hope of its "usurping at no distant period, the place of steam as a mechanical agent, and thus being made, in the most extensive manner, subservient to the uses, and under the control of man."[103]

All manner of electrical devices and processes could be patented. Intriguingly, one patent was taken out by Andrew Crosse, both celebrated and vilified for his production of insect life through electricity.[104] According to the specifications,

> [t]his invention consists in submitting fermentable, fermented, and other liquids to the action of currents of electricity.... The action of the electric currents materially improves the character of the wine, beer, cider, or other liquid, and renders it less

liable to become sour; by this means also, any wine, beer, or other liquid, may be restored from partial acidity, or prevented from becoming more acid; and the impurities of spirituous liquors may be removed by the same process.[105]

Given his past uses of electricity, it is tempting to speculate as to what exactly Crosse hoped to produce by passing electricity through "fermentable, fermented, and other liquids." There is no record of whether he ever profited from this attempt to market electricity as a preventive for hangovers. Another electrical patentee was Joseph Robertson, himself a patent agent as well as editor of the *Mechanics' Magazine*. He patented a process for improving the separation of metals from their ores through the combined use of heat and electricity.[106]

A flurry of patents during the latter half of the 1840s concerned electric lighting. Edward Staite applied for a series of patents detailing improvements to his electric arc light, which used the spark between two points of charcoal as the source of illumination.[107] This was one of the two techniques of electrical illumination in vogue during the period, the other being the use of incandescent wires. No patents were taken out for the latter technique, but William Robert Grove, for example, in 1845, shortly after the notorious Haswell colliery disaster, advocated the use of such "voltaic ignition" for the lighting of mines. He produced a sophisticated design featuring a helical coil of platinum wire as the source of light.[108] Despite Grove's praise for his design's economy, however, the arc light proved more popular with money-minded patentees both in Britain and elsewhere.

Staite was a flamboyant exhibitionist of his new invention. It was soon on show at the galleries and lectured upon at the institutions. He was even to put his light in action in Trafalgar Square:

> The apparatus was so placed . . . as to illumine the whole of Trafalgar-square, the rays reaching as far as Northumberland-house. . . . The rays were continually moved, and as they swept through the foggy atmosphere, they produced the same sort of illumination as the sunlight through atoms of dust. The objects upon which they fell were most brilliantly lighted. The Nelson column, which was selected as the principal point, being frequently as conspicuous as at noonday. If the illumination can be sustained, there is no other means of lighting the streets that can at all be compared with this electric light.[109]

In a city whose dimly lit and crime-ridden rookeries terrified the middle classes as well as offending their sensibilities, Staite's promise of illumination was calculated to draw attention.[110] His great triumph of advertisement was yet to come. In May 1849 a new ballet, *Electra*, opened to the public at Her Majesty's Theatre. The star of the show was Staite's electric arc light. The ballet had been commissioned specifically to show off the brilliance of Staite's invention. The ballet was an instant success, even receiving the accolade of a command performance before Queen Victoria within a few weeks of opening.[111]

The aim of using electricity as a motive power, "usurping the place of steam," was a common one for early Victorian patentees and their promoters. Experi-

6.2  Staite's exhibition of electric light at Trafalgar Square.
*Illustrated London News* 13 (1848): 368.

ments to produce motion by means of electricity had been commonplace
since Oersted's original "twitching needle" experiment of 1820. The best
known probably were Michael Faraday's own experiment in which a wire
rotated around a central magnet, and variations on the theme such as Peter
Barlow's rotating wheel. Despite popular claims to the contrary, these experiments did not provide the foundations for the provision of motive power by
means of electricity.[112] No contemporary ever regarded Faraday's rotating wire
experiment as a primitive electric motor. For contemporary electricians
hopeful of invention, the key to motive power by means of electricity was the
electromagnet.

The electromagnet had been one of the results of William Sturgeon's efforts
during the mid-1820s to build a portable laboratory that would allow the traveling lecturer to reproduce spectacular electrical effects cheaply and reliably. In
its simplest form it consisted of copper wire coiled loosely around a varnished
soft iron core, typically in the shape of a horseshoe.[113] Following its invention,
the new device rapidly underwent dramatic improvement, mainly at the hands
of the American electrician Joseph Henry, who throughout the late 1820s and
early 1830s produced a series of increasingly powerful instruments, capable of
lifting several thousand pounds in weight.[114] The crucial feature of the electromagnet, however, from the perspective of the production of motive power, was
not so much the magnitude of its lifting power as its capacity to produce and cut

off the magnetic force at will. It was this capacity that hopeful inventors aimed to exploit to produce an economical source of motive power.

Henry himself had designed a simple reciprocating electromagnetic motor to show off the potential of the electromagnet to his students at Princeton in the early 1830s. Most designers, however, aimed to use arrays of electromagnets to build rotary engines. By the mid-1830s both William Richie and William Sturgeon had produced such motors on a small scale.[115] The self-publicizing American blacksmith Thomas Davenport patented his rotary electromagnetic engine in the United States, and his promoters put his models on display at the Adelaide Gallery.[116] Enthusiastic discussions of Davenport's engines in the *Mechanics' Magazine* during the latter half of the 1830s frequently took for granted that all that was required for an indefinite increase in power was a correspondingly indefinite increase in the engine's dimensions.[117]

The great hope was that electricity would be viable for locomotion. Steam had barely become established as a source of power for trains and steamboats before predictions were voiced of its imminent replacement by cheaper and safer electromagnetism:

> I am free to confess that I cannot discover any good reason why the power may not be obtained and employed in sufficient abundance for any machinery—why it should not supercede steam, to which it is infinitely preferable on the score of expence, and safety, and simplicity. . . . Half a barrel of blue vitriol, and a hogshead or two of water, would send a ship from New York to Liverpool; and no accident could possibly happen, beyond the breaking of the machinery, which is so simple that any damage could be repaired in half a day.[118]

This was the *Morning Herald* American correspondent's view of Davenport's engine. The view was heartily endorsed by the *Mechanics' Magazine*, which reprinted the encomium.

The first systematic effort by an expert electrician to investigate the potential for electromagnetic engines was carried out by the young brewer's son from Manchester James Prescott Joule.[119] Joule had been an enthusiastic contributor to William Sturgeon's *Annals of Electricity*, and when Sturgeon moved to Manchester as superintendent of the Victoria Gallery of Practical Science, he became an invaluable source of experimental expertise in electricity to Joule. In many respects, Joule's work in establishing the efficiency of electromagnetic motors was similar to Sturgeon's own work in establishing the efficiency of different types of battery.[120] Sturgeon aimed to assess the capacity of different kinds and combinations of batteries for particular types of display. At stake in his experiments was the cost of exhibition.[121] Joule, however, with his hardheaded upbringing in commercial, industrial Manchester, was unambiguously concerned with the *economic* efficiency of electromagnetic engines. As several commentators have noted, Joule quite explicitly adopted the language and concerns of the economist and the engineer in his electrical researches.[122] The economy of nature, the economy of the machine, and the economy of commerce were part and parcel of the same system.[123] These different economies might not necessarily be identical, but they were certainly homologous. The laws of one operated in the others as well.

From the first of his communications to Sturgeon and the *Annals of Electricity*, Joule was primarily concerned with the improvement of electromagnet design, with particular reference to electromagnets to be used in the construction of electromagnetic engines. Economic efficiency could be attained only if all parameters contributing to the magnets' performance were analyzed and means found to maximize their effects. Joule experimented with a whole variety of different designs of magnet. He suggested, for example, that rather than having a solid core, an electromagnet for an electromagnetic engine was best prepared out of a number of separate iron wires tightly bound together. Further experimentation suggested that the wires should be square rather than round, to maximize contact. When wire and solid core magnets were tried on the same machine, they revolved at 162 and 130 revolutions per minute, respectively.[124]

These were not the only issues to be clarified. Joule needed to find out how the length and diameter of the iron cores affected the magnet's performance. Crucially, he also needed a way of distinguishing between the lifting capacity of an electromagnet, defined as the amount of weight it could lift and hold, and what he described as its attraction. *Attraction* (crudely speaking, the pulling power of the magnet) rather than lifting capacity was the crucial parameter to be maximized in improving engine performance. One could measure attraction by suspending a straight steel magnet from the beam of a balance and bringing each of the electromagnets to be investigated to a standard distance of one and a half inches. Joule found that there was no necessary correlation between the lifting capacity and the attraction of an electromagnet. This Joule attributed to practical features of the design. For example, longer magnets had better attraction than short ones of the same diameter, since some of the coiled wire was at a further distance from the core in the short ones.[125]

Another issue Joule considered was the relationship between the quantity of electricity and the design of different magnets. In particular he was interested in the relationship among attraction, quantity, and the sectional area of the magnets. The result was one of his first efforts at quantification: "I think . . . that I have by these experiments discovered a most important law, namely: *The attractive force of the electro-magnet is directly as the square of the electric force to which its iron is exposed: or if E denote the quantity of electricity, M, the magnetic attraction, and W the length of wire,* $M = E^2W^2$."[126] This law had crucial implications that Joule lost no time in pointing out:

> I can scarcely doubt that electro-magnetism will eventually be substituted for steam in propelling machinery. If the power of the engine is in proportion to the attractive force of its magnets, and if the attractive force is as the squares of the electrical force, the economic effect will be in the direct ratio of the quantity of electricity, and the cost of working the engine may be reduced ad infinitum.[127]

The laws governing the forces of electricity translated straightforwardly into those that governed the forces of commerce. Joule hastened to communicate to Sturgeon the design of his engine, "to forestall the monopolizing designs of those who seem to regard this most interesting subject merely in the light of a pecuniary speculation."[128]

Joule quite explicitly adopted the language of the engineer. In his next communications he aimed to ascertain the duty, and the *economic* duty, of his newly designed electromagnetic engine.[129] Duty was the standard term used by contemporary steam engineers to measure the power of their engines. Joule defined duty as the number of pounds his engine could raise per second to a foot's height. The definition of economic duty was crucial to Joule's project: it was "the duty, in pounds raised to the height of one foot by the agency of one pound of zinc."[130] Joule was looking for a way to relate his engine's power—the amount of economic work that it could do—to the cost of that power expressed as the amount of fuel (zinc) expended in its production. Joule might be disdainful of those who wished to monopolize electromagnetic power for personal profit, but he certainly regarded his experimental analysis of his engine's performance as an economic project.

Joule, despite his prognostication that "electro-magnetism will eventually be substituted for steam," clearly did not even at this stage fully share the optimism of many of his electrician contemporaries. After several pages of detailed tables of numerical results, he caustically apologized to his readers: "I have not relieved the tediousness of this paper, by a single brilliant illustration. I have neither propelled vessels, carriages, nor printing presses. My object has been, first to discover correct principles, and then to suggest their practical development."[131] Joule clearly took a different view toward the process of invention from many of his fellow electricians. He was a devotee of discipline rather than flamboyance in experiment. He believed in achieving progress by careful measurement rather than by attracting entrepreneurial attention through impressive displays of electromagnetism's potential. He was one of the first of a new breed of experimenters.[132]

As is well known, Joule's cautious but optimistic predictions soon turned to pessimism. His considerations of the economic relationship between electricity and work soon expanded to embrace the relationship of work and heat. By the mid-1840s he had conclusively abandoned his efforts to develop an economically viable electromagnetic engine, pessimistically denying that such a machine could ever be cheaper to operate than a coal-driven steam engine. He even denied that the electromagnetic engine was ever likely to prove more economically efficient than the animal or human machine.[133] Joule was not the only one who came to see the issue in such stark economic terms. William Robert Grove pointed out pragmatically that since in the production of electricity "we use for fuel manufactured materials in the production of which coals, labour &c., have been expended," while in steam engines "coal and water are used directly,"[134] the electromagnetic engine was unlikely to supersede its steam-driven competitor. Grove maintained, however, that electricity might yet be found to have uses where steam, though more economical, was impractical.[135]

Few paid much attention to pessimism or economic caution. Such voices certainly seemed to do little to discourage hopeful patentees from displaying their engines' capacities. The *Inventor's Advocate* editorialized in 1840 that "there is no object, indeed, to which the attention of the scientific world is now directed, that appears so fraught with important consequences to the interests of mankind, and the development of scientific truth, as the science of electromagnetism."[136] They felt that "[e]nough has already been done to show that the power exists, and that it is capable of being applied. Its extent seems to be unlimited, whilst the cost of its production is comparatively trifling; and we feel confident that human ingenuity will eventually find the means of rendering it subservient to the purposes of man."[137] Human ingenuity throughout the 1840s was indeed making every effort to invent and patent an effective electromagnetic engine.

Enthusiasm for electromagnetic engines continued throughout the 1840s. Regardless of Joule's pessimistic prophecy that electricity would never be a viable source of motive power, the practice during the period was one of optimism. In 1840 at the Leicester Exhibition, Uriah Clarke operated his model electromagnetic locomotive, *Jupiter*, on a circular railway. The model weighed sixty pounds and could run at "considerable speed" for more than two and a half hours at a time.[138] Clarke and Thomas Wright waged battle against one another in the pages of Sturgeon's *Annals of Electricity*, contending for priority in the invention of a reciprocating electromagnetic engine that both asserted to be "the best form of applying the electro-magnetic power."[139] Wright, though, was dismissive of Clarke's prospects: "I have very little expectation however of these engines being applicable to the working of machinery;—at least economically. They are very interesting toys."[140]

In 1842, the *Edinburgh Witness* reported that Robert Davidson, a Glasgow philosophical instrument-maker, was being financed by the directors of the Edinburgh and Glasgow Railway Company to carry out "extensive experiments as to the practicability of applying electro-magnetism for propelling trains along the lines of a railway."[141] The trials showed every sign of being successful. "The ponderous machine, weighing between five and six tons, was instantly set in motion on the immersion of the metallic plates into the troughs containing a solution of sulphuric acid."[142] According to the enthusiastic *Railway Times*, in further trials, the electromagnetic locomotive traveled one and a half miles at a velocity of more than four miles an hour.[143] Confidence was expressed that all that was required to improve on the performance was larger, more powerful batteries, and congratulations were offered to the Railway Company's directors for their "discernment" in financing such a worthy endeavor.

The Edinburgh and Glasgow's directors were not, however, sufficiently discerning or enthusiastic to continue the trials much longer. Before long, Davidson was in London putting his locomotive on show at the Egyptian Hall in Picadilly.[144] Some visitors, at least, were converted. Enthusiastic correspondents to the *Railway Times* announced that the locomotive's "simplicity, economy, safety, and compactness, render it a far more valuable motive power than steam on railways and in navigation."[145] Railway companies were urged to renew their patronage, particularly since Davidson was too poor to afford a patent to defend and develop his invention. Given "skill and capital," the engine's future was assured. One correspondent took Davidson's predicament as the occasion for a blast at the patent laws that "keep the "poor inventor" from becoming rich, and allowing [*sic*] others to pluck him."[146]

Some inventors certainly patented their electromagnetic engines. In November 1839, W. H. Taylor patented an engine that the *Mechanics' Magazine* described as being "the first electro-magnetic engine which has yet been produced, capable of being practically applied as a motive power to machinery."[147] Taylor claimed as his innovation that the electromagnets in his engine were simply switched on and off without any change of polarity. The result was an engine that could run far more quickly and efficiently. The engine was soon on show at the Colosseum, one of London's premier exhibition sites, famous for its panorama of the metropolis as seen from the dome of St. Paul's Cathedral.[148] The *Mechanics' Magazine* was convinced: "Nothing can be more continuously regular or beautiful than the motion imparted to the wheel—the agent (unlike fire, water, or steam) invisible, yet its effects palpable to the senses—capable of being called instantaneously into action (no time lost in getting up the steam,) and of being as instantaneously brought to a full stop."[149] Despite the small size of the Colosseum's model, the *Magazine* felt no doubt that it clearly demonstrated electromagnetism's potential as a motive power. It was only a matter of deciding what type of battery would be most appropriate.

## CONCLUSION

As these examples show, patents were neither necessary nor sufficient for the making of an electrical inventor. They were only one of a range of possibilities available. Indeed, patenting could be quite dangerous for an inventor who also placed great value on the kudos associated with the status of a gentlemanly natural philosopher. Outside the purview of practitioners of "practical science," the subscribers and contributors of journals such as the *Mechanics' Magazine* or the *Inventor's Advocate*, or the gallery exhibitors, their patrons, and their audiences, there was no consensus that the business of invention and that of science were related. Fellows of the Royal Society frowned upon those who by engaging in commerce aimed for personal profit from their philosophy.[150] Those who inhabited the world of patents, exhibitions, and violent epistolary and pamphlet wars stood to one side of the culture of the gentlemanly elite. Theirs, nevertheless, was in all probability the majority view. For those who concerned themselves with such matters, science was "practical science." The editors of the *Inventor's Advocate* were not alone in assuming that what made the science of electromagnetism matter was the electromagnetic engine.[151]

Electricity, through these enterprises in plating or locomotion or lighting, was in some sort entering a culture of commercial, economic calculation. If the dissolution of a zinc plate in a battery was to profitably run an engine or plate a candlestick with silver or illuminate a building, then comparing the zinc's cost with that of the output became an economic imperative. Alfred Smee, for example, included detailed accounting of potential costs in his *Elements of Electrometallurgy*.[152] Even Grove, who did not oppose electricians' participation in consumer culture but only their attempts at personal profit, went out of his way to emphasize the economic viability of his nitric acid battery.[153] In other experiments he sought to maximize the power his battery could produce.[154] His experiments focused on the transformation of natural powers and the economics of their production. In Grove's case, one of the results of this perspective toward his experiments was *On the Correlation of Physical Forces*. In this view, the different physical forces were linked to each other by a principle of correlation. Each force could be exploited to produce any of the others. Correlation was a matter of market economics as much as experimental philosophy.[155] In James Prescott Joule's case, it was the mechanical equivalent of heat. Comparing the practical cost of different engines, the fuel consumed in producing useful work, led to a new law of nature.

If electricity was becoming part of consumer culture when it was put on show at the exhibitions, even more was it a commodity when it was straightforwardly part of manufacturing processes, its cost-effectiveness compared to that of a horse or a sack of coal.[156] The notion of electricity as a source of work had been central to electricians' cosmologies for some time.[157] Much of their experimental endeavor was directed at demonstrating electricity's power by maximizing the power and efficiency of their batteries and machines. This was not just a philosophical principle. Electricity's work was a matter of real economic production as much as a hypothetical abstraction. With electrometallurgy and the electromagnetic engine, putting electricity to work was a way of making nature produce commodities, a way of naturalizing consumer culture. Putting electricity to work was a way of making entrepreneurs of electricians as well.

This chapter implies that all of these things went together. Invention was not an isolated activity. Its negotiation as an identifiable act required a whole range of other negotiations. The business of laying claim to a new device or process required in the first instance the definition of novelty. Invention was a moral matter as well. The inventor's status hinged on whether he could be successful in establishing that his claims to novelty and priority could be properly legitimated. A range of relationships among people, processes, and things were problematized. Work, in particular and in more than one sense, was problematized in the episodes described in this chapter. Invention, for the electricians discussed here, required showing successfully that it was their labor, in one way or another, that was responsible. It also required more often than not that nature should be put to work. Devices such as the electromagnetic engine, or processes such as electroplating, put nature at the service of mankind and made electricity part of the factory system.

# CHAPTER 7

# To Annihilate Time and Space: The Invention of the Telegraph

FOR MANY, the invention of the electric telegraph was the greatest triumph of Victorian ingenuity and invention. It revolutionized communication and had a central role to play in the creation and consolidation of imperial bureaucracy during the second half of the nineteenth century.[1] Many historians have also recently argued for the constitutive role of the telegraph in the development of nineteenth-century physics.[2] Most of these commentators have focused on telegraphy and the production of electrical standards. The spread of telegraph cables fostered a need for accurate measurement and the production of theories that could make sense of the technology's idiosyncrasies. The argument works both ways. Not only is it clear that the imperialist project of expanding the telegraph network was central to nineteenth-century physics, it is just as clear that that physics itself had a crucial role to play in the formation of British empire building.[3] The telegraph was widely celebrated by the Victorians themselves. It seemed to represent the pinnacle of science's application to the arts and hence to the industrial and civilizing progress of the empire.

Accounts of the telegraph frequently waxed lyrical over its potentialities. In 1840, Edward Copleston, bishop of Llandaff, could hardly contain his excitement: "Last night I was hardly able to sleep from the strong impressions made on my mind by the stupendous discoveries and results of experiments by Mr Whetstone [sic] in electricity and his most ingenious mechanical apparatus for an electric telegraph." The new invention seemed to epitomize the virtues of modern rationality:

> It far exceeds even the feats of pretended magic and the wildest fictions of the East. This subjugation of nature and conversion of her powers to the use and will of man actually do, as Lord Bacon predicted it would do, a thousand times more than what all the preternatural powers which men have dreamt of and wished to obtain were ever imagined capable of doing.[4]

Such enthusiasm was far from untypical.

Journalists easily lapsed into rhapsody describing this latest application of electric science. For the *Patent Journal*, the telegraph showed how

> we have trained the electric agent as a dutiful child or obedient servant, to carry our messages through the air by the road we have made for it, and with equal velocity through the earth by a road it makes for itself. Again, traversing the mighty deep in the shape of an angel of peace, bearing the olive branch to countries formerly our most bitter and inveterate foes, it takes another step towards realizing the dream of the poet.[5]

Such accounts frequently emphasized the *domestication* of nature by the telegraph. Through its invention, electricity was no longer "to be admired in proportion to its power or brilliancy, or the skill of the manipulator," but had "been
set to work in good earnest."[6] The analogy to a child or servant spoke volumes.
One of the most frequent assertions was that the telegraph had brought about
"the annihilation of time and space." Like the railways, for which identical
claims were made, the telegraph was seen as a technology that rendered the
world physically smaller and therefore more socially manageable.[7]

This was a point Latimer Clark made while reminiscing before the Society of
Telegraph Engineers concerning his own experiences in the heady days of
telegraphy's early development.[8] He saw himself as spokesman for the generation that had created and lived through a new era through the telegraph, as a
result of which "distance and time have been so changed to our imaginations,
that the globe has been practically reduced in magnitude, and there can be no
doubt that our conception of its dimensions is entirely different to that held by
our fathers."[9] Speculating concerning the origins of the revolutionary technology, Clark pointed to two key factors: the progress of electrical knowledge and
the spread of the railway. By 1837, he asserted, "Scientific men were in possession of every knowledge and appliance necessary for creating a perfect electric
telegraph." It was as a result of the railways, however, that "the world was in
every way ripe and ready for the practical introduction of the telegraph."[10]

Clark was right in pointing out that the early telegraph inventors utilized
preexisting technology. The whole range of the electricians' technology of display was put to work in the form of recording devices.[11] Galvanometer needles,
electromagnets, and electrically decomposed chemicals all could be and were
utilized to make the electric messages visible. Indeed, the telegraph itself
proved a valuable addition to the electricians' repertoire. In Britain, on the
Continent, and in America, demonstrations of electricity's capacity to communicate information at a distance were comparatively commonplace. The frontispiece of Henry Noad's *Lectures in Electricity* gave a view of the electrician's
tools of the trade. Prominent in the background, behind the batteries, electromagnets, and induction coils, was the diamond shape of a Cooke and Wheatstone five-needle telegraph.[12] One of the aims of this chapter is to investigate
what role strategies of exhibitionism had to play in making the telegraph into a
workable technology.

The importance of the railway system in facilitating the successful introduction of the telegraph was another fact upon which contemporaries largely
agreed as they looked back at the beginning of the telegraph's career.[13] Railway
companies provided the financial support that made many of the early experiments possible. In their need for an efficient and reliable signaling technology,
they also provided a potential market for the new product. Moreover, railways
displayed an important resemblance to the telegraph in that both technologies
required possession, or at least use of, continuous stretches of land. Close connection with the railways not only provided the telegraph with a market for its
signals, it also provided it with a space through which to expand its network. It

supplied the "way-leave, protection, and speedy access for repairs in case of defect in the wires" that made the telegraph system a practical projection rather than a visionary dream.[14]

Railways themselves were a new and dramatic technology in the 1830s. The Stockton and Darlington Railway, opened only in 1825, was the first to use steam locomotives on a regular basis, though only for short sections of the line. It was not until Robert Stephenson's triumph at the famous Rainhill trials on the Liverpool and Manchester Railway in 1829 that steam locomotives seemed clearly established.[15] Railways boomed during the 1830s. By 1840 more than a thousand miles of rail was already completed, and many more thousands of miles were projected. Railways were in the process of transforming perceptions of distance, speed, and time. They also had the effect of transforming the rural landscape, bringing the rewards of industry to landscapes that had previously been comparatively unaffected. They provided the opportunity for the engineering feats of Isambard Kingdom Brunel and his cohorts. Embankments, iron bridges, tunnels, and viaducts soon became familiar features of many parts of the countryside. After 1840, those scenes were often also accompanied by the chains of wooden poles and iron wires that carried the electric telegraph.

Wolfgang Schivelbusch has argued that railways differed significantly from previous forms of transportation technology.[16] He points out that whereas earlier means of transport such as roads and canals distinguished between locomotive and track, railways made no such distinction. For example, on eighteenth-century toll roads, any pedestrian, horseman, or carriage was permitted to use the road on payment of the appropriate toll. Similarly, any barge or boat was permitted to use a canal. By the 1830s, however, only the railway company's own steam locomotives could travel on that company's rail. This was not simply a feature of the railway's technology—before steam became the prime means of locomotion, railways had at times been open to all paying users. The crucial point was that railways were projected as a means of *mass* transportation, and that required new disciplines.

Contemporary commentators were aware of this feature of the new transportation technology:

> The wheels, rails, and carriages are only parts of one great machine, on the proper adjustment of which, one to the other, entirely depends the perfect action of the whole. . . . Therefore, from this cause it becomes necessary, in order to secure safety when moving at great speed, to have the parts in contact adjusted to each other in such manner as at all times, and under varying circumstances, to preserve a true relationship one to the other.[17]

Dionysius Lardner also emphasized the unity of the new technology: "A railway, *like a vast machine*, the wheels of which are all connected with each other; and whose movement requires a certain harmony, can not be worked by a number of independent agents."[18] The railways' efficient operation as a system of mass transportation required a network of information collection and distribution. By the late 1830s electricity provided a possible means of constructing that network.

It should be emphasized, however, that the electric telegraph's emergence as the inevitable companion to the railway was by no means a foregone conclusion. As contemporaries themselves noted, it was one thing to put on show the electric fluid's ability to communicate at a distance and quite another to make that ability economically and practically feasible. Telegraph entrepreneurs and inventors had to work hard to find a market for their product that was prepared to provide capital for its realization rather than simply marvel at their ingenuity. It was not at all clear to railway companies either that the electric telegraph was the answer to their need for an efficient system of signaling information. Initially, what seemed to be required was a means of communication over comparatively short distances, from one end of a tunnel to another, for example. Already existing technologies such as pneumatic signaling systems seemed well able to provide for such a requirement.

This chapter seeks to follow two attempts by British telegraph projectors to make their apparatus work and at the same time find and consolidate a market for their product. One of these is the work of the well-known duo of William Fothergill Cooke and Charles Wheatstone, credited with the successful introduction of the electric telegraph as a practical, working technology in Great Britain and the holders of the first British patent for an electric telegraph system. The other comprises the endeavors of the almost unknown Edward Davy, their exact contemporary, whose efforts to succeed practically mirrored those of his successful competitors but ended in complete failure. It will become clear that a whole range of resources, social and technical, had to be marshaled to make the telegraph work. In Latour's idiom, the telegraph had to become an obligatory passage point.[19] Some of the telegraph's features also had to be transformed to to enable it to fulfill its eventual role as a railway signaling technology.[20] The telegraph's technical features as they were put together during the late 1830s and early 1840s were not self-evident utilizations of electricity's technology of display. They were tailored for particular uses. As the railway network became the first market for the new technology, that technology had to be reconfigured so that it fitted its place. The telegraph was mutable. Its successful integration into the marketplace hinged upon its adaptation.

The telegraph was also a remarkably controversy-ridden technology. Disputes abounded concerning priority and rights to specific devices and principles. Particularly notorious was the virulent quarrel between William Fothergill Cooke and Charles Wheatstone, the co-owners of the first telegraphic patent, concerning their respective inventive rights. Both claimed their own contributions as the crucial factors in the achievement of success.[21] The adjudication of telegraphy's origins may have been particularly fraught precisely because of the wide range of resources that went into its making. Not only technical but social planning had been central to the invention's success, and the two could not easily be dissociated. Invention, for the early Victorians, was the property of an individual, and a great deal could hinge on its possession. Invention was imbued with moral properties, particularly when one was charged with having made an illicit claim.

PROJECTION

Cooke, the son of a surgeon and professor of anatomy at Durham University, had joined the East India Company's army in 1825 at the age of nineteen. Resigning his commission in 1833, he took up anatomical modeling as a profession. It was while visiting Heidelberg to improve his skills that Cooke attended a lecture by Muncke at which an electric telegraph designed by Baron Schilling was demonstrated.[22] Struck by the possibilities of Muncke's display, within weeks of the event he had commissioned the construction of a model telegraph and was preparing to return to London to exploit the commercial possibilities of his new invention.[23]

His return was dictated by two concerns. On the one hand, he recognized that neither he nor the local German craftsmen had the skills necessary to produce the parts, and, on the other, he was aware of the necessity of discovering "what others may have done in the same way." With that in mind he anticipated that the different parts of his model telegraph would be "made by different mechanicians for secrecy's sake" and asked his mother to conceal his motives for returning to England.[24] On his return to London he commenced his search for more information and for possible patrons. He was confident of the potential of his invention: "As the commercial and political worlds are equally concerned, I have a choice between Government and the mercantile potentates."[25]

With an eye toward such patronage, Cooke produced a prospectus of his new invention.[26] He was confident that "[t]he national importance of some practicable method, by which the benefits of a rapid telegraphic communication may be extended at a cheap rate to political and mercantile affairs, and in all cases of emergency to the private concerns of individuals is too obvious either to require argument or admit of question."[27] He offered his proposal as cheaper, more efficient, and more secret than any other method of rapid, long-distance communication then in use. The cities and towns of the kingdom would be linked by a network of copper wires buried in the earth along the connecting roads. At each town, one of his instruments, attended by "confidential clerks," would be maintained, keeping the various centers in telegraphic communication.

Cooke detailed his telegraph's utility to government, commerce, and the individual. For the government, "in case of dangerous riots or popular excitement, the earliest intimation thereof should be conveyed to the ear of Government alone, and a check put to the circulation of unnecessary alarm."[28] Such a remark was particularly apt in the latter half of the 1830s as rumors and threats of Chartist uprisings proliferated. Cooke suggested that government agents should have access to their own instruments at each station and be able to cut off the public system if the need arose. In this way, "the Government would be enabled in case of disturbances to transmit their orders to the local authorities, and, if necessary, send troops for their support; whilst all dangerous excitement of the public might be avoided."[29]

Commercial transactions would equally be facilitated if the provinces were placed in timely communication with the metropolis. "An immediate knowledge of the daily state of all the important markets &c., would place the most distant cities of the kingdom on a footing in their mercantile transactions with the capital."[30] The telegraph would provide honest men with "security and confidence" while subverting the efforts of swindlers hoping to profit by circulating false information or forged bills. Individual citizens would benefit also: "The comfort of friends and relations, far distant from each other, would often be materially involved, especially in cases of sickness." In all areas of life the telegraph would contribute toward "the one and only justifiable object of all establishments—the aggregate of comfort and happiness to the nation."[31] Perpetual subscribers would be "admitted to the subscribers' room, where all general, political, and mercantile news would be published as soon as received."[32] Cooke suggested that the telegraph might even be a charitable enterprise, since deaf-mutes, "accustomed from their infancy to abbreviate as far as possible, by signs, their symbolic language," would make excellent clerks for the telegraph stations.

Cooke concluded with a long quotation from Dionysius Lardner's essay on railways, remarking that it was as apt when applied to the telegraph and corresponded "most happily with the united action of both":

> The moral and political consequences of so great a change in the powers of transition of persons and intelligence from place to place are not easily calculated. The two advantages of increased cheapness and speed, besides extending the amount of existing traffic, call into existence new objects of commercial intercourse. The concentration of mind and exertion, which a great metropolis always exhibits, will be extended in a considerable degree to the whole realm. The same effect will be produced as if all distances were lessened in the proportion in which the speed and cheapness of transit are increased. Towns at present removed some stages from the metropolis will become its suburbs; others, now a day's journey, will be removed to its immediate vicinity; and business will be carried on with as much ease, between them and the metropolis, as it is now between distant parts of the metropolis itself.[33]

Cooke had made one other passing reference to the railways in his pamphlet. He suggested that the telegraph might prove useful for short-distance signaling in cases where stationary steam engines were used to provide additional power on steep inclines. There the telegraph might save money by eliminating the need to keep the engine permanently fired, or might save time by preventing delay as the locomotive waited for the engine to heat up before proceeding uphill.[34]

In the meantime, during the summer of 1836, Cooke, aided by his brother, continued to pursue the project. He found mechanics in Clerkenwell and Sudbury to work on his instruments and visited the Adelaide Gallery "to study various scientific instruments, connected more or less with our object in hand."[35] To succeed, Cooke had to find a place for himself in the network of workshops, mechanics, and exhibitions where he would have access to the necessary skills and resources. Even so, making his projection real was not an easy matter. Transforming designs and prototypes into practical, working instruments was not straightforward. By the autumn he was experiencing difficulties with his instrument-makers. He complained that "my clockmaker has again disappointed me." "None but those who have been engaged in such work can imagine the endless series of disappointments and delays to which the inventor of a new and secret instrument is exposed ere he can approach the moment of trial."[36] At the end of November, Cooke was still complaining that Moore (his clockmaker) had "not behaved well." Despite advice from clock- and watchmakers concerning the best form for the mechanism, it still did not work, and Cooke announced himself resigned to failure.[37]

Cooke soon regained his optimism, however, and succeeded in securing an interview with Michael Faraday, the "King of Electro-Magneticians." Faraday politely informed him that the principle of his instrument seemed correct, but declined to speculate as to its practicality.[38] Seeing Faraday was a matter of building the support network. Faraday's name added to the security of the telegraph as a viable instrument. Cooke had clearly hoped that Faraday would take an active role in promoting the invention. He was by now aware that the "only way to ensure success is to part with some of the advantages to be derived from it to those who are to bring it forward, and, as I am entirely unknown, I must get what aid I can, being unable to chuse."[39] A support network and patronage were indispensible requirements for an outsider aiming to sell his product on the London markets. He continued work throughout December and January, conducting experiments on a mile-long length of wire and attempting to interest men of science in his project. He consulted Faraday again, as well as Peter Mark Roget and "Clark, a practical mechanician."[40] Eventually he called on "a Mr. Wheatstone, Professor of Chemistry at the London University." He found to his dismay that Wheatstone "had been employed for months in the construction of a Telegraph, and had actually invented two or three, with the view of bringing them into practical use."[41] After discussion, he and Wheatstone agreed to investigate the possibility of collaboration.

Wheatstone was an experienced electrician, known in philosophical circles mainly for his work on the velocity of electricity, published a few years previously in the *Philosophical Transactions*. These experiments, carried out at the Adelaide Gallery, had established that electricity traveled through a wire at a velocity somewhat in excess of the velocity of light.[42] It was these experiments which proved that electricity could indeed be used to provide practically instantaneous communication between distant points. Wheatstone had continued his experiments through very long wires, investigating the action and effects of electricity through such lengths. He told Cooke that he had already designed several telegraphic instruments in the course of this research.[43]

Cooke and Wheatstone, having agreed to pool their resources, continued to experiment throughout the early months of 1837.[44] Their aim was to take out a joint patent giving themselves exclusive rights to electric telegraphy in the United Kingdom. They were also in negotiation with various commercial entrepreneurs in attempts to acquire financing for their invention. Cooke had been engaged in such negotiations even before his meeting with Wheatstone. In December 1836 he had requested from his father a letter of introduction to Joshua Walker, who had close connections to the Liverpool & Manchester Railway Company. He met the directors of the company in January to discuss the possibility—one that he had foreseen in his pamphlet—of using an electric telegraph to signal through the Lime Street tunnel, which was at a steep incline and required a stationary steam engine to provide additional locomotive power.[45] The directors ultimately chose a reliable pneumatic signaling system rather than risking an untried projection. The episode did, however, provide Cooke with useful future contacts.

Cooke continued to keep in touch with Walker, seeing the railway companies as a potential source of financing for the telegraph. Wheatstone also had his commercial contacts, notably the Enderby brothers, "who rank among the leading men in their line in London—enterprising, determined men."[46] They were rope- and sail-makers and were engaged in manufacturing "iron rope" for Wheatstone's experiments. By the end of May, Cooke was in contact with the Enderbys concerning the possibility of their investment in the telegraph and was also discussing with Walker the potential interest of another railway: the London & Birmingham Railway Company. News of the experiments was slowly leaking out. The May issue of the *Railway Magazine* announced:

> Telegraphic Communication by Galvanic Electricity.—Two gentlemen are employed in experiments on this subject, and anticipate complete success. The apparatus will be laid, if successful, wherever lines of railway are laid. If the method succeeds, it is expected we shall be able to communicate with almost any extremity of the kingdom to which railways extend, in about three minutes.[47]

Cooke and Wheatstone pressed ahead with their experiments, concerned that premature publicity could ruin their prospects of a secure patent. The patent was eventually signed on 10 June 1837, giving Cooke and Wheatstone joint rights to the commercial exploitation of their telegraph.[48]

Negotiations with the London & Birmingham Railway were also proceeding well. By the end of June, Cooke had been introduced to the company's chairman and secretary to describe the invention; following discussions with Robert Stephenson, the company's engineer, arrangements were made for experiments on the telegraph with the prospect of its employment on the railway.[49] By early July experiments over long distances were under way. Preparations were extensive:

By strenuous exertions I succeeded in collecting the . . . vast quantity of wire, cleared the huge workshop of men and lumber, by the constant labour of from 30 to 40 men, and had nearly half a mile of wire arranged by Friday night; proceeding slowly on Saturday morning, having to teach all the men employed—viz., 8 carpenters, 2 wire-workers, and 8 boys—their distinct duties, we got forward more rapidly towards evening, and at 5 o'clock, when the men left off work, I had about four miles of wire well arranged. . . . You may imagine the task when I tell you that 2,888 nails have been put up for the suspension of the wires. The labour can only be conceived by witnessing our proceedings.[50]

Making the telegraph work required more than manipulating the electrical fluid. It required manipulating large amounts of labor and materials. The company had allocated sixty pounds for the prosecution of these exploratory, large-scale experiments.[51]

While Cooke labored on the long lines of wire at the London & Birmingham's London terminal at Euston Square, which soon stretched for almost sixteen miles to Camden Town and back,[52] Wheatstone continued his experiments at King's College, attempting to improve the signaling and recording apparatus. A demonstration of the telegraph's potential was set for Tuesday, 4 July, before the assembled company directors, and Wheatstone worked hard to perfect the apparatus before the crucial exhibition.[53] Cooke and Wheatstone had to convince the hardheaded railway speculators that electricity could indeed be communicated instantaneously through long distances. They had to succeed at the difficult task of making laboratory and lecture theater apparatus work in a new and potentially inhospitable environment.

Demonstrations to the directors and other interested parties continued throughout the summer. Stephenson, the company's engineer, became a close collaborator:

Yesterday Mr. Stevenson [*sic*] witnessed our experiments through 19 miles of wire, extended from Euston Square to Camden Town, and declared himself so satisfied with the result that he begged me to lay down my wires permanently between those two points on my best plan, with a view to extending the communication hereafter, if the Directors approved. . . . He said we must have two or three rehearsals beforehand, that all may go off in "good style." He declared himself quite a "convert to our system," and seemed quite delighted at the correspondence we carried on at so great a distance from each other, requesting me to send the word "Bravo" along the line more than once.[54]

With such a powerful patron in the railway world, the Cooke and Wheatstone telegraph seemed assured a secure future. This did not, however, deter Cooke and Wheatstone from continuing their search for more patronage. Reports were appearing in the press of alternative electric telegraph systems that could still easily supplant their own.

In September, Wheatstone reported to his partner that "Mr. Chadwick, the poor-law commissioner, who is also one of the commissioners for reporting on the establishment of a general police, is very much taken with the electrical telegraph, and intends in the forthcoming report to recommend its adoption by the government; we shall have a visit, for this purpose, from this gentleman, Col[n] Rowan and Mr. Lefevre M.P., the three police commissioners, when the instruments are again in action."[55] In their enthusiasm for railway contracts, the projectors had clearly not forgotten Cooke's suggestion that the telegraph could prove a powerful anti-insurrectionary device. It could easily be presented to Benthamites such as Chadwick as an ideal tool for regulatory surveillance. Throughout September, Cooke, with the aid of his solicitor, Robert Wilson, was also engaged in negotiation with Benjamin Hawes, M.P., in attempts to interest the Admiralty in the electric telegraph.[56]

By December, however, disaster had struck the projected telegraph. The London & Birmingham Railway Company's directors suddenly signaled their decision not to continue their support for Cooke and Wheatstone's telegraphic experiments. In October they resolved that "[t]he Secretary be instructed to inform Mr. Cooke, that it is the intention of the Company to limit the trial of his Magnetic Telegraph for the present to the portion of the Railway between the Euston and Camden Stations."[57] This was a serious setback. By this stage Cooke had been discussing with the company the prospect of their financing the construction of a telegraph, interconnecting London, Liverpool, Manchester, and Holyhead, "for the transmission of all Government & commercial intelligence," as soon as the railway to Holyhead was completed.[58] A link to Holyhead, the gateway to Ireland, was, of course, of major military importance.

The directors made it clear that they regarded the existing telegraph line between Euston and Camden, along with the associated telegraph apparatus, as the company's property, unsurprisingly, since they had financed its construction. They argued that handing the existing apparatus over to Cooke would render the connection between Euston and Camden useless, though they agreed that "Mr. Cooke is at all times at liberty to exhibit it to parties who may desire information on the subject of it, on his giving previous notice to the Secretary."[59] It took several meetings before Cooke could persuade them to return the instruments at half of their original cost.[60] This was a far cry from Stephenson's promises to recommend immediate adoption of the telegraph by the company along the entire length of their line.[61] The long effort to attract railway patronage had been unsuccessful.

At the same time, disturbing reports of rival telegraph systems continued to appear in the press. William Alexander had put his telegraph on show at the Royal Society of Arts.[62] It was soon reported that Alexander and his telegraph had been presented at Kensington Palace to the duke of Sussex, the new queen Victoria's uncle and president of the Royal Society.[63] By January his telegraph was on show at the Adelaide Gallery.[64] Alexander's telegraph remained on show throughout the 1840s and was one of the exhibits at the Great Exhibition of 1851. A more serious competitor was Edward Davy's telegraph, on show at a private exhibition in Exeter Hall. Unlike Alexander, Davy was attempting to acquire a patent (which might seriously interfere with Cooke and Wheatstone's claims) and was also trying to court many of the same patrons from government and the railway companies whose attention they were aiming to capture.

Edward Davy, a London surgeon, had been working on the possibility of telegraphy by means of electricity since the mid-1830s.[65] By 1837 he had working models of telegraphic apparatus, and in January of 1838 in an attempt to raise the necessary finances to patent his telegraph as well as attract attention to its potential, he hired rooms and put his invention on show at Exeter Hall. Wheatstone was immediately concerned: "Davy has advertized an exhibition of an electric telegraph at Exeter Hall which is to be opened Monday next; I am told he employs six wires, by means of which he obtains upwards of two hundred simple and compound signals and that he rings a bell; I scarcely think that he can effect either of these things without infringing our patent; if he has done so I think some step should be taken."[66] He suggested that an injunction should be taken out against both Davy and Alexander's exhibition at the Adelaide Gallery. He instructed Cooke to call on J. Strettell Brickwood, one of the Adelaide's proprietors, to inform him of the consequences of continuing the display.

Davy's exhibition went well, however. Within days of opening he had received visits from the Admiralty as well as from the superintendent of government telegraphs. He anticipated that the show would earn between one and two hundred pounds by the end of the season.[67] As he told his father, "You did not expect to have a son turned showman, but I trust I am merely instrumental in promulgating a useful discovery."[68] Davy was well aware that such showmanship was essential if his telegraph were to be marketed successfully, and wrote enthusiastically to his father of the various eminent potential patrons who had been seen examining his telegraph.

Like his competitors, Davy worked to persuade the railway companies of the utility of the electric telegraph. He was in contact with Isambard Kingdom Brunel, engineer of the Great Western Railway Company, trying to convince him of the merits of his invention. Brunel by this time was also in negotiation with Cooke and Wheatstone. It was a perfect opportunity for Davy to snatch the advantage. He was confident that a telegraph would be laid down on the Great Western, and that all he needed to do was persuade Brunel of the merits of his design. Davy, however, was in the difficult position of having to exhibit his telegraph to potential customers without the protection of a patent. He had to be very careful concerning exactly what was put on show. He was in danger both of losing the right to patent through prior disclosure and of allowing access to potential competitors of the details of his design.

Davy's interactions with Brunel highlight these problems: "the principal difficulty under which I laboured was the impossibility of rendering manifest all the advantages of my mechanism, without entering, more or less, into such explanations as would, more or less, betray my secret—as yet unpatented."[69] Seeing, for example, that Davy's telegraph, like that of his competitors, used six wires, Brunel suspected that Davy's design infringed the Cooke and Wheatstone patent. Davy was not in a position to demonstrate that this was not so. Brunel argued that without a detailed account of the telegraph's principles, as opposed to a simple display of its effects, he could not judge its advantages. Davy was trying to sell him a "pig in a poke," "which, though it might produce very pretty effects, yet, as the *rationale* was not open for canvass, its practicability could not fairly be judged of, nor could [Brunel] confidently assure the company but that it might prove to be an infringement on the others patent."[70] The encounter taught Davy the difficulties of display—of having "to be on the alert to divulge nothing that would impair the security of a future patent-right."[71]

By March, Davy was proceeding to obtain a patent, against strong opposition from Cooke and Wheatstone.[72] Davy's exhibition was one of the issues. It was arguable that having put his telegraph on show was equivalent to publication. Davy argued that he had "exhibited the external appearance or effects only: & that I do not seek a patent for the effects but for the means, or mechanism, which has never been exhibited, but on the contrary kept closely secret."[73] After continued lobbying from Wheatstone, the solicitor general eventually called in an expert witness, no less a personage than Michael Faraday. He concluded that Davy's telegraph was an invention distinct from that of Cooke and Wheatstone and was therefore eligible for a patent. After some delays in raising the necessary capital, Davy was granted a patent for his telegraph in June 1838.[74]

With a safe patent (rendered safer by Faraday's authority on his side) Davy aimed to further publicize his telegraph and acquire financial backing. He even contemplated the possibility of coming to an agreement with "Cooke & C°."[75] Davy was soon reporting to his father that he had "been endeavouring to make connections with some *businessmen*, to assist me in making negociations with the Railway Companies, or in getting up a general Telegraph Comp^y."[76] He expressed confidence that once his invention was more widely known, he would have no further need to solicit patrons. Davy had formed a partnership with "Mr. N.F. Price, of Liverpool St., Bishopgate, & Director of the Arisna[?] Mining Comp^y," to help him in his negotiations. Price was soon in touch with the directors of both the London & Birmingham and the Southampton Railway Companies. Davy was also cultivating a "Cap^t Bellew," who was "intimate with the Engineer of the Birmingham & Gloucester Railway:—& has influence with the Midland Counties Railway:—either of which would be a good step."[77] Another financial contact was "Mr. L. Bunn, a young man, of considerable connection & address."[78]

Davy busied himself over the next several months with promoting the telegraph. His aim was to form a general company as soon as possible, which would then license out the use of the telegraph to the railway companies and other interested parties. Through Price he was in negotiation with both the London & Birmingham and the Southampton Railways, while Bunn in Paris negotiated with the French government concerning the possibility of patenting Davy's telegraph in France.[79] All kinds of strategies could be adopted in the effort to secure patronage. Before an appointment with Charles Fox, resident engineer of the London & Birmingham Railway, "Mr. Price suggested that it would be prudent to hint to that Gentleman, to give him contingent interest in the business, in some shape or other, to secure his best endeavours to forward it with the Directors."[80]

By August 1838, Davy was on the brink of success. John Easthope, director of the Southampton Railway Company and member of Parliament, was enthusiastic. The Southampton line seemed particularly promising since it raised the possibility of a lucrative agreement with the government to carry Admiralty communications between London and the south coast.[81] Price had agreed to purchase a one-quarter share in the patent for £150 and to share the cost of acquiring foreign patents. Davy was convinced that an agreement with the London & Birmingham Railway was immanent. He was confident that the "value of the invention has very greatly increased since what it was six months ago," and was even at the stage of drawing up lists of the trustees and directors of the projected telegraph company. Charles Fox was to be the engineer, while Davy himself would be superintendent of machinery.[82]

By the end of August, however, the situation changed. Price turned out to be unable to afford the £150 for a share in the patent and suddenly announced his skepticism concerning the likelihood of the railway companies' entering into agreements with a telegraph company to use their lines. Throughout this period Davy had been in litigation with his estranged wife and with her creditors. Before the end of the year he fled to Australia, leaving the telegraph in his family's charge. He spent the rest of his life in South Australia.[83] In his absence negotiations with the railway companies failed, and his telegraph apparatus was eventually broken up and taken to his father's home in Devon. Davy's experiences are a good example of the contingencies involved in bringing a new technology out of the workshop and into the market. There was a world of difference between demonstrating to potential patrons that his telegraph worked in the exhibition room at Exeter Hall and persuading them that it would be a practical and viable asset on their railway lines. It is worth noting that after Davy's departure his friend Thomas Watson attempted to keep up the display at Exeter Hall. He found that in Davy's absence the telegraphic apparatus no longer worked.[84] It was still too fragile a technology to survive without the practical know-how of its maker.

During the period when Davy had been courting the railway proprietors, Cooke and Wheatstone had also been similarly engaged. While experiments and negotiations at the London & Birmingham Railway were still continuing, Cooke had already entered into discussions with Isambard Kingdom Brunel, the engineer of the Great Western Railway, projected to run from London to Bristol.[85] Brunel had contacted Cooke himself to discuss the prospect of employing the telegraph on the new railway.[86] Following the London & Birmingham Railway's withdrawal of interest, Cooke courted Brunel assiduously and by the middle of March was on the brink of signing an agreement to build a telegraph for the Great Western. Following some last-minute disagreements concerning the terms of the contract—at one stage Cooke threatened to withdraw completely from the negotiations—the agreement was signed on 3 April 1838.[87]

According to the terms of the agreement, the telegraph was to be put in operation immediately on an experimental basis between the railway's London terminal at Paddington and Drayton. If the trials were deemed successful, the telegraph would be extended to Maidenhead immediately, and on 1 January 1839 the company would determine whether or not to purchase a license to operate the telegraph permanently. Cooke anticipated that if all went well, he would "at once be an independent man."[88] Cooke himself was contracted by the company to lay down the telegraphic wires between Paddington and Drayton at the rate of £165 per mile, for a total of £2,145. He was also given the option of contracting at the same rate were the telegraph to be extended to Maidenhead—at that time the western terminus of the Great Western Railway.[89]

By the end of May, work was ready to proceed. Cooke prepared carefully, training his laborers carefully in the new procedures: "I mean to lay down about a quarter of a mile in the garden over and over again, till each man knows his duty." He wanted "no experimental trying of plans on the permanent line" and "no awkwardness or misunderstanding when under the eye of the engineers and other men."[90] Building a telegraph line in the field was as novel a procedure as the telegraph itself was a novel technology: "Our plans are entirely changed; and I am to lay down the wires in iron pipes. In consequence all the instruments and tools before made are useless; and one fresh set is nearly complete. . . . I have between £600 and £700 worth of goods coming in to-day and to-morrow, and 15 or 20 mechanics at work, constantly requiring direction, so that I have not one moment's rest; and am drawing and making calculations half the night."[91] New ways of working had to be found. Practical experience was required to show the best ways of proceeding.

Work on constructing the line to Drayton took a little over a year to complete. The telegraphic connection was opened in July 1839. The five-needle telegraphic apparatus of Wheatstone's design was installed at Paddington and Drayton.[92] All seemed to be proceeding well. Cooke reported that "[t]he telegraph on the Great Western Railway has given great confidence and satisfaction."[93] Vague rumors of the new invention started to circulate in the press. The *Inventor's Advocate* reported in September that a "Galvanic Telegraph" invented by a "Professor Whitton[!]" was in operation on the Great Western Railway.[94] The same journal provided a more detailed account a few months later:

> This telegraph, which is the useful and scientific invention of Mr. Cooke and Professor Wheatstone, of King's College . . . has been, during two months, constantly worked at the passing of every train between Drayton, Hanwell, and Paddington. . . . As soon as the whole line is completed, the telegraph will extend from the Paddington terminus to Bristol, and it is contemplated that then, information, of any nature, will be conveyed to Bristol, and an answer received in town, in about 20 minutes.[95]

Even Cooke's philanthropic hopes had been realized. The *Advocate* reported that "two of the boys from the Deaf and Dumb Asylum in the Kent-road, have been at the Paddington station for five or six weeks, where they were instructed in the working of the machinery . . . and they are now perfectly competent to superintend the telegraph at any one of the stations."[96]

Another contract was also in the offing. Cooke was contacted in November by Robert Stephenson, who inquired into the possibility of adopting a telegraph on the London & Blackwall Railway. Negotiations were successfully concluded by the end of the month. Cooke was delighted at "having been sought out by my *first Patron*, after having lost sight of him for two years."[97] The Blackwall line was a curiosity and a good example of the continuing flexibility of railway technology at the time. It utilized a stationary steam engine along the entire length of the line rather than the more common locomotive. The engine ran constantly, and carriages carrying passengers were hooked on to an endless rope at the different stations along the line. It was a process that required perfect timing and knowledge of conditions all along the railway. The electric telegraph was well adapted to such constant vigilance. The telegraph's utility was demonstrated a few years later. A correspondent in the *Railway Times* described the telegraph in action following an accident: "The utility of the electric telegraph was, we understand, shown very clearly in this instance, intelligence of the accident having instantaneously been conveyed to the engine house and the engine stopped. The consequences might otherwise have been very serious."[98] By now, the electric telegraph was almost universally regarded as a technology inseparable from the railways. It was an integral part of their workings, designed to signal information and prevent accident.

An excellent opportunity to proselytize in favor of the telegraph's role in the expanding railway network was provided by the Parliamentary Committee on Railways, who in their fifth report, in 1840, included discussions of telegraphy as it related to railway safety. Wheatstone was called to give evidence before the committee, and that section of the report was widely reprinted in the press.[99] He gave detailed accounts of the electric telegraph's mode of operation, emphasizing its advantages over traditional methods of signaling information across long distances. Wheatstone emphasized that despite the telegraph's having been put into operation only over the thirteen miles between Paddington and Drayton, he had no doubt that it could operate just as successfully between Paddington and Bristol.[100]

Charles Saunders, secretary to the Great Western Railroad Company, also gave evidence concerning the uses of the telegraph on the railway.[101] Under questioning as to the advantages of the telegraph he replied, "I think the usefulness to the railway itself is the chief remuneration; it is calculated undoubtedly to simplify the working of the railway, and to diminish the stock of every description, whether of engines or of carriages; to insure greater punctuality, and, in cases of accident, to repair the injury with the least delay, as well as to produce general advantages and greater security in working the railway."[102]

Economy, efficiency, and safety were the key words that the telegraph guaranteed. Saunders detailed the kinds of circumstances under which the telegraph could act to maximize the efficient operation of the railroad.

Unsurprisingly, the parliamentary committee were interested in the telegraph beyond its use in railway signaling. They inquired as to the extent to which possession of an exclusive and rapid means of communication between Bristol (a major shipping port) and London would provide "a great advantage in a commercial point of view over the rest of the public," and "what remedy the public might have under those circumstances." Saunders responded that the telegraph might indeed confer such an advantage, and intimated that the railway company should have the right to exploit their property as they saw fit: "I cannot conceive if a party possesses property, why he should refuse to make it useful to himself, or why he should be called upon to make it as useful to another as to himself."[103]

Of particular concern to the committee was the possibility of a private company's having possession of a means of communication denied to the government.[104] Saunders was keen to emphasize in response that any railway company, if financially recompensed, would be willing either to allow the government to lay down wires on their land or to permit use of their own telegraphic equipment for government dispatches. He was also emphatic, however, that only the government could have power to lay down their own telegraphs on railway land: "it would be a great hardship to make the possessor of the soil give up his right to enable some other party to compete with him."[105] The telegraph raised a number of difficulties once it started being seen as more than railway signaling technology. Questions of ownership and rights to information and the confidentiality of that information became matters of concern.

Cooke and Wheatstone in the meantime continued to sell the telegraph as an integral part of a railway network. They needed to work hard. In 1842 the Great Western Railway Company decided against the hoped-for continuation of the telegraph to Bristol. Cooke agreed to maintain the wires and extend the line a further eighteen miles to Slough at his own expense and to transmit railway messages free of charge.[106] One motive for extending the line to Slough was clear. It was a short distance to Windsor, and the possibility existed that a direct telegraphic link might be negotiated between Buckingham Palace and Windsor Castle.[107] Advertisements appeared in the press, drawing the attention of "Railway Companies, Engineers, and other Parties requiring a certain and instantaneous mode of communication between distant points" to the telegraph. The telegraph was

> peculiarly adapted to the use of Railways, as a portable Telegraph to be carried with every train, and as a self-acting Telegraph for giving notice of the approach of trains to tunnels, level crossings, stations, and inclined planes, whether worked by stationary or assistant engines. It affords a means of working a single line of rails, with perfect safety, in both directions at the same time.[108]

In conjunction with the advertising campaign, Cooke published a pamphlet titled *Telegraphic Railways; or, the Single Way* outlining his vision of the electric telegraph as an integral element of the railway network.[109]

By Cooke's account, the integration of the electric telegraph into railway technology would not only "add to the safety and efficiency of Railway communication" but would facilitate the introduction of a new kind of railway.[110] Most railways of the period, for reasons of safety—in order to prevent the collision of trains traveling in opposite directions—were projected as double lines. Cooke claimed the telegraph as a means of making such double ways redundant, at least as a solution to problems of safety. With the perfect knowledge of conditions throughout the line that would be provided by the electric telegraph, single ways would be as safe as, or safer than, their more expensive counterparts. "It appears ... that the double line is now necessary for safety, only because, absolute punctuality being unattainable, the state of the line cannot with certainty be inferred; and therefore, that it would *not* be necessary for safety if general inferential probability could be superseded by particular and certain knowledge."[111] With the telegraph in attendance, double lines would be required only to deal with bulk of traffic, the same consideration that applied to the widths of common roads.

To demonstrate his point, Cooke followed a hypothetical train on its journey between Derby and Rugby on the Midland Counties Railway, showing how the telegraph guaranteed the absolute safety of even an unexpected, unscheduled locomotive traveling along the track: "*every train might, with the Telegraph, be an express train*, if required."[112] Cooke was emphatic that the telegraph would "give to the single way in perfection, what the double way seeks in vain to obtain inferentially, doubtfully, and at an immense expense."[113] The key to the telegraph's success would, of course, be its enforcement of vigilance, not only over the trains, but over their human operators: "As a further practical benefit conferred by the Telegraph, unremitting vigilance and alertness would be enforced, upon all the officers of the division, by the instant and infallible detection at head-quarters of individual remissness."[114] Just as the mechanical loom imposed uniformity of action in the cotton mill, the telegraph would guarantee such uniformity on the railways.

Cooke's claims concerning the single way and telegraphy's potential were enthusiastically received in several quarters. The editors of the *Railway Times*, noting that Stephenson planned to use the telegraph on the projected Norwich and Yarmouth Railway, were emphatic that "a single line with the electric telegraph is superior in security and efficiency to a double line without it."[115] They had already editorialized vehemently in favor of Cooke's views concerning the single way, warning the promoters of new schemes that their investors' financial interests demanded that they seek every opportunity of minimizing the cost (and maximizing the profit) of future railway construction.[116]

The *Railway Times* was endorsing the view of George Bidder, who, while the bill chartering the construction of the Norwich and Yarmouth Railway was passing through Parliament, had recommended that it should be built as a sin-

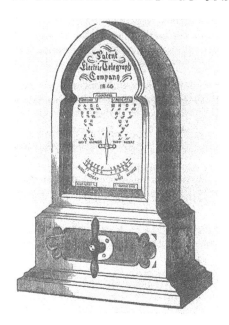

7.1 The *Cooke* and *Wheatstone* single-needle telegraph. Taliaferro Shaffner, *The Telegraph Manual: A Complete History and Description* (New York, 1859), p. 222.

gle way with an electric telegraph.[117] Bidder had worked with Robert Stephenson on the construction of the Blackwall Railway and was a keen advocate of the electric telegraph. He reiterated his views before the Select Committee on Post-Office Communication in 1842 when questioned concerning the proposed railway link between Chester and Holyhead.[118] The link was of concern to the Committee on Post-Office Communication since, when opened, it would be the main conduit for mail between London and Ireland. Bidder claimed that the availability of a telegraph would be particularly useful in such a context: "by means of the telegraph the parties stationed at the look-out at the port would judge a long time before the vessel arrived, as to whether there would be any delay in starting the train . . . and enable, as before stated, preparation to be made for expediting any train that might be detained by adverse winds opposing the passage of the steamer, without interrupting any other communication of the traffic."[119] Yet again, the telegraph could be a means of maximizing regularity on the railroads.

Since Cooke's pamphlet of 1836, and his and Wheatstone's patent of 1837, the telegraph had undergone a transformation. Cooke had projected his invention as a multipurpose device for the instantaneous communication of information at a distance. The telegraph was designed to facilitate "the affairs of Government; of the commercial world; and of the private individual."[120] Cooke in his pamphlet barely mentioned what would rapidly become the primary use of the electric telegraph during its early years, as a system of regulating the movement of stock on the railways. The direction of the telegraph's development during the early 1840s was in large part dictated by the concerns of those willing to finance it. It is noteworthy that Bidder, speculating concerning the telegraph's role on the Holyhead line, saw it as a means of expediting the transmission of government communications by mail rather than as a direct means of conveying information itself.[121] The original aspirations and designs of its makers had very little to do with the telegraph's uses. This was reflected in its construction. The original Wheatstone five-needle telegraph needed no code: combinations of needles simply pointed to different letters of the alphabet, spelling out a message. By the early 1840s the design was obsolete, having been replaced by a two-needle telegraph suited to the transmission of predetermined railway signals. The five-needle telegraph with its distinctive diamond shaped face was ideally suited for exhibiting the instrument's capacities, for putting the technology on show. As the telegraph's cultural place shifted to the railways, the instrument's design had to shift as well. Technical and cultural construction went hand in hand.

ANATOMY OF AN ARGUMENT

The collaboration and business partnership between Cooke and Wheatstone was stormy almost from its inception. Cooke in particular complained bitterly concerning what he regarded as the partisan representation of the telegraph and its origins in the press. Wheatstone, he asserted, was culpable either in himself, in claiming the telegraph as entirely his own invention, or at least in refraining from contradicting such reports when they appeared. The debate continued into the 1850s, well after Wheatstone had relinquished any financial interest in the telegraph patents he had taken out in conjunction with Cooke. Following an article in the *Quarterly Review*, identifying Wheatstone as the inventor, Cooke embarked on a pamphlet war, complaining of Wheatstone's "egotism." Wheatstone's culpability lay in his failure to contradict an account that "contained statements, in his favour and to [Cooke's] prejudice, contrary to the truth and to his own solemn admission . . . by allowing them to remain uncontradicted, he has virtually adopted them."[122]

It is, of course, impossible to provide a definitive historical answer to the question of Cooke's and Wheatstone's relative roles in the invention of the telegraph. As pointed out earlier, any attempt to uniquely identify *the* inventor of any new item is always predicated on a particular account of what kind of activity might count as invention in the first place.[123] Such an account is precisely *what is at stake during the course of disputes such as that between Cooke and Wheatstone in this instance. Not only the identity of the inventor but the nature of inventive activity is at issue* in such episodes. During the early 1840s when the quarrel first emerged, along with challenges to Cooke and Wheatstone's joint claim, the question of who would reap the financial reward from the telegraph's invention was coterminous with the question of who could claim status as its inventor. Even within the partnership between Cooke and Wheatstone, which might be imagined to share a joint interest in the potential

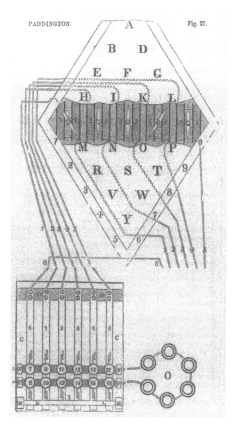

7.2 The Cooke and Wheatstone five-needle telegraph, reading the letter V. From Taliaferro Shaffner, *The Telegraph Manual: A Complete History and Description* (New York, 1859), p. 204.

financial benefits of the telegraph, strong tensions existed concerning the management of the invention, which translated easily into disagreements concerning the nature of invention.

Matters first came seriously to a head between Cooke and Wheatstone during 1840, as Cooke demanded from his partner an explanation of his position with regard to their relative status as inventors and an accounting of his failure to publicly state their joint status as inventors. Wheatstone responded angrily, and Cooke soon demanded a legal arbitration to establish their respective claims. Marc Isambard Brunel, acting for Cooke, and John Frederic Daniell, for Wheatstone, were appointed as arbitrators. Following acrimonious legal dispute, an arbitration statement was finally agreed upon in April 1841.[124] The crucial section of the agreement was the final paragraph:

> Whilst Mr. Cooke is entitled to stand alone, as the gentleman to whom this country is indebted for having practically introduced and carried out the Electric Telegraph as a useful undertaking, promising to be a work of national importance; and Professor Wheatstone is acknowledged as the scientific man, whose profound and successful researches had already prepared the public to receive it as a project capable of practi-

cal application; it is to the united labours of two gentlemen so well qualified for mutual assistance, that we must attribute the rapid progress which this important invention has made during the five years since they have been associated.[125]

The arbitrators had attempted to differentiate between the *science* and the *practical introduction* of the telegraph, placing the two on a par in an attempt to salve both disputants' sensibilities.

It is clear that there were differences between the partners well before 1840. Cooke in his copious correspondence with his family frequently alluded to frustration at the way in which the telegraph was being reported. In September 1837, for example, he complained:

> I saw a most incorrect account of our proceedings in the paper the other day, in which Mr. Wheatstone and Mr. Stephenson are represented as trying the experiments, which have been conducted solely under my direction, *I alone* being in communication with the Directors on the subject, and have been the only person responsible from the commencement. I have good reasons for taking no notice of the newspaper reports for the moment. Do not fear that I shall have my full credit at last.[126]

Significantly, as Cooke was giving vent to such misgivings, he was embroiled in difficult negotiations with Wheatstone concerning the precise details of their business partnership.[127] As early as May, he had announced his intention of describing his "plans and improvements" in postmarked letters to his family as a means of establishing priority should it be required.[128]

The partnership agreement between Cooke and Wheatstone as finally established gave Cooke sole managerial responsibility for exploiting their patent in Great Britain, Ireland, and the colonies. Any profits from the invention were to be divided between the two partners, with 55 percent going to Cooke, the 10 percent above Wheatstone's allotment being his fee as manager.[129] They also agreed that any improvements or modifications in the apparatus developed by either party during the lifetime of the first patent (fourteen years) should be made known to the other party in the agreement. Cooke was later to imply that the prominent position allowed him by this agreement was a tacit endorsement on Wheatstone's part of his superior role in the telegraph's invention. Cooke's "claim to the projectorship was openly advanced throughout the arbitration." He "took the lead in the partnership deed, as in the English, Scotch, and Irish patents, as a matter of right."[130]

The crucial term in Cooke's defense of his rights before the 1840–41 arbitration committee of Brunel and Daniell was *projection*. Cooke asserted that it was the fact of his not only having invented instruments but having also worked out a comprehensive system for their deployment that gave him inventor's rights. He claimed to be

> the Projector of that entire SYSTEM OF PRACTICAL OPERATIONS, which has embodied in the form of a useful, practical invention the idea long floating in the scientific world of an Electric Telegraph; a system which can never be superseded by any improvements in the telegraphic apparatus.[131]

Furthermore, his actual, practical involvement in the detailed, everyday deployment of the telegraph was used as evidence of his superior role in its origins. Wheatstone, by Cooke's accounting, was relegated to the role of an adviser or consultant who had done little more than provide the scientific knowledge required to perfect the apparatus that he had originated. Cooke was interpreting invention in terms of a unitary, individual moment of projection. Once the system had been visualized, the tasks of entering the network, making the instruments work, and acquiring patronage were relegated to the periphery of the inventive progress. Their only significance was the evidence they provided of the preexistence of Cooke's projection.

Wheatstone, unsurprisingly, took a different view. On his showing, it was precisely his science and his superior knowledge of the *principles* involved in the telegraph's construction that made him its inventor. Rebutting Cooke, he asserted:

> Your instrument, however ingenious in its mechanical arrangements, had never been practically applied, and was incapable of being so. On the contrary, the instruments I had proposed were all founded on principles, which I had previously proved by experiments would produce the required effects at great distances.[132]

Completely different accounts of what was required to produce a working telegraph were in play here. For Cooke the crucial features were the design and practical implementation of a system. For Wheatstone the key was the design (not necessarily the construction) of instruments according to scientific principles.

Wheatstone was keen, on the one hand, to emphasize the embeddedness of his work on the telegraph in his philosophical researches in electricity and, on the other, to dissociate the partnership's business arrangements from any claims concerning novelty and priority.

> With respect to my own experiments connected with the telegraph, they are so separate from yours, and so intimately mixed up with other theoretical researches and practical consequences in which you have no interest, that I could not, if I would, associate my [sic] name with them, *particularly since you are unacquainted with the principles on which they are founded.*[133]

The crucial contribution and vital feature of invention on Wheatstone's account was the experimentally based knowledge of *how* the telegraph worked. It was his possession of such knowledge that identified him as the inventor. Commercial considerations had no weight: "When I entered into a commercial speculation with you I had no intention to give up my right to call my own discoveries and inventions my own."[134]

Whereas Cooke based much of his claim to rights and status as an inventor on his having worked out the principles of the telegraph system as a practical—by which he meant a commercially viable—system, Wheatstone sought to base his priority on fitting the telegraph into a succession of experimental researches. He traced his interest in the communication of information at a distance back to his researches on music and sound during the 1820s. His determination of the velocity of electricity in a wire was presented as an attempt to ascertain whether the electric fluid traveled at a speed sufficient to provide effectively instantaneous communication over long distances. The fact that unlike Cooke he had not constructed any specifically telegraphic instruments by 1837 was subordinate to the understanding he had acquired through experiments of the principles whereby such apparatus would have to operate.

Both Cooke and Wheatstone in their submissions to the arbitration committee were engaged in a process of *redescription*. The activities of each over the past few years had to be strategically redeployed and redefined so as to provide evidence of his superior role in the business of inventing the telegraph. The status of their past activities had to be reinscribed so as to match their new claims concerning who had done the inventing and what precisely the invention had been. A whole range of issues, concerning, for example, which instrument did or did not work, what experiments had been successful, and what might or might not be considered a "practical" contribution to the process of building a telegraph, were continually being reconstructed and redeployed by the combatants as each sought to establish his priority. Writing the appropriate history for the instrument was part of the process of establishing its invention and its identity.

Wheatstone's Bakerian Lecture before the Royal Society in 1843 was an exemplar of his expertise in the transmission of electricity at a distance and of the place of the telegraph in his experimental philosophy.[135] He introduced his lecture as an essay on the practical and scientific implications of his telegraphic researches.

> I intend in the present communication to give an account of various instruments and processes which I have devised and employed during several years past for the purpose of investigating the laws of electric currents. The practical object to which my attention has been principally directed, and for which these instruments were originally constructed, was to ascertain the most advantageous conditions for the production of electric effects through circuits of great extent, in order to determine the practicability of communicating signals by means of electric currents to more considerable distances than had hitherto been attempted.[136]

His guide in these researches was the "beautiful and comprehensive theory" of Ohm.[137] Ohm's researches on the theory of electrical circuits were represented as the foundations without which practical telegraphy could not be achieved, and Wheatstone's Bakerian Lecture demonstrated Wheatstone's own expert understanding of Ohm's work, then largely unknown in England.

In many respects, Brunel and Daniell's delicately phrased arbitration was a diplomatic acknowledgment of the very different claims made by the two partners as the grounds for status as the inventor of the telegraph. They agreed with Cooke that he was indeed the first projector of the telegraph as a practical system. They agreed with Wheatstone that his scientific researches were a necessary precondition of the telegraph's success.[138] Unlike both partners they placed the two contributions on an equal footing. The arbitration was as a result susceptible to more than one reading. Since Cooke asserted that projection was the crucial factor, he could claim the arbitration as a vindication of his priority. In just the same way Wheatstone could read the arbitration's recognition of his preeminence in science as a vindication of his claims. This is what happened in the pamphlet war of the mid-1850s.[139]

In the meantime, however, the arbitration could be made public to demonstrate the amity of the Cooke and Wheatstone business partnership. Francis Wishaw, secretary to the Royal Society of Arts, lecturing there, drew attention to the arbitration agreement to explicate the "relative connection of Mr. Cooke and Professor Wheatstone with this invention."[140] Copies of the agreement were made available so that the audience could satisfy themselves that the copatentees of the electric telegraph had been adjudged to have equal rights to kudos as well as cash from its invention. The coinventors had made different contributions befitting their status. The arbitrators had distinguished between the "philosopher's researches into the laws of Nature" and the "labours of the practical man who applies those laws to the purposes of life,"[141] and put them on an equal footing in the business of invention.

Others besides Cooke, however, aimed to dispute Charles Wheatstone's moral status as an inventor. As an adjunct to one of his telegraphic patents, Wheatstone had patented an electromagnetic clock on the same principle. The fact was widely announced, and the clock had been on display at the Royal Society. Alexander Bain, a journeyman clockmaker newly arrived in London from Scotland,[142] however, accused Wheatstone of purloining his invention, asserting that he had himself described the clock to Wheatstone in confidence.[143] Wheatstone's response was terse, if indirect:

> The statement of Alexander Bain, in your number of March 27th does not invalidate the claim of Professor Wheatstone as inventor of the Electro-Magnetic Clock. Professor Wheatstone gave me instructions to make his electro-magnetic telegraphic clock on January 6, 1840, which was more than six months before Mr. Bain says he made his communication.[144]

The letter (clearly) was not from Wheatstone; it was from his instrument-maker, John Lamb.

Bain responded at greater length. He asserted that the mechanism constructed by Lamb for Wheatstone was not part of a clock at all, but of a telegraph, similar to others that Wheatstone had designed before. There was nothing novel in what Lamb had been directed to construct, and he denied that the mechanism could ever be applied to the measurement of time. Bain insisted that he had discussed his own electromagnetic clock, later patented, with Wheatstone, and that at that time Wheatstone "had not the most distant intention of applying electro-magnetism to clocks."[145] Lamb again responded, insisting that "the instructions I received from Professor Wheatstone on January 6th, 1840, related to the construction of an instrument which would show the time of a clock, with which it was to be connected by means of a voltaic circuit."[146]

Bain returned to the attack, complaining that "it would have been more to Professor Wheatstone's credit if he had come manfully forward himself, and stated his claims, instead of directing his workman to do it for him."[147] Wheatstone's own silence, presumably, was quite deliberate, particularly since, as Bain noted, he had been claiming that Bain had also been employed as one of his workmen. By refusing to reply in person, he was asserting the difference in status between himself and a mere mechanic. Bain repeated that the instrument made by Lamb (and patented by Wheatstone) was not a clock but a clockwork telegraphic mechanism designed to "have hands to point to the signals." He denied that he had ever been one of Wheatstone's workmen, and reiterated that previous to the conversation between himself and Wheatstone, the professor "had no intention of employing electro-magnetism to the working of clocks" and had no idea of the advantages of so doing.[148]

The quarrel flared up a year later in the *Literary Gazette*. Following a lecture at the Royal Institution, Bain complained again that Wheatstone was claiming as his own an invention that belonged to him.[149] This time Wheatstone condescended to reply in person. Bain he described as "a working mechanic, who had been employed by me between the months of August and December of the year 1840." In response to the assertion that Bain had described his invention to him three months before his own announcement to the Royal Society, he asserted that

> there is no essential difference between my telegraph-clock and one of the forms of the electro-magnetic telegraph, invented by me, and described in the specification of patent granted to myself and Mr. Cooke in January, 1840; the former is only one of the numerous and obvious applications which I have made, and only requires the idea of telegraphing time to present itself for any workman of ordinary skill to put it into practice. In telegraphing messages, the wheel for making and breaking the circuit is turned round by the finger of the operator, while in telegraphing time it is carried round by the arbor of a clock.[150]

For Wheatstone the essence of the matter was the principle of the mechanism. This was what counted for invention, not the particular application. The electromagnetic clock was simply a telegraph that telegraphed time. He asserted in any case that he had discussed the possibility of telegraphing time with numerous acquaintances several years previously, naming George Bidell Airy, the astronomer royal, among his witnesses.

The full extent of the dispute between Bain and Wheatstone became apparent a year later on the publication of a pamphlet launching a virulent attack on Wheatstone for his behavior in the affair. The author, John Finlaison, represented himself as an unbiased bystander, engaged merely in trying to protect the reputation of his fellow Scotsman.[151] Finlaison detailed Bain's dealings with Wheatstone, alleging that Wheatstone had agreed to purchase models of Bain's apparatus and contracted him to build further models both of his clock and of an electromagnetic printing telegraph of Bain's own invention. According to

Finlaison, Wheatstone had reneged on this agreement and effectively stolen Bain's models. On being challenged, Wheatstone had allegedly come near to physically assaulting the humble mechanic.[152] On this showing, Wheatstone had quite literally stolen Bain's inventions and had behaved in a manner distinctly unbecoming of a gentleman.

Finlaison aimed to thoroughly discredit Wheatstone and therefore to discredit his status as an inventor. Even the dispute between Cooke and his partner, on Finlaison's showing, was no more than a clever marketing ploy:

> Two men embarked in a commercial partnership, with a view to advertise their commodity, pretend to quarrel. To reconcile their imaginary differences, they prevail on an unsuspecting friend to arbitrate between them. . . . The arbitrators are thus innocently induced to sign a manifesto which, under the sanction of their distinguished names, is to be converted into a glowing prologue to every future lecture in praise of the commodity which the partners have thus in the market.[153]

Finlaison was not prepared to allow Wheatstone credit even as an inventor of the telegraph. He and his partner were no more than venal entrepreneurs who had picked up on the insights of others and marketed them as their own. Cooke and Wheatstone's patent contained nothing more than a commercial application of inventions by a long list of predecessors.[154]

Finlaison's assault was enthusiastically welcomed by the *Mechanics' Magazine* who saw the occasion as yet another opportunity to champion the rights of the mechanic-inventor against the corrupt holders of scientific power.[155] Finlaison was applauded for having vindicated the reputation of the "humble watchmaker from Caithness" against his unscrupulous competitor. Wheatstone's defense in this instance was mounted by William Fothergill Cooke, who was perfectly willing to allow his partner the title of inventor against opponents other than himself. Arguing from the point of view of the legal status of the several patents, he asserted that while those held by Bain and his copatentees might allow them rights over particular forms of the electromagnetic clock and the printing telegraph, they did not give them any general claim to the principles involved.[156] He asserted his rights as copatentee to "make any use I please of Mr. Wheatstone's superior forms of the same inventions," which were protected by their own various patents. He repeated Wheatstone's claim that he was entitled to the electromagnetic clock, not only because he had displayed it before the Royal Society previous to Bain's patent, but because the "most superficial examination of the clock proves its mechanism to be absolutely the same with that of the electric telegraph."[157]

It is clear that disputes concerning the identity of the inventors and their inventions in electric telegraphy were particularly fraught. This was not least because by the early 1840s, interest from the railway companies was making it clear that ownership of such patents could be very lucrative indeed. The disputes between Cooke and Wheatstone concerning their own respective claims to invention were inextricably intertwined with their disputes concerning the finances and management of their joint patents. At the same time, to outsiders their public falling out and reconciliation could be represented as nothing more than a ploy to attract attention to the commodity they had for sale. As rival patentees such as Alexander Bain and his business partners appeared on the scene, their main chance of overcoming the already entrenched interests of Cooke and Wheatstone was to find a means of overcoming their moral rights to invention, since the legal rights were protected by patent. Dissecting the anatomy of such arguments provides a means of illustrating the complexities that needed to be negotiated in the establishment of the electric telegraph and its inventors. It is clear that the two issues could not be easily disentangled. Disputes about *who* had invented the telegraph could not be resolved without a problematization of the issue of *what* had been invented. The identity of the inventor and that of the invention were negotiated hand in hand.

COMMUNICATION AS COMMODITY

By the mid-1840s, electric telegraphy was well on the way to being established as a commonplace feature of the railways. The technology was integral to the operation of the busy Blackwall line and was being either constructed or projected in conjunction with the numerous railway schemes that proliferated in mid-decade. On only one line, however, did the telegraph show any clear signs that it might have a function beyond that of signaling the movements of railway stock. This was the telegraph on the Great Western Railway, originally constructed between Paddington and Drayton and later extended to Slough and then to Windsor. In 1842, as discussed above, the Great Western's directors had determined to discontinue their agreement with Cooke and Wheatstone. The upshot had been that Cooke undertook to maintain the telegraph on his own behalf. As well as communicating railway intelligence free of charge, he could also offer his services to others interested in rapid communication.[158]

Rather than take responsibility for the everyday operation of the telegraph himself, Cooke licensed Thomas Home to work the telegraph on the Great Western. Home paid an annual rental of £170 for that right. The telegraph was put on show. One shilling, the standard price for entry into any of London's many exhibitions, was charged for the opportunity to see the "interesting and most extraordinary Apparatus" that comprised the "galvanic and magneto electric telegraph."[159] The public were invited to see the telegraph in action. For their shillings they could also send a message of their own down the line. Messages of all kinds were sent, ranging from the announcement of the birth of Queen Victoria's second son to orders to London grocers for provisions. The most famous message, however, was probably the one that led to the apprehension of John Tawell for the murder in Slough of his former mistress. The press made much of the telegraph's role in his capture and eventual, successful prosecution.[160]

The telegraph on the Great Western line was a technological curiosity comparable to the exhibits on show at the Adelaide Gallery or the Polytechnic. It is significant that the entrance charge to view it in action was the same as that for the majority of London's other shows and exhibitions. Visitors went to marvel at the instrument's ingenuity and the possibilities it offered. Ada Lovelace, for example, enthused about the revolution it would engender in housekeeping:

> The electrical telegraph is laying down on *our* Southampton Rail, & by & bye Wheatstone says that we shall be able to send any message to Town for a shilling, & *get an answer*. . . . Think what a delight Wheatstone says that sometimes friends hold conversations from one terminus to the other; that one can *send* for anyone to speak to one. For instance, I might desire a tradesman to go directly to the Nine Elms' Station, & might discuss an *order* with him about goods &c. Wonderful agent and invention![161]

Like the exhibits at the galleries also, the telegraph on show was a commodity, a technological curiosity to be desired and fetishized. It was on a par with the curious and inexplicable. Like the audience at one of Barnum's shows, the telegraph's audience was tempted to inquire, "How does he do it?"[162]

A significant change in the telegraph's fortunes took place in 1845, when Cooke, in conjunction with some railway entrepreneurs—notably George Bidder, who had worked with Cooke on the Blackwall Railway, among others, and John Lewis Ricardo, nephew of the notorious economist—determined to set up a telegraph company. By this time the partnership between Cooke and Wheatstone had undergone considerable change. Since 1840 Cooke had been given sole management rights over their patents in Great Britain. In 1843, following a further modification to their agreement, Cooke became the sole proprietor of the telegraph in Great Britain, on payment of an annual royalty to Wheatstone for all the telegraph lines constructed during the patents' lifetimes. Having determined to set up a company, Cooke entered into negotiation with Wheatstone and eventually purchased his royalty rights on behalf of the fledgling company for £30,000.[163] Bidder and Ricardo undertook to purchase a $^{23}/_{32}$ share of the patent from Cooke at a price of £115,000, making the total value of the patent £160,000. The final agreement was signed on 23 December 1845, and a bill of incorporation for the Electric Telegraph Company passed before Parliament during 1846.[164] The company's prospectus claimed as its goal "to establish a complete system of Telegraphic communication, connecting the Metropolis with the different Ports and Cities of the Kingdom."[165]

The second half of the 1840s, following the incorporation of the Electric Telegraph Company, witnessed a proliferation of telegraph interest and some attempts to establish new patents. Alexander Bain's various patents were purchased by the Electric Telegraph Company, though never utilized, while Bain himself was briefly a director. The introduction of gutta-percha as a new insulating medium signaled a major breakthrough in making the telegraph work over long distances and considerably lessened the cost of effectively protecting the wires. John Lewis Ricardo in 1849 took out a patent for "improvements in electric telegraphs, and in apparatus connected therewith."[166] The patent protected a particular mode of employing gutta-percha as an insulating medium and a means of suspending the wires from poles so as to protect them from damp. This was not the first proposal to use gutta-percha. Many such schemes were outlined in the *Mechanics' Magazine* in late 1848.[167]

In 1849, George Bachhoffner, previously a professor at the Polytechnic Institution, took out a patent for "improvements in transmitting intelligence by electricity &c." He was closely followed by Frederick Collier Bakewell with a patent for "improvements in making communication from one place to another, by means of electricity."[168] A more serious challenge to the Electric Telegraph Company's monopoly over electric communication was mounted by Alfred Brett and George Little. In 1847 they enrolled a patent for improvements in electric telegraphy, making a number of specific claims to novelty. In particular, they claimed what they described as a hydraulic battery as an improved source of constant current.[169] Within months of the patent's signing, Brett and Little took the Electric Telegraph Company and one of the company's employees, Charles Massi, to court for patent violation concerning the hydraulic battery.[170] With electricity's entrance into the commercial economy, the law courts were becoming a new resource for the resolution of disputes concerning invention, novelty, and priority. Legal, as much as philosophical and technical expertise, was now deemed competent to adjudicate on such matters.

The plaintiffs (Brett and Little) charged that Massi, five days after their patent, had registered an invention of his own, the "percolating galvanic trough," which was practically identical to their battery, and that he had entered into an agreement to sell the rights to the "trough" to the Electric Telegraph Company. Neither side in the case disputed the effective identity of the respective inventions. Massi, however, claimed that since he had entered into an agreement with the company to test his battery before the date of Brett and Little's patent, the matter was already public and could not therefore be the subject of a patent. But in his agreement with the Electric Telegraph Company Massi had stipulated that should they choose not to use his invention, they were bound to secrecy; the judge ruled, therefore, that this could not be considered a public disclosure and allowed the injunction preventing the Electric Telegraph Company from using the device.[171]

A few years later the combatants were in court again. On this occasion, however, the plaintiffs were the Electric Telegraph Company. Presumably as a result of the threat posed by rival companies to their telegraphic monopoly, the Electric Telegraph Company mounted a concerted attack on the Brett and Little patent in its entirety, pleading a breach of their own Cooke and Wheatstone patents.[172] The company alleged that the Brett and Little patent infringed theirs on a number of specific points, including the use of stops to limit the deflection of the needles and the "reciprocating principle, whereby the same wires were used in transmitting backwards and forwards." They alleged that "the invention of the defendants was a mere colourable evasion of the plaintiffs' patent and that they produced the same effect by a process which, although apparently different, was mechanically equivalent." The defendants, on the other hand, claimed that theirs was "an entirely new invention from the same stock of knowledge as the plaintiffs.'"[173]

The jury in large part found for the plaintiffs, listing what they regarded as the differences and the similarities between the two systems. The jury clearly aimed to counter the claim implicit in the Electric Telegraph Company's suit that any system of telegraphy was included in their patent. They ruled that the defendants' "system of the telegraph was a different system from that of the plaintiffs.'" This was the first example of the courts' adjudicating upon a matter of electrical invention. The judge entered the verdict for the plaintiffs and ordered five pounds' costs against the defendants. It seems plausible to suppose that the court's wish to undermine the Electric Telegraph Company's effective monopoly was a factor in the circumspection of the verdict. While finding in principle for the company, the court's ruling established what the limits of the Cooke and Wheatstone patents were. This was a point emphasized by the defendants' attorney.[174] Legal judgments were, however, no more immune from reinterpretation than any other kind. Brett's and Little's attorney insisted that the verdict vindicated his clients as much as it did the Electric Telegraph Company, arguing that the result was only "nominally for the plaintiffs" and protected his clients' rights in most respects.[175]

This was not the only court action in which the Electric Telegraph Company was involved. In 1849, two journalists, Willmer and Smith, Liverpool correspondents for the *Morning Herald*, publicly charged the company with improper use of the information they received for communication.[176] The company was accused of being nothing more than "a monopoly carried out by favouritism the most glaring, blunders and delays the most monstrous and unpardonable, and charges the most exorbitant ever heard of in this or any other country."[177] According to the journalists, the Electric Telegraph Company violated the terms of the act under which it was founded in several respects. The company favored some newspapers (in particular the *Times*) over others, giving them priority in the transmission of news. The company used information given them for transmission for their own purposes. The company both overcharged and charged different rates to different customers for the same services.

Willmer and Smith provided the *Morning Herald* with several examples of the Electric Telegraph Company's perfidy. They cited occasions when telegraph clerks at the Liverpool offices had been instructed from London to interrupt communications from the *Morning Herald*'s correspondents in favor of messages destined for the *Times*. On the occasion of their attempting to send news of U.S. President Taylor's inaugural address to London, following the arrival of a steamer in Liverpool, a communication for which they were charged twenty-one pounds was allegedly delayed for half an hour while the *Times*'s correspondent's news was sent instead. On other occasions, the company had held back the *Morning Herald's* news until they had sent their own version to the newspapers. In short, the implication of Willmer and Smith's charges was that the company made illicit use of the information they received for communication, to their own profit.

One allegation in particular stood out. Willmer and Smith charged that on 27 July 1848, they had received news "of a very astounding description"[178] from Ireland that they immediately attempted to telegraph through to the newspaper's London offices:

> Upon taking it to the Electric Telegraph Office the clerks said that the wires were engaged upon their own message; but, on seeing the news sent to us, they agreed to forward it at once to the *Morning Herald* without charge on the condition that they might *afterwards* use it themselves. This arrangement, to save time, we consented to; the conditions, however, were entirely broken by the company, the intelligence was *first* made known in Capel-court, afterwards to Sir George Grey in Downing-street, and at eleven o'clock was delivered at the *Morning Herald* office, though it had been given to them at Liverpool at a quarter past seven!! So much for good faith.[179]

A little over a month after the accusations had appeared in the pages of the *Morning Herald*, the Electric Telegraph Company took Willmer and Smith before the Bail Court, charging them with libel on the basis of this particular allegation.

The import of the allegation according to the prosecution, and the reason for the bringing of a charge of libel, was that "it was the intention of the writers . . . to insinuate that Mr. Ricardo and the other highly-respectable gentlemen who were directors of that company had made use of the information which they had thus received for stock-jobbing purposes, and had suppressed communications which were intended for the Morning Herald."[180] The attorney general, speaking on behalf of the Electric Telegraph Company, denied the version of events provided by Willmer and Smith, particularly their account of the agreement to communicate the information to London. According to the prosecution, the agreement was that the company "should be at liberty to use the information thus obtained in communicating it to the proprietors as well as all the newspapers. This was agreed to, and the news was sent to the different papers, as well as to the company's news-room."[181] By this account there had been no agreement to inform the *Morning Herald* before all others.

Crucial to the prosecution was the nature of the news that the two journalists had wished to communicate to London. They had received a report that later proved to be false, of revolution in Ireland, the defeat of British troops, and the burning of a railway station. Given the gravity of this information, had it been true, the prosecution claimed that the company's actions were justified. On receiving the news from Liverpool, William Fothergill Cooke, who was in attendance at the London offices, had instructed the clerks there to contact Ricardo with the information before communicating it to the newspapers. Ricardo then did "that which every true and loyal subject was bound to do." He immediately contacted the home secretary (Sir George Grey) with the information and "offered to place the works of the company at the will of the government."[182] There was in fact a clause in the Electric Telegraph Company's charter allowing the government to take over control of their lines in cases of national emergency. That clause had indeed recently been invoked during the course of the Chartist uprisings of 1848.[183] The prosecution asserted emphatically that the directors had "neither on the occasion referred to, nor on any other, ever used the telegraph for their own purposes." The judge allowed the prosecution's case to proceed to a full trial.

In many ways the issue at stake in the dispute between the Electric Telegraph Company and the *Morning Herald*'s irate correspondents was the simple matter of ownership. Who owned the information, the knowledge being communicated along the telegraph lines? It was, after all, the company's business to sell knowledge as much as to transmit it. By this time the Electric Telegraph Company had set up its own newsrooms in many cities where paying subscribers could gain privileged access to the latest information coming down the telegraph lines. The most recent "ship lists, share lists, price currents, Stock Exchange lists, corn markets &c." could all be made available at a price.[184] In these circumstances, news and first access to the news were at a premium. The bursts of electric fluid running through the telegraph wires could translate into a very valuable commodity. On this occasion, the Electric Telegraph Company's rights to that commodity were upheld. The case never proceeded to full trial. Willmer and Smith publicly withdrew their insinuation that the company had been misusing its access to knowledge for financial gain.[185]

The Electric Telegraph Company's relationship with its consumers was, however, often fraught. Complaints of high charges, misconduct, and monopolizing behavior were frequent. Willmer and Smith themselves had pointed to the company's high and inconsistent charges for their services as one cause of complaint.[186] Unfavorable comparisons were drawn between the charges for telegraphy in the United States and those made by the Electric Telegraph Company in the United Kingdom. The press urged that "the incorporated monopolists of this wonderful invention may be brought to see that their speculation and the convenience of the public may be simultaneously consulted with advantage."[187] Correspondents complained that the telegraph was "a means of communication available only to the rich" and threatened "that the British public, which is not accustomed to submit to abuses, will endeavour to frustrate this attempt on their pockets."[188]

Part of the problem was clearly the sheer novelty of the telegraph as a means of communication. Beyond its use as a means of communicating railway signals, it was not clear what role it should play. Other than the railway companies the main users appear to have been the newspapers, intent in particular on receiving foreign news in London as soon as ships arrived in port. Information concerning the fluctuation of stocks and share prices also made up a high percentage of the traffic. In 1847, for example, the telegraph was used in Manchester to inform the corn market of the state of the weather in the surrounding agricultural districts.[189] Private users took advantage of the telegraph for a variety of purposes. An early and not untypical exchange from the Paddington Telegraph Office ran as follows:

> Slough, Nov. 11, 1844, 4.3 p.m.—Send a messenger to Mr. Harris, Duke-street, Manchester-square, and request him to send 6 lbs. of whitebait and 4 lbs. of sausages, by the 5.30 train, to Mr. Finch of Windsor; they must be sent down by 5.30 down train, or not at all.
>
> Paddington, 5.27 p.m.—Messenger returned with articles, which will be sent by 5.30 train, as requested.[190]

The messages still bore the characteristics of communications sent by more traditional means: through the post or by word of mouth. It is worth noting that this example was sent while the cost of communicating by telegraph on the Great Western was a standard one shilling, regardless of length.

At this time the telegraph was still seen as a novelty rather than a mundane instrument expected to operate invisibly and tacitly. The transformations that it would bring about in perceptions of time and space were still nascent and a matter for wry comment: "It appears that directly after the clock had struck 12, on the night of the 31st December last [1844], the superintendent at Paddington signalled his brother at Slough that he wished him a happy new year: an answer was immediately returned, suggesting that the wish was premature, as the new year had not yet arrived at Slough!"[191] This was still an age in which time was a matter for local calculation. Within five years, however, under the aegis of George Bidell Airy, the astronomer royal, plans were afoot to use the telegraph to signal a standard time throughout the kingdom.[192]

In 1845, a game of chess played by telegraph could be an occasion for excited, interested press speculation. The *Morning Herald* announced that "a game of chess will be played this week between two parties nearly 100 miles asunder, at no more perceptible expense of time than would be required if they sat at the same table . . . their only media of communication being the wires of the electric telegraph."[193] The technology's capacities were a matter for showmanship and speculation and an opportunity to engage in the exotic and unprecedented. Thomas Home, the licensee of the Great Western telegraph, was quick to announce that should "any gentlemen feel disposed to avail themselves of the use of the electric telegraph, for the purpose of playing a game of chess, or draughts, I shall feel great pleasure in affording them every facility in my power."[194] The telegraph provided the opportunity for an exercise in difference. Playing chess by telegraph had its particular significance as well. The capacity for such an activity could be regarded by the Victorians as evidence of intelligence. It had been used as a criterion for a machine's capacity for thought. What the game showed was that *intelligence*, as much as electricity, flowed through the telegraph's wires.[195]

As the telegraph's network spread, however, the appearance of unfamiliarity was being replaced by a recognition of commercial potential. Accounts of the telegraph drew attention to its organization and businesslike efficiency:

7.3 Playing chess by electric telegraph. *Illustrated London News* 6 (1845): 233.

The offices are situated at the extremity of a court leading out of the north side of Lothbury, opposite the Bank of England. . . . Entering we pass into a large and lofty hall, with galleries running around supported by pillars. Here the first object that arrests attention is a map of England of colossal dimensions, placed on the wall opposite the entrance,—and covered by a net-work of red lines showing the telegraphic communication at present existing between the metropolis and different towns in the kingdom. Under the galleries are two long counters, over which are the names of the various places to which messages can be sent. Behind the counter are stationed clerks whose business it is to receive the message . . . and pass it to another set of clerks, who transmit it by machinery to the galleries above. Adjoining these are a series of rooms containing the electro-magnetic telegraphs of Messrs. Wheatstone & Cooke. They are placed on desks—and before them are seated the clerks whose province it is to work the apparatus. Each apartment is provided with an electrical clock showing true London railway time—which, as our readers know, is observed throughout the departments.[196]

The electric network was integrated into the facsimile of a typical Victorian countinghouse.

This was not necessarily a virtue. *Punch* in 1854 gleefully lampooned the telegraph's bureaucracy, drawing attention to some other features that Victorians found problematic in the new technology. *Punch* outlined the predicament of a city man, attempting to invite a friend to dinner but wishing to warn his wife beforehand as well. The satire's aim was to show how

the chief object of the officials entrusted with the telegraphs is to discourage the transmission of messages. The plan is to make as much fuss as possible, and to insist upon the observance of details with the same pedantic precision as if a request to your wife at Brighton to secure a bed for Smith, who is coming down with you, was to be registered among the archives of the nation. Then the niggardly, petty-tradesman-like way in which an extra word is made the excuse for an extra demand of money, gives a meanness to the whole affair. Add, that the prices are already far too high, and that, generally speaking, the manners of the electric shopmen impress you with their conviction that they are really doing you a great favour in selling you a pint of the electric fluid, though really the barman who pulls at *his* ivory handle and draws you *his* fluid (when you refresh yourself between the acts of *Norma*) is just as much entitled to give himself airs of importance.[197]

The Victorians were unaccustomed to seeing words as commodities to be carefully weighed and measured before transmission.[198]

In *Punch*'s sketch, the city gentleman was both outraged at the telegraph clerk's translation of his affectionate and verbose message into pounds, shillings and pence and embarrassed at the realization that he had just exposed intimate details of his personal life to the scrutiny of a social inferior. He eventually found it impossible to transform his original epistle into terms sufficiently brief to be affordable while still retaining the necessary information. *Punch* was drawing attention to the difficulties of telegraph communication for its first Victorian users. Norms of privacy and notions of proper behavior were violated. The meaning of communication changed as its components became commodities to be carefully packaged and valued. As Andrew Wynter pointed out, "the telegraphic style banishes all forms of politeness."[199]

As the telegraph turned information into packets of carefully measured words, it in some ways lost its charm for the Victorians. Again, *Punch* expressed the reservations rather well:

> What horrid fibs by that electric wire
> Are flashed about! What falsehoods are its shocks!
> So that, in fact, it is a shocking liar,
> And why? That rogues may gamble in the stocks.
>
> We thought that it was going to diffuse
> Truth o'er the world; instead of which, behold,
> It is employed by speculative Jews,
> That speculative Christians may be sold.
>
> Nations, we fancied, 'twas about to knit,
> Linking in peace, those placed asunder far,
> Whereas those nations are immensely bit
> By its untrue reports about the war.
>
> Oh! let us have the fact that creeps,
> Comparatively, by the Post so slow,
> Than the quick fudge which like the lightning leaps,
> And makes us credit that which is not so.

> The calm philosopher, the quiet sage,
> Fair Science thus abused to see, provokes,
> Especially it puts him in a rage,
> To be, himself, deluded by the hoax.[200]

The telegraph could be seen as a breaker of social protocols, an encouragement to un-English behavior, and a betrayal of scientific purity.

But the telegraph was also seen as a means of binding together the body of the nation. Just as its wires were rapidly recognized as fitting metaphors for the body's nervous system, the nervous system was seen as an apt picture of the telegraph network itself. The central office of the Electric Telegraph Company in Lothbury was "the great brain" at the center of "the nervous system of Britain."

> The physiologist, minutely dissecting the star fish, shows us its nervous system extending to the tip of each limb, and descants upon the beauty of this arrangement, by which the central mouth is informed of the nutriment within its reach. The telegraphic system, already developed in England, has rendered her as sensitive to the utmost extremities as the star fish.[201]

Like the starfish's or the human body's nervous system, what the telegraph was sensitive to was intelligence. The information carried by electricity through its wires was now perceived to be as vital to the state's survival as was the information of its surroundings conveyed by nervous forces to an organism's brain.[202]

This recognition of the telegraph's role in binding together the nation and the empire brought to the fore the question of control. Even in laissez-faire Victorian Britain the issue of who should have control of such a valuable commodity as intelligence was being raised. "Is not telegraphic communication as much a function of Government as the conveyance of letters?"[203] Sir Robert Peel's government had made sure when granting the Electric Telegraph Company's charter that the lines would be at their service, at least in times of national emergency. Like the railways along whose lines the telegraph wires stretched, the telegraph was a means of bringing the periphery closer to the center and under its more direct surveillance. As the early- and mid-Victorian state multiplied its own instruments of surveillance, both these technologies were seen as having a role to play,[204] and electricity became big business. Electricians found themselves at the center of engineering the nineteenth-century state.

This chapter, like the previous one, has emphasized the contingencies of invention. Cooke and Wheatstone's successful efforts, as well as Edward Davy's failure, underline the technical and social problems of getting the telegraph to work and of persuading others that it was an invention that mattered. A range of resources was required for such a task. The dispute between Cooke and Wheatstone concerning their own instrument's origins and their respective claims to priority in invention makes explicit the fragilities of establishing moral as well as financial claims to invention. Making such claims brought together the establishment of technical skill in electrical experimentation, the details of financial partnership, and the managerial skills of overseeing the technology's large-scale implemantation. Finally, the chapter emphasizes the way in which such processes made electricity part of a burgeoning commercial, consumer culture. Through technologies such as the telegraph, electricity was reified as a commodity, to be manufactured, bought, and sold. As such, it acquired new meanings and significances as it helped transform the ways in which Victorians communicated with each other.[205]

# CHAPTER 8

# Under Medical Direction: The Regulation of Electrotherapy

LATE-NINETEENTH-CENTURY commentators, looking back at the second quarter of their century, located the new respectability of electricity as a cure for disease during this period. No one denied that electricity had frequently been used for therapeutic purposes for at least a century before this time. That past practice was dismissed, however, as largely empiric-ridden and unsystematic. Only in the early Victorian period by these accounts did electricity become a tool of orthodox medicine. A correspondent in the *Lancet*, for example, citing a French military report and applying it to the British situation, identified improved technology and a new understanding of the nature and function of the nervous system as the keys to the successful use of electricity as therapy.[1] This assessment of electrotherapy's progress was shared by one of the first historians of the technology.[2] For Hector Colwell, as he himself admitted, electrotherapy's history was the progress of its technology. His essay was a survey of the machinery that made possible electricity's application to the human body. The most comprehensive recent survey takes a similar view.[3] Developments in electrical technology clearly were a significant factor in the rise of electrotherapy. Such developments cannot be divorced, however, from their cultural contexts. This chapter aims to place the new respectability of early Victorian electrotherapeutics in the wider context of contemporary electrical and medical entrepreneurship. The market for electrotherapy and its technologies was part of a far wider market for electricity.

As is well-known, the British medical profession during the first half of the nineteenth century was in a state of increasing flux. Many commentators argued that the traditional structures governing medicine and inherited from the last century or before were inappropriate for the new century. The vast social changes associated with the industrial revolution had greatly increased the number of practitioners and the demand for their services.[4] The profession remained dominated, however, by the "three estates" represented by the Royal College of Physicians, the Royal College of Surgeons, and the Society of Apothecaries. These three ancient corporations still governed medicine, in the metropolis at least, with an iron hand, protecting privileges for their members while denying them to the rapidly expanding band of general practitioners.[5]

The three corporations of medicine divided the profession (*vocation* would be a more appropriate word) into distinct social groups. Their division was also a division of labor in treating the body and its ailments. The Royal College of

Physicians was the oldest of the corporations, founded in 1518. In principle its members had a monopoly over the practice of "physic" or internal medicine.[6] The physician's task was to diagnose illness and prescribe medication. A university degree from the ancient universities of Cambridge or Oxford was required for full fellowship in the corporation. Without that privilege a budding physician could aspire only to the affiliations of licentiate or extra-licentiate. The Royal College of Surgeons was formally of more recent origin, having been established only in 1800. London's surgeons had, however, been organized as a separate city company since they parted from the barbers in 1745. Their redesignation as a college rather than a city guild signaled efforts to equalize the social standing of surgeons and physicians.[7] The surgeon's avocation was the treatment of external disease, performing operations, setting broken bones, and in general physically manipulating the body. The Society of Apothecaries had parted company with the Grocers' Company in 1617, obtaining their own royal charter. Their responsibilities were to supply, compound, and sell drugs in the City of London. They were also permitted to prescribe medicines but not to charge for that service. The license of the Society of Apothecaries, authorized by the Apothecaries' Act of 1815, permitted medical practice. It was easily the most popular early-nineteenth-century medical qualification.[8]

In principle, membership in any of these corporations precluded the practitioner's engaging in the business associated with another group. A fellow of the Royal College of Surgeons, for example, was not permitted to practice internal medicine or to prescribe or sell drugs. In practice, by the early nineteenth century, most practitioners were not in a position to observe such distinctions. It was increasingly the norm for medical practitioners to hold both the license of the Society of Apothecaries and membership in the Royal College of Surgeons. Limiting their practices to only one of the three functions defined by the three corporations was financially unacceptable to practitioners in an increasingly competitive marketplace. Neither could most patients afford the cost of three separate doctors for their ailments. The "general" practitioner, engaging in both medicine and surgery and frequently selling his own drugs, became more and more commonplace.

A gap was rapidly developing between surgeons and physicians employed by the large London hospitals, mainly fellows of the Colleges of Physicians and Surgeons, and this broad mass of general practitioners. Hospitals were coming to play an increasingly important role in London medical life.[9] These hospitals were private institutions, funded by a combination of endowments and charity. Their governors and trustees, drawn from the aristocracy and London's commercial elite, typically regarded their positions as opportunities to combine philanthropy with monetary reward.[10] Appointments at the hospitals were limited. While they provided their incumbents with little in the way of direct financial gain in terms of salary, they allowed them to develop the experience and reputation they needed to acquire a large and lucrative private practice. This was particularly so at a time of increasing competition for patients.[11] Hospital medical schools also provided a substantial income for hospital incumbents since students were required to pay their lecturers for the privilege of attendance.[12] Hospital appointments usually required the patronage of senior fellows of the Royal Colleges. The vast majority of London practitioners, as well as those who pursued their profession in the provinces, felt that the old corporations did not adequately represent their interests.

As a result the first half of the century saw the foundation of a number of medical societies characterized by political outlooks from radical to moderate, but all calling for medical reform. These ranged from the British Medical Association and the National Association of General Practitioners, who lobbied for the creation of a new College of General Practitioners, to the more ameliorist Provincial Medical and Surgical Association, focused on reforming the corporations from within.[13] Parliament was continually lobbied by groups demanding reform and recognition. Reform of medicine's institutions, like reform in political or scientific institutions, could be seen as part of the agenda of newly powerful social groups following the end of the Napoleonic Wars. Calls for an end to old corruption in all these areas made similar claims, used much the same vocabulary, and came from much the same constituencies.[14]

In 1823, Thomas Wakley founded the *Lancet*, which for the next several decades was to be a radical thorn in the flesh of the medical establishment.[15] Wakley was fierce in his denunciations of the corporations. He condemned the leadership of the Royal Colleges as "crafty, intriguing, corrupt, avaricious, cowardly, plundering, rapacious, soul-betraying, dirty-minded BATS."[16] The corporations were monopolies that deliberately excluded the mass of practitioners from the rights to which they were entitled. A convinced meritocrat, Wakley lobbied for recognition of medical men as expert professionals rather than college dilettantes. In 1828 the dissenting London University opened its doors, offering to medical students an alternative to the expensive medical schools and challenging orthodox methods of teaching. Students at the new university were exposed to the latest theories from France and the German lands. Geoffroy St. Hilaire or Lamarck could prove useful bludgeons with which to attack medicine's old aristocracy.[17]

New practices could flourish in such an atmosphere. Budding practitioners were looking for new ways to ply their trade. Mesmerism, for example, revived briefly at the London University during the late 1830s.[18] Electrotherapeutics was one resource that a medical man could draw upon to found a reputation. Innovation could also be seen as dangerous, however, as the case of mesmerism demonstrates. Like mesmerism, electrotherapeutics could easily be tarred with the brush of heterodox and subversive materialism.[19] It could also be dismissed as mere quackery, which medical radicals despised as much as they hated the old corporations. New practices had to be provided with genealogies and social settings that insulated them from such attacks. A variety of intellectual and social resources were available to neophyte medical practitioners during this period. Success, however, would be contingent upon convincing others, particularly the powerful medical elite, of the validity and social acceptability of one's practices.

The first section of this chapter focuses on the activities of the young physician Golding Bird at Guy's Hospital in London and his efforts to establish electrotherapeutics as a viable form of medical practice within a hospital context. His success depended upon his ability to acquire patrons and the extent to which he could divorce the medical administration of electricity from its contemporary unsavory context. The second section examines electrotherapeutics with particular reference to its use as a cure for hysteria, focusing on the ways in which notions of the hysterical female body were put together during the second quarter of the nineteenth century, and the ways in which electrical models and practices could be adapted to this context. The final section outlines the material technology of electrotherapeutics, its origins, and its adaptations to the needs of a medical market.

ELECTROTHERAPEUTICS AT GUY'S HOSPITAL

Golding Bird became a pupil at Guy's Hospital, London, in 1832, having been apprenticed to the London apothecary William Pretty since the end of 1829. Guy's was one of the more recent charity hospitals, founded in 1721. Like other such institutions established during the eighteenth century, the hospital's purpose was to care for the sick poor. As was common for London hospitals by the early nineteenth century, it also housed a medical school. By the 1830s, Guy's was one of medical London's more prestigious establishments with many influential and renowned practitioners holding positions there.[20] During the course of Bird's studies there, he succeeded in attracting the patronage of both Astley Cooper, then the senior surgeon at the hospital, and Thomas Addison, then assistant physician. He was licensed by the Society of Apothecaries in 1836. In 1838 he took the degree of M.D. at St. Andrews and became a licentiate of the Royal College of Physicians in 1840, becoming a fellow four years later. Bird was appointed to lecture on natural philosophy at the Guy's Medical School in 1836. He was an enthusiastic chemist and had already produced a number of papers, particularly on the chemical analysis of blood and urine.[21] He joined the London Electrical Society shortly after its foundation in 1837. Bird was physician to the Finsbury Dispensary from 1838 to 1843 and lectured also at the Aldersgate Medical School. In short he was engaged in a wide range of the activities of a typical, if ambitious, young medical practitioner of the period attempting to make his way in a difficult and competitive climate.[22]

In October 1836 an "electrifying room" was fitted up at the hospital at the instruction of Benjamin Harrison, the treasurer of Guy's. Bird was placed in charge of the establishment. It is now impossible to recover the details of the decision to establish this addition to the hospital's resources. The surviving minute books of the hospital's various committees are very patchy in their coverage of the early nineteenth century. The only direct reference to the electrifying room is in a report by the treasurer to the Court of Committees in August 1838, almost two years after the facility had been first set up. "The Treasurer informed the Court that in consequence of the number of patients in the Hospital to whom electricity was applied, he had found it expedient to appropriate a room for the purpose where the apparatus could be kept in constant order and readiness & a person who has other occupation in the Hospital is in regular attendance; a senior physicians pupil regulates the application and keeps records of all cases where electricity is applied."[23] Harrison's support was presumably vital in the setting up of the electrifying room. During his long tenure as treasurer of Guy's, Harrison had virtually run the hospital as his own private fief. No appointments could be made without his patronage, and no new initiative could be carried through without his support.[24] Harrison was treasurer of Guy's from 1797 until 1848. His influence was such that he was popularly known as "King Harrison." It is indicative of his power that Harrison apparently felt no need to inform the hospital's governing body of the electrifying room's existence until two years after its first establishment.

The electrifying room was in more senses than one an experimental project. This was certainly not the first time that electricity had been used at a hospital. John Birch had used electricity at St. Thomas's Hospital before the end of the eighteenth century, and he was certainly not the only example.[25] It is worth remembering Giovanni Aldini's visit to London in the 1810s and his electrical experiments on human corpses.[26] In the case of Birch at St. Thomas's, however, the use of electricity was discontinued after his death in 1815, suggesting that it had been at Birch's own initiative and had not been fully integrated into the hospital's routine. Birch himself claimed that he had found it difficult to persuade others to take up electricity as a therapy.[27]

At Guy's Hospital, on the other hand, the use of electricity as a therapy under Bird's direction was carefully systematized and integrated into the hospital's everyday routine. Outpatients and inmates of the hospital who had been prescribed electricity as part of their regimen attended the electrical room at three o'clock daily to be submitted to various forms of the electrical treatment. A student acted as clinical clerk, making notes on all the cases and registering their progress, the aim being to ascertain the precise therapeutic power of electricity and the range of diseases for which it was applicable.[28] This was an attempt to regulate the medical application of electricity, circumscribing its use and bringing it inside orthodox practice. It certainly appears that the regular use of electricity at Guy's was a matter of public record and note. The hospital was singled out in the satirical "New Frankenstein" as a place of electricity.[29]

The various forms of treatment utilized at the hospital were drawn from the large repertoire of techniques and instruments developed by the popular electrician's culture of which Bird himself, as a member of the London Electrical Society, was a part.[30] Many of the techniques in question had been in use for electrotherapeutic purposes since the mid–eighteenth century.[31] Broadly speaking, two types of electricity were employed: ordinary static electricity derived from a frictional electrical machine or a Leyden jar, and galvanic electricity, from a battery or, increasingly, from some form of electromagnetic apparatus. Ordinary static electricity was typically employed in one of three ways.

Patients could be seated on an insulated stool and connected to the prime conductor of the electrical machine so that their skin surfaces became electrified. This form of treatment was known as the "electric bath." Another closely related and very common form of treatment was to give the patient an electric bath and then draw sparks from the body, usually from the spine, with a conductor. Finally, patients could be placed in the circuit of a Leyden jar so that they underwent a momentary electric shock. More often, however, galvanic or electromagnetic apparatus was used in electric shock treatment, particularly when a large dose of electricity was deemed desirable.[32]

An important feature of the experimental treatments carried out at Guy's Hospital was that electrotherapeutics was confined to the treatment of a range of specific nervous disorders, most frequently chorea and hysterical paralysis. Bird and his associates were keen to emphasize that, unlike the majority of past practitioners, they did not employ electricity either as a universal panacea or as a last resort when all other forms of treatment had proven ineffective. By circumscribing electricity's use in this manner, they aimed to present their use of the therapy as a rational practice, immune from the charges of empiricism that had often been leveled at past practitioners and were a frequent accusation made against contemporary quacks. "Empiricism" as used by early-nineteenth-century medical commentators invariably had derogatory connotations. An empirical therapy was one that was not based on a rational understanding of the method whereby the therapy affected the human body. "Rational" therapy did not necessarily imply a theoretical commitment to any particular account of the human body; it might, for example, be based on a systematic reading of past case histories and the treatments employed there. It was in this sense that the experiments in the electrifying room were aimed at providing electrotherapeutics with a rational base.

This point was emphatically made in the first paper that publicly discussed the use of electricity at Guy's. In a contribution to *Guy's Hospital Reports*, Bird's patron Thomas Addison represented his past practice as a failure to realize the efficacy of electrotherapeutics:

> It is, nevertheless, much to be feared, that many persons, like myself, have been led greatly to underrate its efficiency, either in consequence of its vague and indiscriminate recommendation, or from the inefficient and careless manner in which it has been applied. Certain it is, that, although I have often ordered it myself, and have more frequently witnessed its employment by others, I had never for a moment entertained the belief that it possessed the power over the disorders alluded to, which I am now inclined to concede it.[33]

Addison's implication was that while previous uses of electricity had been empirical and indiscriminate, the work done in the electrifying room was a carefully regulated and therefore rational application of the therapy.

In his first contribution on electrotherapeutics to *Guy's Hospital Reports*, Bird amassed an impressive list of case studies to support his claims concerning electrotherapy's efficacy. In the section dealing with the electrical treatment of chorea, he listed thirty-six cases that had been treated since the electrifying room's establishment. In only one of them had the treatment had no beneficial effect. He also provided a table of patients treated for amenorrhea, listing twenty-four cases, of which only the first four were unsuccessful. Bird's contribution was carefully systematic. He specified the age, occupation, and (where necessary) the sex of each patient as well as their symptoms and the particular electrical therapy they had received. Thomas Addison in his report went into careful detail as well concerning the physical appearance of the patient.[34] Every effort was made to ascertain and make explicit precisely the circumstances in which electricity provided a viable cure.

The role of the patient in this context was unambiguously that of the experimental object. Patients at the London hospitals were invariably drawn from the poorer sections of the working classes. Only those who had nowhere else to go usually turned to hospitalization for the cure of illness. Their perception of the hospital was as a place of death where at best their bodies would be manipulated for the amusement of others, and at worst they would end up on the dissecting table.[35] Their position as objects of charity at the hospitals meant that once within the walls they rarely if ever had a say in the treatment and use of their own bodies. Indeed, one of the advantages of a hospital appointment for the budding practitioner was that it provided a ready supply of docile bodies upon which the doctor could hone his skills and develop new techniques of treatment.

Electricity had an advantage over some other forms of therapy in this respect, as its effects could, so to speak, be read directly from the patient's body without any need for the patient's own testimony. The need for patient cooperation was a problem with mesmerism, for example, which many medical practitioners argued handed over too much control to the patient in the interpretation of its effects. Mesmerized bodies were not necessarily docile.[36] An interesting insight into the attitude of men like Bird toward their patients may be gleaned from the following letter: "Dr. Cholmely wishes to assure Mr. Bird that he never entertained the thought of making any complaint of Galvanic neglect. He is sorry to learn that Mr. Bird is ill, & he upon no account wishes the Clinical female to undergo the Voltaic process for the cure of aphasia until the reestablishment of his health."[37] Patients such as the "Clinical female" were regarded as objects who would "undergo . . . the process," as much as they were perceived as people to be treated.

By far the most common symptoms exhibited by patients presented at the electrifying room were paralysis and the involuntary movement of limbs. Such symptoms were often accompanied in female patients (the majority) by irregular menstruation. The onset of the symptoms was frequently attributed to particular traumatic events in the patient's recent past. Sarah Wheeler, for example, admitted on 5 November 1839 suffering from involuntary motion of the right arm and shoulder, attributed her condition to "fear, produced by the threats of her schoolmistress."[38] Electricity was rarely the therapy of first resort, or the only therapy in a regime of treatment. Typically the patient would initially be treated by a regimen of chemicals, sulfate of zinc being the most common antispasmodic drug administered. Electricity in the form of a spark drawn from the spine or passed through the affected limb would be used only if the drugs failed to bring about improvement in the patient's condition. Once electricity had been prescribed, patients attended the electrifying room on a daily basis until their symptoms subsided. Cessation of the involuntary movement and in the case of female patients the resumption of normal menstruation were taken as signs that the therapy had been successful, and the patient would be discharged.

Accounts in *Guy's Hospital Reports* pinpointed chorea in particular as a disease in which the employment of electricity could lead to spectacular results. Again, however, the accounts were clear that electricity was by no means a panacea—any such claim would carry overtones of quackery. Benjamin Guy Babington, physician at Guy's, emphasized in his survey of cases treated at Guy's that the electrifying room's work was on the way to demonstrating the limitations, as well as the possibilities, of electricity's medical employment.[39] The procedures at Guy's were unprecedented; there had "never been so complete an opportunity for trying the medical effects of electricity and galvanism, in all their varieties, and in all diseases, as in the Electrical Apartment at Guy's Hospital."[40] The steady accumulation of evidence under hospital conditions was what was required to make electricity part of proper practice. Electricity was presented as being a component of a routine regimen of treatment. It was introduced as circumstances and symptoms indicated and discontinued when no longer required or when shown to be ineffective. It was part of a standardized armory of therapies, a part of everyday practice. It was divorced from its aura of the marvelous and spectacular.

The emphasis on electricity's routinized utility was present a few years later in Henry Marshall Hughes's survey of chorea cases treated at Guy's.[41] Hughes presented extensive tables of the cases dealt with, providing details of the age and sex of the patient, the duration and cause of the disease, the treatment, and its eventual outcome. In his conclusions he emphasized that while electricity was not invariably successful as a therapy for chorea, it was particularly effective in those cases for which it was the appropriate treatment: "it produces its effects more rapidly than any other remedy with which I am acquainted; but it is a remedy which is not of universal application in chorea."

> The cases in which it appears to be more especially applicable are those occurring in young women, in whom the disease assumes somewhat of a hysterical character, and those protracted cases in boys in whom other remedies have been tried ineffectually, and in whom the disease is dependent upon no obvious source of irritation, and has an injurious effect upon the general health instead of being affected by it.[42]

Electricity had its place at Guy's, both literally in the electrifying room and as a standard, limited, and understood part of regimented treatment. The importance of defining that place was underlined by William Withey Gull, who complained bitterly of the hospital's inability "to enforce upon our patients a more regular attendance . . . after having carefully recorded the details of their history, progress and present state, our labour has in more than one half the cases been lost by their non-attendance, or by their attending so irregularly, or for so short a time, as that no practical inference could be drawn."[43]

Men such as Golding Bird wanted not only to acquire large and successful medical practices, they were intent on making their reputations in the new discipline of physiology. Indeed, a "scientific" reputation was rapidly becoming a useful component in a practitioner's career capital. Bird had already used his chemical expertise at Guy's to good effect, attracting the attention of Sir Astley Cooper. He had taken advantage of the Pupils' Physical Society's and later the Senior Physical Society's meetings to publicize his chemical endeavors. As a result he was asked to contribute an account of the chemistry of milk to Astley Cooper's *Anatomy of the Breast*.[44] Another good example was Marshall Hall, who used his physiological expertise as the foundation of a highly lucrative medical practice, despite his lack of hospital affiliation.

Hall's work on the physiology of the nervous system moved easily from discussions of electrical vivisections on frogs and rabbits to accounts of experiments carried out on human patients. Hall's language was unambiguous:

> I waited with anxiety for opportunity of submitting this question to the decision of experiment. . . . A little child, aged two years, was perfectly paralytic of the left arm. The slightest shock of galvanism was directed to be applied which should produce an obvious effect. It was uniformly observed that the paralytic limb was agitated by a degree of galvanic energy which produced no effect on the healthy limb.[45]

The use of electricity in such a case was presented as both therapy and experiment. While it cured the patient, it also allowed Hall to generalize from experimental observations made on animals to human subjects. Hall's work provided resources for other electrically minded medical practitioners as well. Thomas Williams, a radical Welsh doctor trained at Guy's, drew on Hall's doctrines to explicate his own electrical experiments.[46] Williams carried out experiments on dogs and rabbits, demonstrating the existence of what he termed (following Hall) an "electro-genic" state between the motor and sensitive roots of the spinal nerves. He speculated that his experiments suggested how muscular contractions might take place by induction in a similar manner.[47]

Golding Bird's close links with other London electricians—unusual for a medical practitioner aiming at elite status—were a crucial factor in his success. Members of the London Electrical Society, such as Bird himself, could have access to a variety of sources that could be exploited by medical electricians. A good example would be the work of Henry Letheby, who contributed to the society numerous descriptions of dissections of the *Gymnotus electricus* and was explicit in his claim that the gymnotus provided direct and conclusive evidence that electricity was the nervous force.[48] Many other examples could be cited from chapter 5. This interest was not all that Bird might gain from membership in the society, however. The members' emphasis on the technology of display had led to the proliferation of electrical apparatus that could be exploited for a variety of purposes.[49] Bird through this association was therefore familiar with the skills and practices needed for the competent manipulation of a wide range of electrical apparatus. He had the competences needed to make electrotherapeutics work as a practical technology.[50]

These practical skills and technologies were, however, embedded in a culture which was far removed from that of London's medical elite. The electricians' displays of shocks and sparks took place at venues such as the Adelaide Gallery or the Royal Polytechnic Institution rather than in the prestigious surroundings of the Royal Society, the Royal Institution, or the Royal Colleges of Physicians or Surgeons. The administration of electric shocks for therapeutic purposes was a routine part of this popular culture.[51] The electricians also had links to such irregular practitioners of medical electricity as William Hooper Halse. In order to attract the patronage and support of the medical elite, Bird needed to divorce his practices from such an unsavory context. The establishment of the electrifying room at Guy's was an attempt to shift physically the social context of electrotherapeutic practice. Rather than being performed in the dubiously popular context of the Adelaide Gallery or the private house of an irregular practitioner, or even within a hospital on an irregular basis and at the whim of an individual practitioner, electrotherapeutic treatment could take place within a circumscribed space where its application could be carefully and systematically policed. The hospital apparatus of clinical clerks and assistants could be brought to bear on the practice, rendering it safe for elite medical consumption.

## DEFINING THE HYSTERICAL BODY

The main targets for electrotherapeutics during this period were, unsurprisingly, women. A perusal of Golding Bird's or Thomas Addison's case histories of patients treated at Guy's Hospital makes it clear that the majority were female. Women, it was argued, were more prone to the kinds of nervous disorders that electricity could cure. Those disorders were commonly grouped together under the label of "hysteria." Hysteria during the first half of the nineteenth century was almost universally regarded as a disease that affected women and to which all women were to some extent susceptible. Almost as universally it was regarded as having some link to the dysfunction of the female reproductive organs.[52] These were far from new claims. Contemporary commentators almost invariably drew attention to the long history of the connection and to the Greek origins of the term *hysteria* itself. Showing that the exception proved the rule, those few cases in which instances of male hysteria were discussed made explicit the ambiguous status of the sufferer. In his discussion of hysteria and its links with the sexual organs and emotions, Edward John Tilt mentioned more than one case of male hysteria but suggested that these by no means disproved the case for a concrete connection between the disease and female sexual dysfunction. On the contrary he argued that men who suffered from hysteria possessed a predisposed nervous system that rendered them susceptible to the disorder. The strong implication was that men who were at risk from hysteria were men who acted like, or were constituted like, women.[53] Hysteria was regarded as a mark of femininity.

Beyond this consensus, however, accounts of hysteria varied extremely widely. There was no coherent definition of the hysterical body during the early Victorian period. Different writers gave varying accounts of the symptoms that might be regarded as indicative of hysterical disorders, and of the extent to which diseases of the sexual organs in general should be regarded as related to hysteria. Neither was there any across-the-board agreement among doctors concerning what form of therapy might be appropriate. Electric shock treatments of various kinds were among a wide range of remedies that were on offer. Accounts of the hysterical body were locally constituted in case histories of patients, descriptions of their physical appearance and behavior, and their responses to treatment. A common feature in all such accounts was the attribution of machinelike qualities to the female body and following from that an insistence that the body required regulation and surveillance. Upon the failure of the woman to control her physical self, the doctor was required to supply a means of restoring that control. Electricity provided a physical means for so doing.[54]

A good example of such a claim may be found in the pronouncements of Thomas Laycock. His work is of particular relevance since he was widely and approvingly cited by his contemporaries as having placed the study and treatment of hysteria on a new and more philosophical basis.[55] Laycock, born in 1812, received his education at the dissenting and radical London University. During 1834 he was in Paris where he studied anatomy and physiology. He became a member of the Royal College of Surgeons in 1835 and was awarded his M.D. from the University of Göttingen in 1839. In later life he achieved fame as one of the founding fathers of mental physiology.[56] During the early 1840s he was a frequent contributor to the reformist *British and Foreign Medico-Chirurgical Review*. In 1846 he was appointed a lecturer in clinical medicine at the York School of Medicine and in 1856 secured an appointment as a professor at the University of Edinburgh. During the late 1830s Laycock published a series of papers in the *Edinburgh Medical and Surgical Review* on hysteria in women and its relationship to disorders of the ovaries in particular.[57] In 1840 he published a comprehensive treatise on the nervous disorders of women.[58] His early career adheres closely to that of the nonconformist, radical, or reformist doctors discussed in Desmond's *Politics of Evolution*, who were educated at the London University and trained in French and German physiology.[59] His focus on the diseases of women can be understood as a strategy for forging a place for himself in the competitive medical market facing such men.

Both in his papers and in his later treatise, Laycock presented women in general as being particularly prone to nervous disorders. Hysteria was described as "affecting in some of its varied forms, almost every female."[60] This was the result of the relationship among the nervous system in women, their reproductive organs—in particular the ovaries—and the encephalon, described as "the organ of the instinctive faculties subservient to the reproductive process."[61] It was also the case that the female nervous system itself was particularly prone to disorder. Discussing the mental and corporeal peculiarities of women, Laycock straightforwardly asserted, "Without preface it may be stated that by universal consent the nervous system of the human female is allowed to be sooner affected by all stimuli, whether corporeal or mental, than that of the male."[62] This was a common trope in early Victorian medical literature.[63]

Edward John Tilt, for example, gave a specific account of the anatomical differences between the nervous systems of the female and the male and of the consequences arising from that difference. In his view "the ganglionic system of nerves is more developed in woman than in man." An understanding of nervous diseases in women and an understanding of the more gross physical differences between men and women could come only from an understanding of that fact:

> The difference in size in the two sexes depends on the greater development of the organs of *animal* life in the male, and as the nerves and ganglia of the ganglionic system in the trunk are in relation with the organs of *vegetative* life, these nerves and ganglia are proportionally larger in women; physiology and pathology likewise show that there is a greater amount of vegetative power in woman, for while the proper development of the testicles at once immutably imparts its characteristic effects to man,—the noblest of created beings,—in woman, the corresponding organs react more strongly on her system during the reproductive period of life, subjecting it to incessant vicissitudes of health and disease.[64]

The female's proneness to nervous disease was a result of her nervous system's design for a passive life suitable for reproduction.

Thomas Laqueur has remarked that in both medical and popular discourse on the body, the male body is typically regarded as unproblematic while the female body is rendered alien and in need of explication as a result of its difference from that norm. He has also suggested that the late eighteenth and early nineteenth centuries saw the emergence of new ways of representing that alienness.[65] Briefly, he suggests that until the early eighteenth century men and women were regarded as being structurally identical, the female simply being a less perfect or less developed form of the male. Female genitalia, for example, were simply inverted male genitalia. He then charts the emergence during the eighteenth century of a two-sex model, in which men and women were regarded as being irreducibly dissimilar, having fundamentally different anatomies and physiologies. Laqueur also notes, however, the coexistence of the two models well into the nineteenth century.

Laycock's account of the female, hysterical body is certainly consistent with this view. While women's bodies were fundamentally different from those of men, particularly in terms of their respective nervous systems, it could be argued that traces of the earlier single-sex model remain in the representation of women's bodies as being similar to those of prepubescent children, less developed than those of men. What matters about these different representations, however, is the way in which they were strategically deployed to place the female patient in a particular relationship with her male doctor. The interesting point about Laycock's description of the female body is its portrayal as inherently unstable and constantly in need of professional, medical intervention. Laycock's strategy was to show that women could not be trusted to control their own bodies. As wayward machines their bodies needed direct intervention by a professional, trained in the science of managing the body, in order to guarantee its proper functioning. As a child, the woman needed moral guidance.

The peculiar physiology of the hysterical woman had visible consequences, giving her an identifiable, typical appearance. Laycock suggested that in distinguishing cases of hysteria, the "very plump mammae [breasts], prominent nipples, and dark-tinted aureolae are so frequently present in young hysterical females, as to aid with other symptoms in forming a diagnosis."[66] The physical appearance of hysteria—"the embonpoint of the Parisian prostitutes, many of whom are perfect models of symmetry and grace"—went together with behavioral and dietary features of the disease:

> This hysterical embonpoint is a most constant symptom. It is somewhat analogous to the development of the mammae before noticed, and, like the latter, may assist materially in forming a diagnosis between hysteria and diseases resembling it. With this state of the surface the appetite is much impaired or abolished, the most minute portion of food, and that only farinaecous, being taken for months altogether. Yet to the great surprise of every one, the limbs, mammae, and trunk, continue round and plump. Indeed the most common subjects of hysteria are those endowed with this brilliant plumpness of the surface and delicacy of finish.[67]

The inherent instabilities of the female physiological machine endangered the patient such that the need for direct, external intervention and control could not be denied.

Morality as much as physical well-being was at stake in hysteria. The bad habits consequent upon an absence of proper regulation and supervision of female life were a major factor in the development of hysteria:

> Young females of the same age, and influenced by the same novel feelings towards the opposite sex, cannot associate together in public schools without serious risk of exciting the passions, and of being led to indulge in practices injurious to both body and mind. . . . Frequently, too, the daily exercise is little more than a lounging walk in two and two file, and consequently the sensory system becomes charged (as it were) with excitability, for nothing diminishes the affectability of this system so much as indolence. The consequence of all this is, that the young female returns from school to her home a hysterical wayward, capricious girl; imbecile in mind, habits, and pursuits; prone to hysteric paroxysms upon any unusual mental excitement.[68]

Even within the security of the parental home, the influences of fashionable life continued to contribute to this process of moral and physiological delinquency, unless the "unfortunate young lady" was subject to proper discipline. Laycock made starkly clear the consequences of ignoring such dangers and underlined the need for constant, medical supervision. What he meant by "practices injurious to both body and mind" was clarified in an anonymous review of his own work a decade later, when he remarked upon the dangers that "vicious habits have been already acquired, and the ovaria have been unduly excited by lesbian pleasures."[69]

Women's sexuality, as well as their sexual organs, were implicated in the production of hysteria:

> The power of the sexual passion in exciting hysteria is evident from the general fact, that it frequently follows disappointments, and affects unmarried females. Self-pollution [masturbation] is mentioned by Villermay as a cause of hysteria, and I believe with great justness; other writers also refer to it; I think those case marked by irregular arterial action originate in this practice. Strong sexuality is another very evident cause, especially when combined with continence.[70]

Laycock was not at all alone in making such a claim. Marshall Hall, similarly engaged in making his career as an expert on nervous disorders, remarked that in "the female sex, it is at the moment when disappointments in love are most apt to take place; and in the male, when disappointments in projects of ambition are most apt to occur, that we most frequently observe insanity."[71]

Actual sexual misconduct was, of course, itself to be regarded as a hysterical symptom. Laycock remarked that "young ladies in a hysteric paroxysm will sometimes utter expressions which one would think it impossible for them to know." He warned that "[i]n some cases of aggravated hysteria there is a slighter degree of nymphomania, but the display of it is usually confined to lascivious glances only, although two cases have been mentioned to me, in which the medical attendant was solicited, and I have heard of others. This depraved feeling leads the patient not unfrequently to feign retention of urine, that catherism may be performed."[72] While suppressed sexuality was to be seen as a cause of hysteria, explicit expressions of sexual desire or actual intercourse was to be regarded as a potential symptom of the disease.

These extensive descriptive passages have been reproduced to illustrate the range of images that could be strategically deployed to put together a picture of the hysterical woman and to show the necessity of medical intervention. Laycock and many other writers were quite explicit that the social mores of early Victorian society made it impossible for young women of the middle classes to express their sexuality in a physical manner. While it was acceptable (if not condonable) for men to indulge in premarital sexual activity, this was most certainly not the case for young women. While it was taken for granted that women did have legitimate sexual desires, it was also taken for granted that they could not be allowed to fulfill those desires. Doctors such as Laycock presented themselves, therefore, as mediators who could manage the fraught and delicate relationship between what woman's nature dictated concerning female sexual behavior and what society demanded. Edward John Tilt was only one of many who recommended early marriage as a means of avoiding such difficulties and the consequent dangers of hysteria.[73]

Proper management of the body and of the passions was a key issue in early Victorian discussions of sexuality. Commentators took it for granted that there was more to sex than procreation. Sexual desire was a legitimate emotion and its gratification with enjoyment a legitimate activity for men and women. In fact a common popular belief supported by many medical practitioners was that female orgasm was essential for conception to take place. Enjoyment was therefore central to the sexual act's procreative function.[74] In an age of Malthusian fears, however, overindulgence was dangerous. Uncontrolled sexual behavior was held to be a physical as much as a social evil. The result in men was spermatorrhoea and impotence. The result in women was hysteria. Continence outside marriage and temperance within were the ideal solutions. Physical relationships needed to be managed for the protection of bodily health and social standing. Where personal management failed, however, doctors could offer means by which proper regulation could be reestablished.

Laycock in particular identified electricity as the means whereby medical experts could mediate in the physical and therefore moral and social problems of hysteria. Electricity not only provided a dynamical physical mechanism that explicated the link between mind and matter, morality and nature, but also served as a tool that allowed the doctor to intervene directly and correctly regulate the imbalances in the female physiological machine. He was not the only writer on female diseases to liken, at least by analogy, the workings of the female body to those of an electrical machine. Thomas William Nunn, for example, drew attention to the interrelations of the female breasts and reproductive organs in such terms:

> The ovaria, uterus and mammae form, as it were, a reproductive pile, the circuit being completed by the nervous system. If there be no antecedent ovarian excitement, no impulse is transmitted to the breast. As in the galvanic battery, if no chemical action takes place at one pole, no electric current traverses the wire and no sign is elicited at the other.[75]

Like the battery in a completed circuit, the female body was a closed system in which changes in one part would inevitably affect others. Using this model Nunn could suggest, for example, that a failure to breastfeed after childbirth could lead to uterine or ovarian disorders (including hysteria) since it broke the usual action of this "circuit."

Thomas Laycock's construction of an electrical body was more complex. His account of the nervous system's action drew extensively on Marshall Hall's controversial theory of reflex action, which sought to explain how nervous stimulation at one point in the body could lead to a response at some other point without the action's being mediated by the conscious will. Hall claimed that the spinal marrow formed a distinct organ, at the center of a system of

nerves that he termed the "excito-motory" system. Sensory impressions received by one of these nerves could cause a current of nervous influence to be
conveyed to the spinal marrow where it would be reflected to some other portion of the body. He described the excito-motory system as the "seat or nervous
agent of the appetites and passions."[76]

Laycock invoked Hall's account of the excito-motory system and its extension to the brain as being crucial to the understanding of hysteria. On the one
hand, he suggested that disorders of the ovaries could impinge upon the excito-
motory system, causing a current of nervous force to be conveyed to the
encephalon (which was taken to be that part of the brain responsible for the
instinctive faculties of reproduction), where it would cause a disturbance that
would lead to hysteria. Intense sexual emotions or passions aroused in the encephalon could also react on the cerebrum through the excito-motory system in
the same way, causing hysterical paroxysms.

The mechanism by which the excito-motory system operated was represented as being at least analogous to electricity: "Sensation is not a perception
of the qualities of bodies, but of the changes which these excite in the terminat-
ing molecules of the sensitive nerves; and which changes appear to be propa-
gated to the brain from molecule to molecule, just as the (so-called) electric or
galvanic fluid."[77] The principles by which hysteria was excited were thus
claimed by Laycock to have far-reaching consequences:

> They form the connecting link between the phenomena of consciousness and the
> molecular changes in organic matter upon which the phenomena of heat, electricity,
> galvanism, and magnetism depend. They point out a new path of experimental in-
> quiry into the phenomena of life and thought, and, if traced out in all their relations,
> cannot fail to change the whole aspect of mental philosophy.[78]

Laycock was demonstrating how the whole apparatus of contemporary natural
philosophy could be brought to bear in regulating the female body through the
intervention of experts such as himself. Electricity was a solution to a moral and
social problem. At the same time, of course, the hysterical female body, as a site
for systematic experiment, was to become an obligatory passage point for the
new natural philosophy.[79]

The mode of application of electricity as a cure for hysteria was quite
straightforward. The passing of an electric current through the sexual organs
was a cure for the disease. The aim of the therapy was to check the "utero-
ovarian disease" that caused hysteria. Without such intervention "there is little
hope that the cerebral or cerebro-spinal disorder thence resulting, will be
cured, or even palliated."[80] Electricity would restore the disordered body, and
therefore the disordered mind:

> In one case under my care, in which there was the morbid condition of the will, I have
> described in a most marked form, and a concomitant painful alteration of the charac-
> ter, the remedy was eminently successful; in three weeks the bowels were moved
> daily without purgatives or enemata, although these had been necessary for three or

four years previously; the patient gained considerably in weight, improved daily in mind, and in four months the menses re-appeared, after having been wholly suspended *for five years*. The latter result was the more remarkable, inasmuch as the patient presented certain external characteristics which usually indicate atrophy of the ovaria.[81]

Doctors were alerted to watch for the warning signs: "an undefinable anxiety and restlessness, usually the most intense early in the morning. The bowels are generally much constipated, the tongue scabrous, presenting a rough, *fissured*, appearance, the complexion muddy, the countenance grave, or peevish, or anxious." If these symptoms appeared, then "the daily local application of electrogalvanism . . . will be the best remedy."[82]

Doctors were well aware of the delicacy of their situation in pointing to physical manipulation of the sexual organs as a means of both diagnosing and curing hysterical conditions in women. Their therapies advocated an invasion of bodily privacy that ran directly counter to the mainstream of early Victorian morality. "Feelings of delicacy, deserving of respect even when carried beyond the bounds of discretion, raise a barrier between the female patient and the practitioner." Given the necessity of such intervention, authorities offered protocols for the budding practitioner that would allow him to overcome his patient's feminine modesty. Tilt advocated that "as a General always puts the sun in the face of the enemy, so should the practitioner always let the light fall upon the face of his patient, where, as on a map, is often traced the outline and the character of disease, one look often better enabling him to unravel its manifold complications than many a prolonged inquiry."[83] He advised doctors to avoid looking the patient, unnervingly, directly in the eye. He should not show signs of impatience at the woman's "long, rambling unconnected tale" of her symptoms. Discussion of specifically sexual disorders should be approached obliquely through questions concerning the health of other, less intimate organs. When physical manipulation was required, the doctor was advised to make clear his sympathy at such a painful invasion of modesty. The "weaker sex" had to be persuaded "as they submit with patience to the disease itself, so is it incumbent on them to submit to an examination, distressing no doubt, but necessary for the recovery of their health."[84]

Not all experts on female diseases were as sanguine concerning the delicacy and modesty of their patients. One danger of physical examination and treatment was that the doctor might be duped into pandering to the depraved sexual urges of hysterical women. Marshall Hall inveighed against the increasingly prevalent use of the speculum in vaginal examination. Such a practice threatened to be the cause of a "new and lamentable form of hysteria." Its use put in jeopardy the "spotless dignity of our profession, with its well-deserved character for purity of morals," by the "dulling of the edge of virgin modesty, and the degradation of pure minds, of the daughters of England." He pointed to "the most revolting attachment, on the part of such patients, to the practice and to the practitioner."[85] He was not alone in his concern. Robert Brudenell Carter was equally apprehensive that such physical intimacy could be as much a cause as a cure of hysterical diseases.

Carter was unusual among contemporary commentators on female nervous diseases in denying a direct causal link between disorders of the reproductive organs and hysteria. He insisted that the cause of hysteria lay in the emotions and that there was no physical basis to the disease. He divided the condition into three stages of development: the primary, caused by some sudden and intense emotion; the secondary, where ideomotor reflexes caused unconscious links between one emotion and another; and the tertiary, which was entirely conscious and self-induced.[86] The only cure for hysteria and the only means of preventing the willful tertiary stage was to subdue the patient's will to that of the doctor. Anything said by the patient herself in the course of interrogation was to be routinely dismissed. Humiliation was advocated as a means of returning the recalcitrant "girl" to her senses. Carter emphasized that physical examination of the patient's sexual organs was simply pandering to the desires of a hysteric: "I have, more than once, seen young unmarried women, of the middle-classes of society, reduced, by the constant use of the speculum, to the mental and moral condition of prostitutes; seeking to give themselves the same indulgence by the practice of solitary vice; and asking every medical practitioner, under whose care they fell, to institute an examination of the sexual organs."[87]

Despite his marked divergence from Thomas Laycock's views concerning the causes of hysteria, Carter, like his contemporary, sought to present himself as an expert whose medical knowledge could be invoked to control the unruly female body. The victim of hysteria was not the female patient but the Victorian family. Standards of proper behavior and decorum were at risk from the inability of Victorian women to comport themselves in a socially acceptable manner and to control their bodies' sexuality. Doctors such as Carter and Laycock put themselves forward as mediators who could regulate what the woman was unable to regulate herself. Laycock painted a nightmare vision of the possible consequences of hysteria:

> The gentle, truthful, and self-denying woman has unaccountably become cunning, quarrelsome, selfish; piety has degenerated into hypocrisy, or even vice, and there is no regard for appearances, or for the feelings of others, except in so far as they may minister to the vanity or selfishness of the patient. . . . It . . . invests the patient's character with an unfeminine character, perverting all the feelings and sentiments connected with the sexual functions, either directly or indirectly, and so exciting insane cunning, destructiveness, infanticidal impulses, morbid appetites &c.[88]

Hysteria turned valued Victorian ideals of femininity, and particularly of woman's role in the family, on their heads. Laycock offered electricity as a means of restoring the female body and therefore the Victorian family to its proper state of equilibrium. Along with electricity came a whole technology of intervention: machines that could regulate the female machine.

MARKETING THE MACHINE

In promoting the use of electricity in medicine, men such as Golding Bird had a distinct advantage over the mass of medical practitioners in that they combined their medical skills with membership in the electricians' cultural network. Bird's membership in institutions such as the London Electrical Society was a crucial feature of his success since it provided him with the tacit skills and knowledge required to enable him to use complex electrical apparatus successfully. The overlap between the electrical and medical communities was not, however, a large one. The medical contingent may have formed a significant proportion of the London Electrical Society's membership, but that was small and they constituted only a tiny percentage of the London medical community as a whole. If electrotherapeutics was to be marketed successfully to this broader community of medical practitioners, it was essential that a new technology be developed that made it easy to apply the treatment without detailed technical knowledge. Early voltaic batteries were expensive and cumbrous. They required constant attention to guarantee their continued proper functioning. Making electrotherapeutics easy to apply would also deprive electrical quacks of their claims to expertise in the application of electricity.

Bird was certainly aware of the dangers posed to his practice by the activities of quack electricians. In the *Lancet* in 1846 he fulminated against the activities of self-styled medical galvanists, blaming their success at least partially on the ignorance of his fellow medical practitioners who refused to inform themselves of electricity's virtues or the proper means for its application.[89] He presented his case as being as much a moral as a medical crusade. Shameless charlatans were duping and cheating the public: "These medical galvanists lay heavy contributions on the public, and one of them, in the west end of the metropolis, regularly pockets some thousands per year by his practice. There is scarcely a part of London where these irregulars do not exist—the small suburb of Pentonville alone rejoices in two, if not three."[90] One solution, he suggested, was to educate the medical profession in natural philosophy, for without such education the medical practitioner would be unable to beat the unprincipled quack at his own game. Such rhetoric was of course well placed in the pages of the *Lancet* during the 1840s. Thomas Wakley was engaged in a relentless campaign against the quacks, defending the interests of his fellow hopeful professionals.

Bird emphasized that very little training was required for proficiency in electrotherapeutics:

> Every practitioner should feel it his duty to be as conversant with the application of electricity and its modifications as of a cupping-glass; and there is no more necessity for increasing the "specialities" of our profession by the electrician or galvanist, than by a qualified administerer of glysters. Largely as I have employed the remedy in question, I have never needed any other electrician than the wife or mother—often the footman or maid-servant.[91]

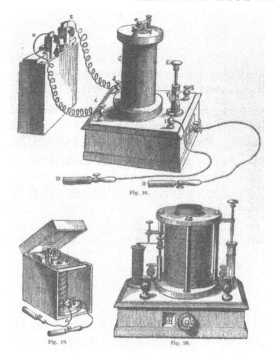

8.1 Medical electrical apparatus. From Alfred
Smee, *Elements of Electro-biology* (London, 1849),
p. 99.

Such a claim was, to say the least, more than a little disingenuous. Bird himself
went on to emphasize that it was only recent improvements in electrical technology, in particular the introduction of the induction coil to replace "the large
and cumbrous voltaic battery," that rendered electricity at all easy to apply
therapeutically.[92] The point of his remark was that electrotherapeutics should
remain firmly under the control of professionally trained medical practitioners.
It should be the preserve of the general doctor, not the specialty of a mere
electrician.

Golding Bird had himself been involved in attempts to improve the induction coil's ease of use in medical practice. His main innovation was a device that
made the repeated switching on and off of the electricity supply, which was
essential to the induction coil's action, an automatic process.[93] His contact-
breaker operated by means of a spring-and-magnet device. When the circuit
was complete, a small electromagnet became magnetized, attracting the contact-breaker and thus breaking the circuit. This then demagnetized the electromagnet so that the contact-breaker would be drawn back by the spring, completing the circuit once more.

He was not the only electrically informed medical practitioner to attempt the
development of apparatus specifically designed for therapeutic purposes.
Henry Letheby, lecturer in chemistry at the London Hospital and medical offi-

cer of health to the City of London, designed a version of the induction coil in which the electric fluid traveled in only one direction.[94] In ordinary induction coils, of course, the direction of the fluid's flow changed each time with the making or the breaking of the primary current. Letheby solved the problem using a pair of spoked wheels, arranged such that only the current from either the "make" or the "break" of the circuit could pass. As a result only a succession of shocks in a single direction would be felt. This was an important therapeutic feature, as Letheby pointed out:

> Another principle to be kept in view is, to pass the current in the route of the vis nervosa, that is from the centre to the periphery, in motor paralysis, but from the extremities to the centres when the sensitive nerves are affected. If this important feature is not borne in mind, we cannot expect ever to do much good in the application of electricity or galvanism; indeed, there is much reason for believing that the uncertainty of their therapeutic powers may have arisen from a want of observance of this principle.[95]

The contrivance had the added advantage of dividing the current into two parts. The shock from the break current was known to be more severe than that from the make current, which suggested that the former could be used for afflictions of the sensitive nerves while the lesser shock was used for motor nerve disorders.

Golding Bird's efforts were applauded by the surgeon John Crowch Christophers in the *Lancet*. Like Bird, Christophers expressed concern that electrical quackery could undermine the position of practitioners of electrotherapeutics who wished to remain within the orthodox medical profession. He was fully in favor of Bird's "attempt to wrest a most important agent from the grasp of an unprincipled host of quacks, self-styled 'Medical Galvanists' and 'Medical Electricians' whose ignorance and effrontery eminently tend to bring it into discredit and disuse."[96] Like Bird, Christophers emphasized that with the proper equipment, and minimal training, any person could apply the electrical treatment. He thanked Bird for his efforts in "putting it in the power of all to obtain an efficient apparatus at an outlay so trivial." As a result of the efforts of men like Bird, he concluded, "it is to be hoped, that when the public is made to see that footmen and maidservants, under medical direction, make excellent galvanists and electricians, we may find the self-dubbed 'medical galvanist' and 'medical electrician' fast returning to more useful, and though, no doubt, less lucrative, more becoming, occupations; for when simplicity and truth begin, mystery and quackery end."[97] The quacks threatened the orthodox practitioners of electrotherapeutics on two fronts. Not only did their continued existence deny the regulars' claim to legitimacy, they were also in open competition with the orthodox practitioners in their search for lucrative practices. The irregulars, of course, were also in a position to employ means of advertisement, such as notices in the press and popular broadsheets, that were unavailable to the (in principle) gentlemen of orthodox medicine. Thus they posed a direct threat to their livelihoods and status that had to be opposed.

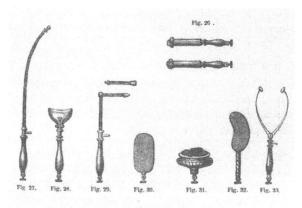

8.2 Electrodes for medical electrical apparatus. From
Alfred Smee, *Elements of Electro-biology* (London, 1849),
p. 111.

By the mid-1840s a wide variety of medical equipment, specifically tailored
for medical use, had appeared. The induction coil apparatus designed by Golding Bird was being produced and sold by William Neeve, a scientific instrument-maker working in Holborn. Other instrument-makers like E. M. Clarke
also produced induction coils and electromagnetic machines specifically for the
medical market.[98] Such instruments were small, portable, and relatively cheap.
They were easy to operate without detailed knowledge of electricity. All medical practitioners could now practice electrotherapeutics. Even electrodes could
be designed specifically for their particular role. Alfred Smee in 1849 could
illustrate a whole range of examples developed to facilitate application on and
insertion in different parts of the body.[99]

Kemp & Co. of Edinburgh throughout the 1840s manufactured and sold an
electromagnetic coil machine that appears to have been particularly popular.
The apparatus was a simple induction coil with the addition of a magnetic
contrivance similar to that devised by Golding Bird to allow the machine to be
self-acting. A core of iron wires could be inserted into the coil to regulate the
intensity of the electric current.[100] Simple instructions made the machine easy
to operate. All the medical practitioner needed to do was fill the battery jar with
a mixture of one part sulfuric acid to eight parts of water, and place the silver
and zinc plates of a Smee battery in the solution. The coil's electrodes were
provided with sponges that, when soaked in salt water, could be used to convey
the galvanism to the relevant part of the body. The practitioner could vary the
intensity of the current simply by inserting the bundle of iron wires into the coil
or withdrawing them.

Kemp & Co. had acquired a long list of testimonials affirming the efficacy of
their electromagnetic coil machine. Henry Bence Jones of St. George's Hospi-

tal in London reported that "during the two months' trial we have given your Electrical apparatus, we have found it much more convenient than the Plate Machine, which has hitherto been exclusively used in the Hospital." The principal medical officer at Chatham reported that the apparatus was "very well adapted for common use, and very efficient as compared with the machine previously in use."[101] Most of the testimonials came from Edinburgh practitioners, however. Thomas Stewart Traill, professor of medical jurisprudence at the university, described the machine as "by far the most efficient and economical apparatus hitherto employed for the Medical application of Voltaic Electricity." David Craigie, physician to the Royal Infirmary, concurred: "[C]omparing it with other instruments intended for the same purpose, it possessed over them the great advantage of being much more energetic, much more uniform, more easily kept in order and applied, and infinitely less likely to go wrong, than other instruments of the same nature."[102]

A considerable number of distinguished Edinburgh medical men agreed, expressing themselves well satisfied with Kemp & Co.'s machine's performance. There were testimonials from such luminaries as William Alison, professor of the practice of physic, John Goodsir, professor of anatomy, William Gregory, professor of chemistry, and Allen Thomson, professor of anatomy at Glasgow University. It is clear that Kemp & Co. had gone to some trouble to acquire such valuable advertising material. From some of the testimonials (particularly those of Henry Bence Jones and Sir James McGregor, director-general of the Army Medical Department) it is apparent that the writers had received samples of the electromagnetic coil machine for testing.[103] Other manufacturers and vendors of electrogalvanic machines tried different tactics to sell their wares. Horne, Thornthwaite and Wood, of London, published a whole booklet accompanying their machine, explaining its mode of operation and detailing at length the cures it could bring about.[104]

The Horne, Thornthwaite and Wood machine was in essence similar to that manufactured by Kemp & Co. In fact, such apparatus designed for medical use reached a more or less standard form quite quickly. It consisted of an induction coil and a Smee (silver and zinc) battery, with directors (electrodes) for applying the electricity to particular parts of the body. Horne in his guide cataloged the many diseases that could be cured by electricity, providing detailed hints as to the best ways of curing each particular set of ailments. Deafness, amaurosis, afflictions of the liver, paralysis, tic douloureux, and asphyxia, to name but a few, were amenable to treatment by the electric fluid flowing through their electrogalvanic machine.[105] The therapy was recommended on the grounds of its cheapness and convenience as much as anything else:

> From practical experience it is found that this very desirable assistance can be derived from the aid and power of the electro-galvanic fluid, without inconvenience and very little expense to the sufferer—a no small desideratum, at a time when the efforts of the medical adviser are daily thwarted—when he finds it necessary to order change

of air and scene, &c., and the necessary consequences—the withdrawal from business, thereby rendering the recommendation useless to nine-tenths of the suffering applicants and community.

With this powerful medical agent the same drawback does not exist, being almost as available to the poor as to the rich; neither does it withdraw them from the place that their daily avocations necessarily bind them to.[106]

The electrogalvanic machine provided the appropriate therapeutic technology for a busy commercial culture. The new coil machines were certainly cheap and convenient in comparison to their predecessors. Plate electrical machines had typically been used for electrotherapy. The smallest useful size had a glass plate of twelve inches in diameter, costing £4.10s. At least one Leyden jar costing about 10s. would also be required.[107] Kemp & Co.'s electromagnetic coil machine, on the other hand, cost £2.13s.[108]

Simpler galvanic cures were also on the market. The mid-1840s witnessed a fad for galvanic rings. These rings made of zinc with a band of copper inlaid on the inner side were worn on the finger as protection against a variety of painful afflictions. They would "alleviate the excruciating and agonizing pains of tic-dolereux and nervous headache, painful rheumatic affections, spasms, cramps, pains, &c. &c.—All that is required is to include any finger in the galvanic ring so as to fit close, but not too tight."[109] Another handbill assured the potential purchaser that the "great influence it exerts on the nervous system is truly astonishing; it must be witnessed to be believed."[110] According to Charles Walker, the skeptical editor of the *Electrical Magazine*, it was impossible to walk into a coffee shop without seeing "two or three out of every half-dozen of those present glory in these absurd circlets."[111] Such devices, of course, were part of that "quack" culture from which men such as Golding Bird and his cohorts were keen to distance their own practice of electrotherapeutics. Galvanic rings, along with galvanic belts and bracelets, formed the popular base, however, of the technology upon which they aimed to found their specialist expertise.

The new technology had the virtue of being invisible. As Christopher Lawrence has noted, technology had a rather dubious status for many nineteenth-century medical practitioners.[112] While some sought to enhance their status by forging links with scientific expertise, others, particularly in London, sought to present themselves as gentlemen trained in the art, rather than the science, of medicine. By this account, it was the ineffable, personal craft of the gentlemanly practitioner, rather than the technical skill and knowledge of the scientific expert, that made a doctor. Such men were unlikely to embrace the new electrotherapeutics if its use was to carry with it the taint of the technician. The new electrical technology ironically could overcome this difficulty by placing the actual manipulation of the machine in the hands of servants while the practitioners might retain control over the regulation of the therapy.

CONCLUSION

Through the introduction of new technology and the routinization of practice at places like the electrifying room at Guy's Hospital, electrotherapeutics became less of a performance and more of a process. Descriptions of patients, their symptoms, and their therapies in Golding Bird's case histories or Thomas Laycock's treatises were literally clinical. They did not share with William Hooper Halse, for example, the emphasis on the dramatic and the spectacular.[113] The aim of electrotherapeutics was to regulate the body, in particular the female body. To do so successfully required a body that was amenable to such control. As a result, electrotherapeutics and increasingly electrical, machine-like accounts of the body and its nervous system went hand in hand.[114] Practitioners such as Bird or Laycock, aiming to find a niche for themselves in a competitive medical market, were as much electrical entrepreneurs as William Fothergill Cooke or the Elkington cousins.[115]

Some historians have already pointed toward the relationship of medicine to Victorian commercial and consumer culture. M. Jeanne Peterson, for example, points to the foundation of specialist hospitals during the second half of the century as being as much an example of medical entrepreneurship as doctors' advertising their skills in the press.[116] Discussing the construction of Victorian sexual mores, Michael Mason notes the role a mass market may have played in putting together notions of Victorian gentility.[117] New medical technologies of the body as they were put together emerged out of and played a role in constructing Victorian consumer culture. The body came to be something that could be consumed. Ambitious doctors keen to make a living and a name found places for themselves in a rapidly expanding market. In order to succeed, men such as Bird or Laycock needed to find a product. The electrical body and its attendant technology constituted the product they needed to sell to make their careers.

# CODA

*THE DISCIPLINING OF EXPERIMENTAL LIFE*

BY THE 1850s, electricity had successfully emerged from laboratories, exhibition halls, and lecture theaters to became part of fully public, commercial life. In so doing, it had not, of course, left its old contexts behind. Rather, these places of experiment took on new significances themselves as they became adjuncts to railway networks, manufactories, and hospitals. Electrical experimenters acquired new skills as they attempted to find places for themselves in an industrial world. The second half of this book surveyed the ways in which electricity took its place in commercial culture during the 1840s. What should be clear is the contingency of this process. Entrepreneurs and electricians had to struggle to make electricity's technology of display into a stable and reliable feature of the Victorian world. The result was the opening of new spaces for electricity and for electricians. There was a new market in which hopeful experimenters and inventors could hope to sell their wares and make themselves.

Charles Vincent Walker's example is instructive, particularly since he had a finger in much of what happened to electricity during the 1840s. In the late 1830s, Walker was a self-confessed "tyro" member of the London Electrical Society, assisting in William Sturgeon's experiments.[1] By the early 1840s he was the society's treasurer and secretary, responsible for editing its proceedings. He had embarked on his own publishing career with his pamphlets on the new industry and science of electrometallurgy. Following the London Electrical Society's demise he briefly attempted to sustain its legacy, editing and publishing the short-lived *Electrical Magazine*. As an engineer for the South Eastern Railway Company, he was soon involved in telegraphic endeavors. By the early 1850s, as the railway company's telegraphic superintendent, he was collaborating with George Bidell Airy, the astronomer royal, to link the Royal Observatory at Greenwich to the telegraph system, to aid astronomical observation and provide the railway companies with a standard time signal. His career was a success. He managed the transfer of his skills from the laboratory to the marketplace. The former electrical "tyro" and humble railway engineer became a fellow of the Royal Society in 1855. A few years after the founding of the Society of Telegraph Engineers in 1871, Walker became its president.

The new electrical arts such as electrometallurgy made a world of goods available to a new generation of Victorian consumers, eager for the trappings of luxury. Some of the exhibits at the Great Exhibition of 1851 highlighted the means by which electricity could be used in this way, and could serve as a celebration of Victorian aspirations. The Elkingtons had contributed numerous pieces to the exhibition, displaying the versatility of the new technology. Medals, sculptures, tea services, and fruit bowls, all finished by electroplating, were on show. The highlight of the Elkingtons' exhibits was a mock Elizabethan vase, celebrating the "triumph of science and the industrial arts." At the vase's four corners stood the figures of Francis Bacon, Isaac Newton, William Shakespeare, and James Watt, representing English prowess in the various departments of science and arts. The whole edifice was crowned by the figure of Albert, the prince consort and moving force behind the exhibition. The vase was, of course, completed in electroplate.[2] Its own mode of production, as much as the subject matter it portrayed, was a "triumph of science and the industrial arts."

Electricity facilitated new forms of control and discipline in a variety of ways. Commentators celebrated the power that electricity gave them over nature. Many regarded the proper understanding of electricity as the key to unlocking nature's secrets. The electric force governed the universe. As William Sturgeon remarked, "Nature's laboratory" was well equipped with an "exhaustless source" of "electric streams" to govern the natural economy.[3] Other, less poetically inclined commentators concurred. James Joule considered electricity to be central to the transformation of force from one manifestation to another. It was the "grand agent for carrying, arranging and converting chemical heat." Similarly, William Robert Grove regarded electricity, pragmatically if nothing else, as the best means of managing the correlation of physical forces.[4] Electricians aimed to demonstrate electricity's omnipresence by displaying the variety of its powers.

Understanding and knowing how to manipulate electricity was therefore a way of controlling nature. The "vast apparatus of nature" could be at human disposal. Disciplining nature was a way of disciplining the body and society as well. Through electrometallurgy, not only could consumer goods be made available to the masses, but a potentially recalcitrant workforce could be imbued with new values. The mass production made possible by the large-scale deployment of the electric fluid provided the worker with objects of contemplation that would elevate his moral status and render him docile.[5] By providing the amateur hobbyist with an edifying recreation, electrometallurgy could even contribute to the domestic tranquillity, as well as the material trappings, of the English household. A future of electric ships crossing the Atlantic and electric locomotives on the railroads was hailed as providing new standards of "simplicity, economy, safety, and compactness."

The most clear example of electricity's disciplinary powers for the early Victorians was the telegraph. The power to "annihilate time and space" that this new application of electricity provided seemed unprecedented. Its early promoters were certainly well aware of the electric telegraph's capacities for the regulation of social life. As well as drawing attention to its considerable commercial possibilities, William Fothergill Cooke in his projection of his new system's potential drew attention to the role it might play "in case of dangerous riots or popular excitement" in alerting the government and allowing the swift deployment of troops.[6] The Admiralty and the police commissioners, as much as railway entrepreneurs, were early targets of telegraph inventors' efforts to market their product. The government was certainly sufficiently well aware of the telegraph's potential uses to insist on the addition of a clause to the Electric Telegraph Company's bill of incorporation, putting the telegraph at their service in case of emergency. That clause was indeed invoked during the Chartist uprisings of 1848.[7] This was not, of course, the only way in which the telegraph disciplined. It created new forms of commercial discourse and interaction.

Control, discipline, and regulation were virtues (or vices) that many Victorian commentators associated with machinery. Men such as Babbage, Barlow, and Ure eulogized the machine's capacity not only to control itself but to make its capacity for regulation universal. Others bemoaned the loss of personal autonomy and self-control that such a system would bring about. Machinery operated at a constant, unvarying pace that in turn communicated itself to the labor of the workforce using the machine. The result was to be a disciplined, ordered worker. Electricity fitted well into such discourses. Many commentators argued that electricity was the force that regulated nature. The universe was regulated and ordered by electric power. That power made subservient to the power of man could be a force for regulating society as well. Electricity and its technologies provided tools for disciplining the body and humans' interactions with each other. Thus it might well be regarded as the epitome of machine culture.

Ironically, however, at least one interested commentator looking back from the late Victorian era saw electricity's cultural role in a different light. Lord Salisbury, the Tory prime minister, talking before the Society of Telegraph Engineers in 1889 under its new guise as the Institution of Electrical Engineers, saw electricity as preparing the end of nineteenth-century industrial culture. He drew attention to the leveling features of electricity, particularly the electric telegraph. The telegraph was "a discovery which operates . . . immediately on the moral and intellectual nature and action of mankind." Its discovery had "assembled all mankind upon one great plane, where they can see everything that is done, and hear everything that is said, and judge of every policy that is pursued at the very moment those events take place. And you have by the action of the electric telegraph combined together almost at one moment, and acting at one moment upon the agencies which govern mankind, the opinions of the whole of the intelligent world with respect to everything that is passing at that time upon the face of the globe."[8] The new culture of mass information that the telegraph had ushered in was central to the running of the late Victorian state. As Salisbury emphasized, it was the telegraph that controlled the "gigantic armies held in leash by the various governments of the world" and therefore controlled the balance of power in Europe.

According to Salisbury, however, this was not to be electricity's greatest cultural contribution. New systems of electrical power would have the effect of replacing the steam engine, and with it the cultural conditions of the industrial revolution. In Salisbury's view, the technical limitations of steam power imposed social limitations as well: "the nature of the steam-engine was such that the force which it produced could only act in its own immediate neighbourhood, and therefore those who were to utilise that force and translate it into practical work were compelled to gather round the steam-engine in vast factories, in great manufacturing towns, in great establishments where men were collected together in unnatural and often unwholesome aggregation."[9] The steam engine, despite its massive contributions to the country's wealth and imperial power, was also responsible for the social and political evils that had dogged the first part of the century.

The availability of electrical power, however, would have the effect of reversing the congregation of working men in large cities and the dangerous potential of their gathering. "If ever it shall happen that in the house of the artisan you can turn on power as now you can turn on gas . . . you will then see men and women able to pursue in their own houses many industries which now require the aggregation of the factory. You may, above all, see women and children pursue those industries without that disruption of the family which is one of the most unhappy results of the present requirements of the industries."[10] Electricity possessed the key characteristic of being able to act at a distance. Just as the electric telegraph made possible the instantaneous transmission of information over vast distances, so Salisbury anticipated that power, too, could be distributed in such a way. The result would be to "sustain that unity, that integrity of the family, upon which rests the moral hopes of our race, and the strength of the community to which we belong."

In Salisbury's fantasy, the widespread adoption of electrical power would have the result of transforming society. Just as the technical limitations of the steam engine had brought about the factory system, the possibilities of electrical technology would bring about the return of a previous system of manufacture in which the artisan family, working from its own home, was the central unit of production. Rather than being obliged to gather around the source of power in the steam engine, and being prey to the social and political temptations of the large industrial town, artisans could return to a rural idyll with the power being brought to them instead. Electricity, ironically enough, could be seen as offering an antidote to the "galvanic" life that Carlyle had identified with machine culture. The nature of the technology made the centralization characteristic of steam industry redundant and therefore made possible a return to older values without loss of the economic power of industrial culture.

Interestingly, at least one more recent commentator has concurred with the Tory prime minister in seeing the advent of electrical technologies as marking the end of machine culture. Marshall McLuhan remarked also on a "peculiarity about the electric form, that it ends the machine age of individual steps and specialist functions."[11] According to McLuhan, the key characteristic of electrical technologies—he had in mind the telegraph in particular—was their organicism. Whereas machine technologies had acted as means of extending particular parts of the human body, electrical technologies acted as extensions of the nervous system itself. Such an analysis would not have surprised many Victorian commentators who themselves drew attention to the homology between the nervous system and the new telegraph technology.[12] They, however, saw electricity as being part of, and not separate from, the machine culture of their era. The telegraph was central to the operation of the railway system. Electrical locomotives would run on those railways just like their steam counterparts. In the Elkingtons' factories, galvanic batteries were machines producing consumer goods just like any others.

This book has focused throughout on electricity as part of the world of things, more often than not as things to be bought and sold, as commodities. Historians of science sometimes forget how important things, material artifacts, can be. For most Victorians who paid attention to such matters, electricity had little immediate to do with fields of force or even a luminiferous ether. Electricity was a matter of spectacular shows of light and sparks, amusing galvanic shocks to the unwary, electric telegraphs communicating at a distance, and fancy cutlery electroplated in silver. The dominant features of science were its material products and the performances and processes surrounding them. Those products often existed in an uneasy space where workshop, laboratory, and exhibition overlapped. It was here that electricity as a commodity to be displayed and admired was put together. This book has tried to examine and understand at least some features of that strange space.

# NOTES

## INTRODUCTION TO PART ONE

1. John Heilbron, *Electricity in the Seventeenth and Eighteenth Centuries: A Study of Early Modern Physics* (Berkeley and Los Angeles: University of California Press, 1979); Simon Schaffer, "Natural Philosophy and Public Spectacle in the Eighteenth Century," *History of Science* 21 (1983): 1–43.

2. William Robert Grove, *On the Progress of Physical Science since the Opening of the London Institution* (London, 1842), p. 24. For Grove, see Iwan Rhys Morus, "Correlation and Control: William Robert Grove and the Construction of a New Philosophy of Scientific Reform," *Studies in History and Philosophy of Science* 22 (1991): 589–621. For the London Institution, see John Hays, "Science in the City: The London Institution, 1819–40," *British Journal for the History of Science* 7 (1974): 146–62.

3. Sidney Ross, "Scientist: the Story of a Word," *Annals of Science* 18 (1962): 65–85; Jack Morrell and Arnold Thackray, *Gentlemen of Science: The Early Years of the British Association for the Advancement of Science* (Oxford: Oxford University Press, 1981).

4. James A. Secord, *Controversy in Victorian Geology: The Cambrian-Silurian Dispute* (Princeton: Princeton University Press, 1986); Martin Rudwick, *The Great Devonian Controversy: The Shaping of Gentlemanly Knowledge among Gentlemanly Specialists* (Chicago: University of Chicago Press, 1985); Maxine Berg, *The Machinery Question and the Making of Political Economy, 1815–1848* (Cambridge: Cambridge University Press, 1980); Morrell and Thackray, *Gentlemen of Science*, are a representative sample.

5. David P. Miller, "The Royal Society of London, 1800–1835" (Ph.D. diss., University of Pennsylvania, 1981); "Between Hostile Camps: Sir Humphry Davy's Presidency of the Royal Society of London, 1820–1827," *British Journal for the History of Science* 16 (1983): 1–47; "The Revival of the Physical Sciences in Britain, 1815–1840," *Osiris* 2 (1986): 107–34; Iwan Rhys Morus, "The Politics of Power: Reform and Regulation in the Work of William Robert Grove" (Ph.D. diss., University of Cambridge, 1989).

6. Adrian Desmond, "Artisan Resistance and Evolution in Britain, 1819–1848," *Osiris* 3 (1987): 77–110; "Lamarckism and Democracy: Corporations, Corruption and Comparative Anatomy in the 1830s," in *History, Humanity and Evolution*, ed. James Moore (Cambridge: Cambridge University Press, 1989), pp. 99–130; *The Politics of Evolution* (Chicago: University of Chicago Press, 1989); Roger Cooter, *The Cultural Meaning of Popular Science: Phrenology and the Organization of Consent in Nineteenth Century Britain* (Cambridge: Cambridge University Press, 1984); James A. Secord, "Extraordinary Experiment: Electricity and the Creation of Life in Victorian England," in *The Uses of Experiment*, ed. David Gooding, Trevor Pinch, and Simon Schaffer (Cambridge: Cambridge University Press, 1989), pp. 337–83; Iwan Rhys Morus, "Currents from the Underworld: Electricity and the Technology of Display in Early Victorian England," *Isis* 84 (1993): 50–69.

7. Mary Shelley, *Frankenstein, or the Modern Prometheus* (London, 1818), ed. with intro. by Marilyn Butler (London: Pickering & Chatto, 1993).

8. Butler argues convincingly that Shelley drew heavily on the contemporary dispute between William Lawrence and John Abernethy in portraying Frankenstein's science. Butler, introduction to Ibid., pp. ix–lvii.

9. For an exemplary account of the ways in which experiments could construct careers, see Steven Shapin and Simon Schaffer, *Leviathan and the Air-Pump: Hobbes, Boyle and the Experimental Life* (Princeton: Princeton University Press, 1985).

10. There is a large literature on the relation of natural and political economies during this period. Of particular relevance is Norton Wise (with the collaboration of Crosbie Smith), "Work and Waste: Political Economy and Natural Philosophy in Nineteenth-Century Britain," *History of Science* 27 (1989): 263–301, 391–449; 28 (1990): 221–61.

11. The recent bicentenary appears to have given birth to a nascent Faraday industry. Prod-

ucts include Geoffrey Cantor, *Michael Faraday: Scientist and Sandemanian* (London: Macmillan, 1991); David Gooding, *Experiment and the Making of Meaning: Human Agency in Scientific Observation* (Dordrecht: Kluwer, 1990); Frank A.J.L. James, ed., *The Correspondence of Michael Faraday*, vol. 1 (London: Peregrinus Press, 1991).

12. The most notable example is John Tyndall, *Faraday as a Discoverer* (London, 1868). Also Henry Bence Jones, *The Life and Letters of Faraday*, 2 vols. (London, 1870); John H. Gladstone, *Michael Faraday* (London, 1872).

13. Silvanus P. Thompson, *Michael Faraday: His Life and Work* (London, 1898).

14. Even one of Faraday's nineteenth-century biographers acknowledged this point. See ibid. Also Silvanus P. Thompson, *The Electromagnet, and Electromagnetic Mechanism* (London, 1891).

15. For London's shows, see Richard Altick, *The Shows of London* (Cambridge: Harvard University Press, 1978).

16. Morus, "Currents."

17. Donald Cardwell, *The Organization of Science in England*, 2d ed. (London: Heinemann, 1972); Colin Russell, *Science and Social Change, 1700–1900* (London: Macmillan, 1983).

18. For London's lecturers, see John N. Hays, "The London Lecturing Empire, 1800–50," in *Metropolis and Province: Science in British Culture, 1780–1850*, ed. Ian Inkster and Jack Morrell (London: Hutchinson, 1983), pp. 91–119.

19. Particularly important were Harry Collins, "The TEA Set: Tacit Knowledge and Scientific Networks," *Science Studies* 4 (1974): 165–86; "The Seven Sexes: A Study in the Sociology of a Phenomenon or the Replication of Experiments in Physics," *Sociology* 9 (1975): 205–24.

20. Harry Collins, ed., *Knowledge and Controversy: Studies of Modern Natural Science*, Special Issue, *Social Studies of Science* 11, no. 1 (1981).

21. Shapin and Schaffer, *Leviathan and the Air-Pump*; Steven Shapin, "History of Science and Its Sociological Reconstructions," *History of Science* 20 (1983): 157–211, also contains a cogent argument in favor of the historical uses of sociology. For recent overviews, see Jan Go-

linski, "The Theory of Practice and the Practice of Theory: Sociological Approaches in the History of Science," *Isis* 81 (1990): 492–505; Peter Dear, "Cultural History of Science: An Overview with Reflections," *Science, Technology and Human Values* 20 (1995): 150–70.

22. Harry Collins, *Changing Order: Replication and Induction in Scientific Practice* (London: Sage, 1985).

23. Simon Schaffer, "Glass Works: Newton's Prisms and the Uses of Experiment," in Gooding, Pinch, and Schaffer, *The Uses of Experiment*, pp. 67–104; "Late Victorian Metrology and Its Instrumentation: A Manufactory of Ohms," in *Invisible Connections: Instruments, Institutions and Science*, ed. Robert Bud and Susan Cozzens (Bellingham: SPIE Optical Engineering Press, 1992), pp. 23–56; Jeff Hughes, "The Radioactivists: Community, Controversy and the Rise of Nuclear Physics" (Ph.D. diss., University of Cambridge, 1993).

24. Collins, *Changing Order*.

25. For an argument against this perspective, see Peter Galison, *How Experiments End* (Chicago: University of Chicago Press, 1987).

26. In particular, Bruno Latour, *Science in Action: How to Follow Scientists and Engineers through Society* (Milton Keynes: Open University Press, 1987).

27. Bruno Latour, *We Have Never Been Modern* (Hemel Hempstead: Harvester Wheatsheaf, 1993); Andrew Pickering, *The Mangle of Practice: Time, Agency and Science* (Chicago: University of Chicago Press, 1995).

28. For an overview of neo-Marxist approaches to the history and sociology of science and technology, see Donald MacKenzie and Judy Wacjman, eds., *The Social Shaping of Technology: How the Refrigerator Got Its Hum* (Milton Keynes: Open University Press, 1985).

29. David Gooding, "In Nature's School: Faraday as an Experimentalist," in *Faraday Rediscovered: Essays in the Life and Work of Michael Faraday, 1791–1867*, ed. David Gooding and Frank A.J.L. James (London: Macmillan, 1985), pp. 105–35, on p. 107.

30. Karl Marx, *Capital*, 3 vols. (Harmondsworth: Penguin, 1976), 1:164–65.

31. Thomas Richards, *The Commodity Culture of Victorian England* (London: Verso, 1991).

32. For similar arguments, see Ludwik Fleck, *Genesis and Development of a Scientific*

*Fact* (Chicago: University of Chicago Press, 1979); Jan Golinski, *Science as Public Culture: Chemistry and Enlightenment in Britain, 1760–1820* (Cambridge: Cambridge University Press, 1992).

CHAPTER 1
THE ERRORS OF A FASHIONABLE MAN

1. Joseph Agassi, *Faraday as a Natural Philosopher* (Chicago: University of Chicago Press, 1971).

2. For a recent, exemplary account of a philosopher's self-fashioning, see Mario Biagioli, *Galileo Courtier: The Practice of Science in the Culture of Absolutism* (Chicago: University of Chicago Press, 1993).

3. Morris Berman, *Social Change and Scientific Organization: The Royal Institution, 1799–1844* (London: Heinemann, 1978).

4. George Ellis, *Memoir of Sir Benjamin Thompson, Count Rumford* (Boston, 1871).

5. Berman, *Social Change*.

6. Ibid., p. 25; Henry Bence Jones, *The Royal Institution. Its Founder and First Professors* (London, 1871), p. 194.

7. Donovan Chilton and Noel G. Coley, "The Laboratories of the Royal Institution in the Nineteenth Century," *Ambix* 27 (1980): 173–203.

8. Ibid., quoted on p. 177 from William T. Brande, *Manual of Chemistry* (London, 1819).

9. Bence Jones, *Royal Institution*, pp. 207–8.

10. Berman, *Social Change*, chap. 2. See also David Knight, *Humphry Davy: Science and Power* (Oxford: Blackwells, 1992).

11. Golinski, *Science as Public Culture*, chap. 7. See also Jan Golinski, "Humphry Davy and the Lever of Experiment," in *Experimental Enquiries*, ed. Homer Le Grand (Dordrecht: Reidel, 1990), pp. 99–136.

12. For an example of galvanism's unsavory context, see *The Sceptic* (London, 1800), p. 6.

13. June Z. Fullmer, "Humphry Davy's Adversaries," *Chymia* 8 (1962): 147–64.

14. June Z. Fullmer, "Humphry Davy and the Gunpowder Manufactory," *Annals of Science* 20 (1964): 165–94.

15. Quoted in Berman, *Social Change*, p. 133.

16. The *Chemist* lasted less than two years, being founded in 1824 and folding before the end of 1825. See Golinski, *Science as Public*

*Culture*, pp. 243–45; Russell, *Science and Social Change*, pp. 139–46.

17. [Thomas Hodgskin], "Apology and Preface," *Chemist* 1 (1824): v–viii, on p. vii.

18. [Thomas Hodgskin], "Humphry Davy's Copper," *Chemist* 2 (1825): 78–79, on p. 78.

19. Quoted in George Foote, "Sir Humphry Davy and His Audience at the Royal Institution," *Isis* 43 (1952): 6–12, on p. 12.

20. There were several nineteenth-century versions: Bence Jones, *Life and Letters*; Gladstone, *Faraday*; Thompson, *Faraday*; Tyndall, *Discoverer*; along with numerous obituaries and short notices. Relevant twentieth-century accounts are L. Pearce Williams, *Michael Faraday* (London: Chapman & Hall, 1965), and Cantor, *Faraday*. Much of the best recent scholarship on Faraday may be sampled in Gooding and James, *Faraday Rediscovered*.

21. Williams, *Faraday*, pp. 7–8.

22. For London lecturing at this time, see Hays, "London Lecturing Empire."

23. James Nasmyth did this when seeking work from Henry Maudslay. See James Nasmyth, *Autobiography*, ed. Samuel Smiles (London, 1883), pp. 142–43.

24. Williams, *Faraday*, p. 29.

25. Gladstone, *Faraday*, p. 19.

26. Michael Faraday to Benjamin Abbott, 8 March 1813. Published in James, *The Correspondence of Michael Faraday* (henceforward *Correspondence*), pp. 44–46.

27. Gladstone, *Faraday*, pp. 20–21.

28. Brian Bowers and Lenore Symons, eds., *Curiosity Perfectly Satisfyed: Faraday's Travels in Europe, 1813–1815* (London: Peregrinus Press, 1991), p. 1.

29. Michael Faraday to Benjamin Abbott, 25 January 1815, ibid., pp. 144–52; *Correspondence*, pp. 115–23.

30. This should be separated from his position at the Royal Institution, where he was clearly the institution's servant. For an alternative reading of Faraday's relationship with Davy, see David Knight, "Davy and Faraday: Father and Son," in Gooding and James, *Faraday Rediscovered*, pp. 33–49.

31. Bowers and Symons, *Curiosity*, p. 146; *Correspondence*, p. 117.

32. Introduction to *Correspondence*, p. xxxii.

33. Isaac Watts, *The Improvement of the Mind*, new ed. (London, 1809).

34. Ibid., pp. 24–25.

35. Williams, *Faraday*, pp. 12–13, emphasizes the way in which Faraday systematically followed Watts's prescriptions.

36. For the City Philosophical Society, see Frank A.J.L. James, "Michael Faraday, the City Philosophical Society and the Society of Arts," *RSA Journal* 41 (1992): 192–99. Faraday's close relationship with his fellow member Benjamin Abbott is discussed in Frank A.J.L. James, "The Tales of Benjamin Abbott: A Source for the Early Life of Michael Faraday," *British Journal for the History of Science* 25 (1992): 229–40.

37. Michael Faraday to Benjamin Abbott, 1 June 1813, *Correspondence*, p. 57.

38. Michael Faraday to Benjamin Abbott, 11 June 1813, *Correspondence*, p. 60.

39. Ibid.

40. James, "City Philosophical Society," p. 195.

41. Bence Jones, *Life and Letters*, 2:451; Gladstone, *Faraday*, p. 140.

42. Thompson, *Faraday*, p. 231.

43. Ibid., quoting from the recollections of Faraday's niece, Constance Reid.

44. Benjamin H. Smart, *A Practical Grammar of English Pronunciation* (London, 1810); *The Practice of Elocution* (London, 1819). For an overview of class differentiation in language use, see K. C. Phillipps, *Language and Class in Victorian England* (Oxford: Blackwells, 1984).

45. Smart, *Practice of Elocution*, p. 1.

46. Ibid., p. 2.

47. Ibid., p. 63.

48. Ibid., p. 64.

49. Marquis of Ridolfi, "Method of Separating Platina from Other Metallic Substances Which Are Found with It in the State of Ore," *Quarterly Journal of Science* 1 (1816): 259–64.

50. Michael Faraday, "Some Experiments and Observations on a New Acid Substance," *Quarterly Journal of Science* 3 (1817): 77–81.

51. See, for example, Michael Faraday, "An Account of Some Experiments on the Escape of Gases through Capillary Tubes," *Quarterly Journal of Science* 3 (1817). 354–55.

52. See, for example, Humphry Davy to Michael Faraday, 3 August 1815, *Correspondence*, pp. 133–34.

53. Michael Faraday, "On Sirium," *Quarterly Journal of Science* 6 (1819): 112–15, on p. 115.

54. Michael Faraday, "On Sirium, or Vestium," *Quarterly Journal of Science* 7 (1819): 291–93.

55. Humphry Davy to Michael Faraday, October 1818, *Correspondence*, p. 171.

56. Humphry Davy, "New Experiments on Some of the Combinations of Phosphorus," *Philosophical Transactions* 108 (1818): 316–37, on p. 337.

57. Quoted in Williams, *Faraday*, p. 46.

58. [Michael Faraday], "Historical Sketch of Electro-magnetism," *Annals of Philosophy* 2 (1821): 195–200, 274–90; 3 (1822): 1078–1121.

59. Michael Faraday, "On Some New Electro-magnetical Motions, and on the Theory of Magnetism," *Quarterly Journal of Science* 12 (1821): 74–96.

60. David Gooding, "Experiment and Concept-Formation in Electromagnetic Science and Technology in England, 1820–1830," *History and Technology* 2 (1985): 151–76; "In Nature's School"; "Magnetic Curves and the Magnetic Field: Experimentation and Representation in the History of a Theory," in Gooding, Pinch, and Schaffer, *The Uses of Experiment*, pp. 183–223; *Experiment and the Making of Meaning*.

61. Michael Faraday to James Stodart, 8 October 1821, *Correspondence*, p. 228.

62. Ibid.

63. For a discussion of this issue, see Iwan Rhys Morus, "Different Experimental Lives: Michael Faraday and William Sturgeon," *History of Science* 30 (1992): 1–28.

64. William Wollaston to Michael Faraday, ca. 1 November 1821, *Correspondence*, p. 235.

65. Michael Faraday, "On Fluid Chlorine," *Philosophical Transactions* 113 (1823): 160–64. Davy insisted that Faraday add the sentence "The President of the Royal Society having honoured me by looking at these conclusions, and suggested that an exposure to heat under pressure would lead to interesting results, the following experiments were commenced at his request." Michael Faraday, "On the History of the Condensation of Gases, in Reply to Dr. Davy, Introduced by Some Remarks on That of Electro-magnetic Rotation," *Philosophical Magazine* 8 (1836): 521–29; reprinted in Michael Faraday, *Experimental Researches in Chemistry and Physics* (London, 1859), pp. 135–41, on p. 137.

66. Henry Warburton to Michael Faraday, 8 July 1823, *Correspondence*, pp. 322–23.

67. Managers' Minutes (henceforward MM RI), 7 February 1825, in *Archives of the Royal*

*Institution,* ed. Frank Greenaway, 15 vols. (London: Royal Institution, 1971–76), vol. 7.

68. Michael Faraday, "Reply to Dr. John Davy's Remarks on Certain Statements of Mr. Faraday Contained in His 'Researches in Electricity,'" *Philosophical Magazine* 7 (1835): 337–42, reprinted in Michael Faraday, *Experimental Researches in Electricity,* vol. 2 (London, 1844), pp. 211–17, on p. 216.

69. Berman, *Social Change.*

70. MM RI, 13 June 1825.

71. Michael Faraday to William Hosking, 5 March 1829, *Correspondence,* p. 478. Hosking was an architect and civil engineer: *Dictionary of National Biography* (henceforward *DNB*), 9:1289–90. In January 1829 he delivered a course of six lectures at the Western Literary Institution on the modern architecture of London. This topic may have been his offering for the Royal Institution.

72. MM RI, 22 January 1827.

73. Ibid., 1 February 1830.

74. Ibid., 17 November 1828.

75. Ibid., 13 April 1829.

76. Both previous porters, including Slatter, who had replaced Faraday as laboratory assistant during his tour with Davy, resigned, complaining that their pay was insufficient to compensate for their new duties. Ibid., 18 May 1829.

77. Ibid., 15 March 1830.

78. See John M. Thomas, *Michael Faraday and the Royal Institution* (London: Adam Hilger, 1991), app. 1, p. 212. They were "Recent Experimental Investigation of Volta-electric or Magneto-electric Induction" (17 February); "Magnetic-electric Induction and the Explanation It Affords of Arago's Phenomena of Magnetism Exhibited by Moving Metals" (2 March); "Evolution of Electricity Naturally and Artificially by the Inductive Action of the Earth's Magnetism" (30 March).

79. From Mrs. Owen's Diary, 7 June 1837, quoted in *The Life of Richard Owen,* 2 vols. (London, 1894), 1:112–13. The date must be misprinted since the discourse was actually presented on 9 June 1837.

80. Ibid., p. 153.

81. Thompson, *Faraday,* p. 232.

82. Bence Jones, *Life and Letters,* 2:447.

83. Michael Faraday to William Jerdan, 12 March 1828; 30 May 1828, *Correspondence,* p. 452; pp. 453–54.

84. Michael Faraday to Walter Calverley

Trevelyan, 7 February 1827, *Correspondence,* p. 422.

85. Thomas, *Royal Institution,* app. 3, p. 217.

86. Berman, *Social Change,* p. 126.

87. MM RI, 9 January 1826.

88. Ibid.

89. Ibid.

90. See MM RI, passim.

91. Michael Faraday to Edward Magrath, 23 July 1826, *Correspondence,* pp. 417–18.

92. "Shewing How the Tories and the Whigs Extend Their Patronage to Science and Literature," *Fraser's Magazine* 12 (1835): 703–9.

93. Ibid., p. 708.

94. *Fraser's Magazine* 13 (1836): 66–67.

95. "Michael Faraday," *Fraser's Magazine* 13 (1836): 224.

96. The *Researches* were republished in three volumes between 1839 and 1855. All quotations will be from this source: Michael Faraday, *Experimental Researches in Electricity* (henceforward *ERE*), 3 vols. (London, 1839–55).

97. Chilton and Coley, "Laboratories."

98. Thomas Martin, ed., *Faraday's Diary* (henceforward *Diary*), 7 vols. (London: G. Bell & Sons, 1932), 1:367 .

99. Ibid. p. 369.

100. Ibid., pp. 369–70.

101. Ibid., p. 373.

102. Ibid.

103. Ibid., p. 376.

104. Ibid., pp. 380–85; pp. 385–89.

105. Faraday had been engaged to teach chemistry at the Woolwich Royal Military Academy since 1829.

106. For the central (and invisible) role of technicians, see Steven Shapin, "The House of Experiment in Seventeenth Century England," *Isis* 79 (1988): 373–404.

107. For replicability and publicity, see Collins, *Changing Order.*

108. *ERE,* 1st ser., 1:3.

109. Ibid., p. 4.

110. Ibid., p. 16.

111. Ibid., pp. 34–35.

112. Michael Faraday to Richard Phillips, 29 November 1831. Published in *The Selected Correspondence of Michael Faraday,* ed. L. Pearce Williams, 2 vols. (Cambridge: Cambridge University Press, 1971), 1:209–11.

113. The episode is detailed in Brian Gee, "Faraday's Plight and the Origins of the Magneto-electric Spark," *Nuncius* 5 (1990): 43–68.

114. Jean Nicolas Hachette to Michael Faraday, 9 July 1832, in Williams, *Selected Correspondence*, 1:229.

115. Leopoldo Nobili and Vincenzo Antinori, "On the Electro-motive Force of Magnetism: With Notes by Michael Faraday," *Philosophical Magazine* 11 (1832): 401–13, on p. 405.

116. Jean Nicolas Hachette to Michael Faraday, 30 August 1833, in Williams, *Selected Correspondence*, 1:259.

117. Quoted in Thompson, *Faraday*, p. 241.

118. Tyndall, *Discoverer*, p. 41.

119. *ERE*, 3:1–26. The paper was read to the Royal Society on 20 November 1845.

120. Ibid., pp. 1–2.

121. Ibid., 1:285–86. The experiments were briefly discussed in his eighth series, first published in 1834.

122. *Diary*, 4:256–57.

123. Frank A.J.L. James, "Time, Tide and Michael Faraday," *History Today*, September 1991, 28–34.

124. *Diary*, 4:264.

125. Ibid., p. 267.

126. Ibid., p. 272.

127. Ibid., p. 277.

128. Ibid., p. 294.

129. "Weekly Gossip," *Athenaeum*, 8 November 1845, p. 1080. Quoted in David Gooding, "He Who Proves Discovers: John Herschel, William Pepys and the Faraday Effect," *Notes and Records of the Royal Society* 39 (1985): 229–44, on p. 229.

130. William Robert Grove, *On the Correlation of Physical Forces* (London, 1846), based on lectures delivered at the London Institution. For Grove and correlation, see Geoffrey Cantor, "William Robert Grove, the Correlation of Forces, and the Conservation of Energy," *Centaurus* 19 (1976): 273–90; Morus, "Correlation and Control."

131. *ERE*, 3:6. The law, as expressed here by Faraday, is incomprehensible.

132. Ibid., p. 14.

133. Ibid., pp. 14–15.

134. Ibid., pp. 19–20.

135. Ibid., p. 20.

136. Ibid.

137. For a discussion of this aspect of Faraday's presentation of his work, see Morus, "Different Experimental Lives," pp. 7–13.

138. *ERE*, 3:21.

139. Michael Faraday, "A Speculation Touching Electric Conduction and the Nature

of Matter," *Philosophical Magazine* 24 (1844): 136–44; *ERE*, 2:284–93. The discourse was delivered at the Royal Institution on 19 January 1844.

140. For an interesting account of Faraday's views on the interaction of light and matter, see Frank A.J.L. James, "The Optical Mode of Investigation: Light and Matter in Faraday's Natural Philosophy," in Gooding and James, *Faraday Rediscovered*, pp. 137–61.

141. See Gooding, "He Who Proves Discovers."

142. John Herschel to Michael Faraday, 9 November 1845, in Williams, *Selected Correspondence*, 1:459–60.

143. Michael Faraday to John Herschel, 13 November 1845, in Williams, *Selected Correspondence*, 1:465.

144. William Whewell to Michael Faraday, 20 November 1845, and Michael Faraday to William Whewell, 22 November 1845, in Williams, *Selected Correspondence*, 1:465, 466.

145. Michael Faraday to John Herschel, 22 December 1845, in Williams, *Selected Correspondence*, 1:475.

146. John Herschel to Robert Hunt, 5 January 1846, quoted in Gooding, "He Who Proves Discovers," p. 239.

147. Ibid.

148. William Whewell to James D. Forbes, 5 May 1846, quoted in Simon Schaffer, "The History and Geography of the Intellectual World: Whewell's Politics of Language," in *William Whewell: A Composite Portrait*, ed. Menachem Fisch and Simon Schaffer (Oxford: Clarendon Press, 1991), pp. 201–31, on p. 228.

149. Ibid., pp. 226–30.

150. Tyndall, *Discoverer*, p. 208.

151. Ibid., p. 45.

152. Fleck, *Genesis and Development*, concurs on the epistemological priority of public space.

153. Martin Rudwick, "Charles Darwin in London: The Integration of Public and Private Science," *Isis* 73 (1982): 186–206; Robert Iliffe, "In the Warehouse: Privacy, Property and Priority in the Early Royal Society," *History of Science* 30 (1992): 30–68.

154. See, for example, Cornelia Crosse, "Science and Society in the Fifties," *Temple Bar* 93 (1891): 33–51.

155. Steven Shapin, "The Mind Is Its Own Place: Science and Solitude in Seventeenth Century England," *Science in Context* 4 (1991):

191–218, argues the importance of such cultural distancing in the figuring of the scholar.

156. Simon Schaffer, "Self-Evidence," *Critical Inquiry* 18 (1992): 327–62, charts the rise of the cult of natural philosopher as disembodied genius.

CHAPTER 2
THE VAST LABORATORY OF NATURE

1. Thompson, *Faraday*, p. 226.

2. For "black boxes," see Latour, *Science in Action*; Schaffer, "Glass Works."

3. For the "core set," see Collins, *Changing Order*.

4. Sophie Forgan, "Faraday—From Servant to Savant: The Institutional Context," in Gooding and James, *Faraday Rediscovered*, pp. 51–67.

5. For a similar example of the spatial dimension to an experiment's transparency, see Schaffer, "Glass Works."

6. Collins, *Changing Order*.

7. The most famous example of such a loss is, of course, that of N-rays. Mary Jo Nye, "N-Rays: An Episode in the History and Psychology of Science," *Historical Studies in the Physical Sciences* 11 (1980): 125–56.

8. Biographical details may be found in James P. Joule, "A Short Account of the Life and Writings of the Late William Sturgeon," *Memoirs of the Manchester Literary and Philosophical Society* 14 (1857): 53–83; William W. Haldane Gee, "William Sturgeon," *Electrician* 35 (1895): 632–35; Thompson, *Electromagnet*, app. A: "William Sturgeon," pp. 412–18.

9. Joule, "A Short Account," p. 54.

10. Ibid., p. 77.

11. Nathan Reingold, ed., *The Papers of Joseph Henry* (henceforward *Henry Papers*), 5 vols. (Washington, DC: Smithsonian Institution Press, 1972–), 3:307.

12. Ibid., 4:152. Joseph Henry to Asa Gray, 1 November 1838. It is not entirely clear here whether Henry was offering Sturgeon a backhanded compliment or insulting the entire London community!

13. The London Electrical Society is discussed in detail in chapter 4. See also Morus, "Currents."

14. Joule, "A Short Account," p. 80.

15. Donald Cardwell, *James Joule: A Biography* (Manchester: Manchester University Press, 1989), pp. 25–28.

16. Gooding, "In Nature's School."

17. Peter Barlow, "A Curious Electro-magnetic Experiment," *Philosophical Magazine* 59 (1822): 241–42.

18. Ibid., p. 241.

19. See, for example, Peter Barlow, "An Account of Some Electro-magnetic Combinations, for Exhibiting Thermo-electric Phenomena, Invented by James Marsh of Woolwich," *Philosophical Magazine* 62 (1823): 321–27.

20. "Electro-magnetic Rotation," *Philosophical Magazine* 62 (1823): 237. Sturgeon was described here as "a pensioned artilleryman of Woolwich, who has successfully devoted himself to scientific pursuits."

21. William Sturgeon, "Electro-magnetical Experiments," *Philosophical Magazine* 63 (1824): 95–100; quotation on p. 95.

22. Ibid., p. 98.

23. William Sturgeon, "Electro- and Thermo-magnetical Experiments," *Philosophical Magazine* 63 (1824): 266–68; "Description of a Rotative Thermo-magnetical Experiment," *Philosophical Magazine* 63 (1824): 269–71.

24. Sturgeon, "Thermo-magnetical Experiments," p. 271. For Halley's theories of terrestrial magnetism, see Patricia Fara, *Sympathetic Attractions: Magnetic Practices, Beliefs, and Symbolism in Eighteenth-Century England* (Princeton: Princeton University Press, 1996), esp. pp. 152–55.

25. William Sturgeon, "On Electro-magnetism," *Philosophical Magazine* 64 (1824): 242–49.

26. Ibid., p. 243.

27. Compare with Peter Galison and Alexi Assmus, "Artificial Clouds, Real Particles," in Gooding, Pinch, and Schaffer, *The Uses of Experiment*, pp. 225–74.

28. Sturgeon, "Electro-magnetism," pp. 246–47.

29. William Sturgeon, "Account of an Improved Electro-magnetic Apparatus," *Annals of Philosophy* 12 (1826): 357–61. Reprinted in William Sturgeon, *Scientific Researches, Experimental and Theoretical in Electricity, Magnetism, Galvanism, Electromagnetism and Electrochemistry* (Bury, 1850), pp. 103–12.

30. "Portable Electro-magnetic Apparatus," *Transactions of the Society of Arts* 41 (1823): 47–52.

31. Ibid., p. 49.

32. Ibid., p. 48.

33. Ibid., p. 104.

34. Ibid., p. 105.

35. Quoted from an unpublished lecture by Sturgeon in Joule, "A Short Account," p. 78.

36. See above and Sturgeon, "Thermo-magnetical Experiment."

37. David Gooding, "Final Steps to the Field Theory," *Historical Studies in the Physical Sciences* 11 (1981): 231–75; "Experiment and Concept-Formation"; Williams, *Faraday*.

38. William Sturgeon, "On Electro-pulsations and Electro-momentum," *Philosophical Magazine*, n.s., 9 (1836): 132–36, on p. 132.

39. Ibid., p. 134.

40. For a discussion of some of the cultural resonances of scholarly disengagement, see Shapin, "The Mind Is Its Own Place."

41. For an instructive account of the alarm caused by controversy between gentlemen, see Secord, *Controversy*, pp. 231–33. For Faraday's refusal to engage in dispute, see Cantor, *Faraday*, p. 141. Faraday concurred with the common perception that controversy had no place in philosophical discourse. This is not to suggest that gentlemen did not as a matter of fact engage in controversy. The point is that such engagements were regarded as deviations from an ideal.

42. Iwan Rhys Morus, "The Sociology of Sparks: An Episode in the History and Meaning of Electricity," *Social Studies of Science* 18 (1988): 387–417. The following paragraphs are largely based on material from this essay.

43. *ERE*, 3d ser., 1:102.

44. Ibid., p. 104.

45. Ibid., p. 103.

46. Ibid., p.105.

47. Ibid., pp. 105–6.

48. Ibid., p. 106.

49. Ibid., p. 106.

50. Ibid., p. 107.

51. Ibid.

52. Ibid., 7th ser., p. 207. Italics in the original.

53. Ibid., pp. 207–9.

54. Ibid., p. 209.

55. Ibid., p. 211.

56. Ibid., p. 217.

57. William Sturgeon, "On the Relation by Measure of Common and Voltaic Electricity," *Annals of Electricity* 1 (1836–37): 52–65. Faraday's researches are reprinted on pp. 61–65.

58. Ibid., p. 53.

59. Ibid.

60. Ibid., p. 54.

61. Ibid.

62. Ibid., p. 57.

63. Ibid., p. 59.

64. Ibid., p. 60.

65. William Sturgeon, "Remarks on the Preceeding Paper, with Experiments," *Annals of Electricity* 1 (1836–37): 367–76.

66. Ibid., p. 371.

67. Ibid.

68. William Sturgeon, "Fifth Memoir on Experimental and Theoretical Researches in Electricity, Magnetism &c.," *Annals of Electricity* 5 (1840): 121–35; 293–301.

69. Ibid., p. 293.

70. Michael Faraday to Edward Solly, 23 November 1836, in *The Correspondence of Michael Faraday*, ed. Frank A.J.L. James, vol. 2 (London: Institution of Electrical Engineers, 1993), p. 388.

71. *Diary*, 2:330; *ERE*, 9th ser., 1:322.

72. *Diary* p. 331: "It is I think evidently dependant on induction." Note, however, that he continues, "See the paper I have written to the Phil. Mag. Though the experiments were made on the day of the date given [15 October], the rough notes were not entered here until the 20th Octr. at Brighton." It is possible, therefore, that the decision to treat the phenomenon as one of induction was not immediate. The passage is a salutary reminder of the retrospective nature of Faraday's *Diary*.

73. *ERE*, 9th ser., 1:324–28; *Diary*, pp. 333–340.

74. *ERE*, 9th ser., 1:330–31.

75. For the notion of "surrogates," see Collins, *Changing Order*, pp. 100–6, 125–27.

76. *ERE*, 9th ser., 1:332.

77. Ibid., p. 334.

78. For the notion of "evidential context," see Trevor Pinch, *Confronting Nature: The Sociology of Solar-Neutrino Detection* (Dordrecht: Reidel, 1987).

79. Quoted in Bence Jones, *Life and Letters*, 2:45n.

80. William Sturgeon, "On the Electric Shock from a Single Pair of Voltaic Plates," *Annals of Electricity* 1 (1836–37): 67–75. Henry's experiments were published in Joseph Henry, "Facts in Reference to the Sparks &c. from a Long Conductor Uniting the Poles of a Galvanic Battery," *Journal of the Franklin Institute* 15 (1835): 169–70.

81. Sturgeon, "On the Electric Shock," p. 67.

82. Ibid., p. 70.

83. Ibid., p. 72.

84. Ibid., p. 74.

85. Ibid., p. 75.

86. William Sturgeon, "Remarks on Mr. Faraday's 'Ninth Series of Experimental Researches in Electricity,' with Experiments," *Annals of Electricity* 1 (1836–37): 186–91; quotation on p. 186.

87. Ibid., p. 186.

88. Ibid., p. 187.

89. Ibid.

90. Sturgeon, "Relation by Measure"; Morus, "Sociology of Sparks."

91. William Sturgeon, "An Experimental Investigation of the Laws Which Govern the Production of Electric Shocks &c. from a Single Voltaic Pair," *Annals of Electricity* 1 (1836–37): 192–98.

92. Ibid., p. 195.

93. Ibid., p. 196.

94. Ibid.

95. Ibid., p. 198.

96. For another example of the relationship between audience and the aims of experimentation, see Mario Biagioli, "Galileo's System of Patronage," *History of Science* 28 (1990): 1–62.

97. Collins, "The TEA Set"; Collins, *Changing Order*.

98. For an instructive account of the strategic nature of the categories of public and private, see Iliffe, "In the Warehouse."

99. Joseph Henry to Asa Gray, 1 November 1838, *Henry Papers*, 4:152.

100. Henry's European Diary, 8 April 1837, *Henry Papers*, 3:250.

101. William Sturgeon to Joseph Henry, 23 September 1839, *Henry Papers*, 4:263.

102. Ibid.

CHAPTER 3
BLENDING INSTRUCTION WITH AMUSEMENT

1. Zeta, "The Scientific Amusements of London," *London Polytechnic Magazine and Journal* 1 (1844): 225–34, on pp. 229–30.

2. Berg, *Machinery Question*.

3. Clive Behagg, "Secrecy, Ritual and Folk Violence: The Opacity of the Workplace in the First Half of the Nineteenth Century," in *Popular Culture and Custom in Nineteenth-Century England*, ed. Robert Storch (London: Croom Helm, 1982), pp. 154–79.

4. Iorwerth Prothero, *Artisans and Politics in Early Nineteenth-Century London: John Gast and His Times* (London: Methuen, 1981); Edward P. Thompson, *The Making of the English Working Class* (Harmondsworth: Penguin, 1968).

5. Prothero, *Artisans and Politics*; John Rule, "The Property of Skill in the Period of Manufacture," in *The Historical Meanings of Work*, ed. Patrick Joyce (Cambridge: Cambridge University Press, 1987), pp. 99–118.

6. Iwan Rhys Morus, "Manufacturing Nature: Science, Technology and Victorian Consumer Culture," *British Journal for the History of Science* 29 (1996): 403–34.

7. [Joseph Robertson?], "Preface," *Mechanics' Magazine* 1 (1823): iii–iv, on p. iii.

8. Prothero, *Artisans and Politics*, pp. 191–203.

9. Ibid.

10. Clive Behagg, "The Democracy of Work, 1820–1850," in *British Trade Unionism, 1780–1850: The Formative Years*, ed. John Rule (London: Longmans, 1988), pp. 162–77.

11. Thomas Hodgskin, *Popular Political Economy* (London, 1827), p. 251. Quoted in Berg, *Machinery Question*, p. 172.

12. Behagg, "Secrecy"; Rule, "Property of Skill."

13. Altick, *Shows of London*, p. 376.

14. "National Repository," *Mechanics' Magazine* 11 (1829): 45; 58–60, on pp. 58–59.

15. John F. W. Herschel, *A Preliminary Discourse on the Study of Natural Philosophy* (London, 1830), pp. 70, 71–72.

16. Behagg, "Democracy of Work," p. 163.

17. James A. Bennett, "Instrument Makers and the 'Decline of Science in England': The Effect of Institutional Change on the Elite Makers of the Early Nineteenth Century," in *Nineteenth-Century Scientific Instruments and Their Makers*, ed. P. R. deClercq (Amsterdam: Editions Rendopi B.V., 1985), pp. 13–27.

18. William Ginn, "Philosophers and Artisans: The Relationship between Men of Science and Instrument Makers in London 1820–1860" (Ph.D. diss., University of Kent at Canterbury, 1991).

19. Berman, *Social Change*.

20. Dorothy Bathe and Greville Bathe, *Jacob Perkins: His Inventions, His Times, and His Contemporaries* (Philadelphia, 1943).

21. M. Veronica Stokes, "The Lowther Arcade in the Strand," *London Topographical Record* 23 (1974): 119–28.

22. Altick, *Shows of London*.

23. "New Exhibition Room, Adelaide Street, Strand," *Literary Gazette* 16 (1832): 378.

24. "Exhibition of Works of Popular Science," *Mechanics Magazine* 17 (1832): 158–60, on pp. 158–59.

25. "A Revival and Removal," *Mechanics' Magazine* 19 (1833): 336.

26. Adelaide Gallery Handbill, Adelaide Gallery File, Theatre Museum, London.

27. See John W. Dodds, *The Age of Paradox* (New York: Rinehart & Co., 1952), p. 138.

28. Quoted in Altick, *Shows of London*, p. 377.

29. *Society for the Illustration and Encouragement of Practical Science, Gallery for the Exhibition of Objects Blending Instruction with Amusement, Adelaide Street, and Lowther Arcade, West Strand. Catalogue for May, 1836*, 14th ed. (London, 1836).

30. Ibid., p. 37.

31. L. F. Menebrea, "Sketch of the Analytical Engine Invented by Charles Babbage," trans. Ada Lovelace, *Taylor's Scientific Memoirs* 3 (1843): 666–731, on p. 706.

32. Simon Schaffer, "Babbage's Intelligence: Calculating Engines and the Factory System," *Critical Inquiry* 21 (1994): 203–27.

33. Adelaide Gallery Handbill, Adelaide Gallery File, Theatre Museum, London.

34. Neil Harris, *Humbug: The Art of P. T. Barnum* (Boston: Little & Brown, 1973), p. 98.

35. "Gallery of Practical Science," *Literary Gazette* 18 (1834): 578.

36. "Gallery of Practical Science," *Literary Gazette* 19 (1835): 715.

37. Henry's European Diary, 20 March 1837, *Henry Papers*, 3:179.

38. William Fothergill Cooke, *Extracts from the Private Letters of the Late Sir William Fothergill Cooke, 1836–39, relating to the Invention and Development of the Electric Telegraph* (London, 1895), p. 8.

39. *Henry Papers*, pp. 180–81.

40. Quoted in Toshio Kusamitsu, "Great Exhibitions before 1851," *History Workshop* 9 (1980): 70–89, on p. 80, from Manchester Mechanics' Institution, *Annual Report*, 1839, p. 60.

41. J. L. Pritchard, *Sir George Cayley, the Inventor of the Aeroplane* (London: Max Parrish, 1961), pp. 124–39.

42. "Our Weekly Gossip," *Athenaeum* 11 (1838): 554–55, on p. 554.

43. Ibid., p. 555.

44. E. M. Wood, *A History of the Polytechnic* (London: Macdonald, 1965).

45. Peter Cunningham, *Handbook of London, Past and Present*, 2d ed. (London, 1850), p. 403; *Punch* 5 (1843): 91.

46. "Our Weekly Gossip."

47. Royal Polytechnic Institution Handbill, Polytechnic File, Theatre Museum, London. For Armstrong and his machines, see Willem Hackmann, "Electricity from Steam: Armstrong's Hydroelectric Machine in the 1840s," in *Making Instruments Count: Essays on Historical Scientific Instruments Presented to Gerard L'Estrange Turner*, ed. R.G.W. Anderson, J. A. Bennett, and W. F. Ryan (Aldershot: Variorum Press, 1993), pp. 146–73.

48. Edmund Yates, *Recollections and Experiences*, 4th ed. (London, 1885), p. 92.

49. William Leithead to Joseph Henry, 8 October 1839, *Henry Papers*, pp. 268–69.

50. Zeta, "Scientific Amusements."

51. For Sturgeon's departure to Manchester and his activities there, see Robert Kargon, *Science in Victorian Manchester: Enterprise and Expertise* (Manchester: Manchester University Press, 1977), pp. 36–41.

52. Yates, *Recollections*, p. 92.

53. See, for example, Cunningham, *Handbook*.

54. Jehangeer Nowrojee and Hirjeebhoy Merwanjee, *Journal of Residence of Two Years and a Half in Great Britain* (London, 1841), p. 120.

55. Richard Beamish, *Memoir of the Life of Sir Marc Isambard Brunel* (London, 1862); David Lampe, *The Tunnel* (London: George Harrap, 1964).

56. "National Repository," p. 60.

57. For an account of the role of exhibitions in promoting consumer culture, see Richards, *Commodity Culture*. Obviously, my account suggests that the link was forged well before 1851.

58. For Saxton's background, see Arthur H. Frazier, "Joseph Saxton's First Sojourn in Philadelphia, 1818–1831," *Smithsonian Journal of History* 3 (1968): 45–76; Joseph Henry, "Memoir of Joseph Saxton, 1799–1873," *Biographical Memoirs of the National Academy of Sciences* 1 (1877): 219–316. Lukens's magnets are described in Franklin Peale, "Notice of a Large Magnet," *Journal of the Franklin Institute* 6 (1830): 284. For the Philadelphia context, see Bruce Sinclair, *Philadelphia's Philosopher Me-*

*chanics: A History of the Franklin Institute, 1824–1865* (Baltimore: Johns Hopkins University Press, 1974).

59. Eugene S. Ferguson, ed., *Early Engineering Reminiscences (1815–40) of George Escol Sellars* (Washington, DC: Smithsonian Institution Press, 1965), pp. 131–32. Sellars described Saxton as Jacob Perkins's "right hand man."

60. *Henry Papers*, p. 179.

61. For a fascinating discussion of the role of space in defining reputation, see Shapin, "House of Experiment."

62. Smithsonian Institution Archives, Record Unit 7056. Joseph Saxton Papers (henceforward SIAJSP), entries for 28 October 1832, passim.

63. *ERE*, 1st ser. 1:1–41. Republished from *Philosophical Transactions* 122 (1832): 125–62. The experiments were first announced in November 1831.

64. SIAJSP, 2 May 1832.

65. Ibid., 4 May 1832.

66. Ibid., 4 June 1832.

67. Ibid., 18 May 1832. The Paradox Head is described in *Catalogue*, p. 37 n. 37.

68. The fountain pen is mentioned in SIAJSP, 9 September 1832. It did not write very well.

69. Ibid., 4 June 1832. Saxton records passing cheering crowds as he returned home from the gallery at one o'clock in the morning.

70. Ibid., 7 June 1832. For Watkins, see E.G.R. Taylor, *Mathematical Practitioners of Hanoverian England* (Cambridge: Cambridge University Press, 1966), p. 438.

71. SIAJSP, 17 June 1832. The Oxford gathering was the second annual meeting of the newly founded British Association for the Advancement of Science. See Morrell and Thackray, *Gentlemen of Science*.

72. SIAJSP, 5 June 1832.

73. Ibid., 6 June, 13 June, 15 June, 20 June 1832.

74. Ibid., 27 October, 7 December, 8 December 1832.

75. For Faraday's privacy in the laboratory, see Gooding, "In Nature's School."

76. Barlow, "Curious Electro-magnetic Experiment."

77. During most of October 1832, for example, Saxton was engaged in building a model diving bell for the gallery: SIAJSP, 13 October 1832, passim.

78. Ibid., 6 December 1832.

79. Ibid., 20 June 1833.

80. Ibid., 26 June 1833.

81. Henry, "Memoir."

82. "The British Association—Journal of a Week at Cambridge. Conclusion," *Literary Gazette* 17 (1833): 490–93, on p. 490.

83. Ibid. For Faraday, Nobili and Antinori, and the spark, see chap. 1. Also Gee, "Faraday's Plight."

84. "To Correspondents," *Literary Gazette* 17 (1833): 510.

85. "National Gallery of Practical Science. Adelaide Street," *Literary Gazette* 17 (1833): 730.

86. Ibid.

87. Ibid.

88. Ibid.

89. For the importance of witnesses in establishing matters of fact about nature, see Steven Shapin, "Pump and Circumstance: Robert Boyle's Literary Technology," *Social Studies of Science* 14 (1984): 481–520; Shapin and Schaffer, *Leviathan and the Air-Pump*.

90. For some account of contemporary English reading habits, see Richard Altick, *The English Common Reader: A Social History of the Mass Reading Public, 1800–1900* (Chicago: University of Chicago Press, 1957).

91. The publicity was obviously useful to Faraday as well since he claimed the discovery of the "principle" upon which Saxton's machine was based.

92. SIAJSP, 1 January 1835.

93. Ibid.

94. SIAJSP, 6 January 1835.

95. William Sturgeon, "An Account of Some Experiments Made with the Large Magnet at the Exhibition Rooms, Adelaide-street," *Philosophical Magazine* 5 (1834): 376–77.

96. Ibid., p. 377.

97. Francis Watkins, "Observations on Mr. Sturgeon's Letter," *Philosophical Magazine* 6 (1835): 239.

98. Ibid.

99. William Sturgeon, "Explanatory Facts," *Philosophical Magazine* 7 (1835): 231–34.

100. Ibid., p. 231.

101. Ibid., p. 232.

102. Ibid. Watkins's paper was "On Magneto-electric Induction," *Philosophical Magazine* 7 (1835): 107–13.

103. Sturgeon, "Explanatory Facts," p. 232.

104. Ibid., p. 233.

105. Francis Watkins, "Letter from Mr. Watkins," *Philosophical Magazine* 7 (1835): 335.

106. For another account of this dispute, see Brian Gee, "The Early Development of the Magneto-electric Machine," *Annals of Science* 50 (1993): 101–33; Arthur Frazier, "Joseph Saxton at London and His Magneto-electric Devices" (unpublished manuscript, Smithsonian Institution).

107. J. E. Burnett and A. D. Morrison-Low, *"Vulgar and Mechanick": The Scientific Instrument Trade in Ireland, 1650–1921* (Dublin: Royal Dublin Society, 1989), pp. 49–50.

108. Edward M. Clarke, "On a New Phenomenon in Magneto-electricity," *Philosophical Magazine* 6 (1835): 169–70; "On Certain Optical Effects of the Magnetic-electrical Machine, and on an Apparatus for Decomposing Water by Its Means," *Philosophical Magazine* 6 (1835): 427–28.

109. Watkins, "On Magneto-electric Induction," p. 107.

110. Edward M. Clarke, "Remarks on a Peculiar State of Polarity Induced in Soft Iron by Voltaic Magnetism," *Philosophical Magazine* 7 (1835): 422–23, on p. 423.

111. Edward M. Clarke, "Description of E. M. Clarke's Magnetic Electric Machine," *Philosphical Magazine* 9 (1836): 262–66.

112. Ibid., p. 262.

113. Ibid., p. 263.

114. Ibid., p. 264.

115. Ibid., pp. 265–66.

116. Joseph Saxton, "Mr. J. Saxton on His Magneto-electric Machine; with Remarks on Mr. E. M. Clarke's Paper in the Preceding Number," *Philosophical Magazine* 9 (1836): 360–65.

117. Ibid., p. 364.

118. Ibid., p. 365.

119. This shows how the status of spaces could change with the epistemological status of the work carried out there. See Iliffe, "In the Warehouse."

120. Morus, "Sociology of Sparks."

121. E. M. Clarke, *Philosphical Magazine* 10 (1837): 455.

122. Frederick Boase, *Modern English Biography* (London, 1865), 7 vols. 4:674–75; "The Panopticon of Science and Arts," *Year-book of Facts in Science and Arts* 17 (1855): 9–11.

123. Saxton, "Remarks," p. 365.

124. E. M. Clarke, *List of Prices of Magnetical, Philosophical, Optical, and Chemical Instruments and Apparatus, Manufactured by Edward M. Clarke, Magnetician* (London, 1837?).

125. Hodgskin, *Popular Political Economy*.

126. Rule, "Property of Skill."

127. SIAJSP, 18 November 1832.

128. Berg, *Machinery Question*; Schaffer, "Babbage's Intelligence."

129. This was the *Mechanics' Magazine*'s manifesto: "Now the entire end and aim of knowledge—considered in its physical relations—is to multiply the means and resources of the human race—to give man such an absolute mastery over the treasures of nature and art, that he shall have always at his command plenty of everything calculated to make life comfortable and pleasant. . . . It not only enables the manufacturer and workman to make the most of every useful art, but instructs them how to strike out new lines of profitable employment for themselves when those to which they have been accustomed fail them." "Preface," *Mechanics' Magazine* 17 (1832): iii–vi, on p. iv.

130. Morus, "Manufacturing Nature"; Richards, *Commodity Culture*.

CHAPTER 4
A SCIENCE OF EXPERIMENT
AND OBSERVATION

1. Susan F. Cannon, *Science in Culture: The Early Victorian Period* (New York: Science History Publications, 1978); Miller, "The Royal Society of London, 1800–1835."

2. Charles Babbage, *Reflexions on the Decline of Science in England* (London, 1830); David P. Miller, "Method and the Micropolitics of Science: The Early Years of the Geological and Astronomical Societies," in *The Politics and Rhetoric of Scientific Method*, ed. John Schuster and Richard Yeo (Dordrecht: Reidel, 1986), pp. 227–57; Miller, "Between Hostile Camps."

3. For the society, see William H. Brock, "The London Chemical Society, 1824," *Ambix* 14 (1967): 133–39; Russell, *Science and Social Change*, pp. 139–46.

4. A. W., [later corrected to A. M.], "Chemical Society," *Chemist* 1 (1824): 167.

5. J. G., "Chemical Society," *Chemist* 1 (1824): 206.

6. Behagg, "Democracy of Work." Anne Secord, "Science in the Pub: Artisan Botanists in Early Nineteenth Century Lancashire," *His-*

*tory of Science* 32 (1994): 269–315, gives a brilliant analysis of the institutional practices of such artisan organizations.

7. Minute Book of the London Electrical Society (henceforward MBLES), entry for 16 May 1837 (unpaginated), Special Collections MS 42, Institute of Electrical Engineers.

8. MBLES, 16 May 1837.

9. "Electrical Society of London," *Annals of Electricity* 1 (1836–37): 415–17.

10. "Electrical Society of London," *Annals of Electricity* 2 (1838): 62–64.

11. Ibid., p. 63.

12. William Sturgeon, "Address, Delivered by W. Sturgeon, Esq., Lecturer on Experimental Philosophy at the Honourable East India Company's Military Seminary Addiscombe &c. &c. at a General Meeting of the London Electrical Society," *Annals of Electricity* 2 (1838): 64–72, on p. 70.

13. Ibid.

14. Ibid., p. 71.

15. William J. Ashworth, "The Calculating Eye: Baily, Herschel, Babbage and the Business of Astronomy," *British Journal for the History of Science* 27 (1994): 409–41.

16. [Hodgskin], "Apology and Preface."

17. J. G., "Chemical Society."

18. Charles V. Walker, "Inaugural Address," *Journal of the Society of Telegraph Engineers* 5 (1876): 1–22, on p. 17.

19. The president was the Whig reformer George Birkbeck. "London Chemical Society," *Chemist* 2 (1825): 162–68.

20. Secord, "Science in the Pub."

21. Joseph Priestley, *The History and Present State of Electricity* (London, 1767), preface. Reprinted in *Joseph Priestley, Selections from His Writings*, ed. Ira V. Brown (University Park: Pennsylvania State University Press, 1962), pp. 191–209, on p. 203.

22. Sturgeon, "Address," p. 68.

23. For Priestley, see Simon Schaffer, "Priestley and the Politics of Spirit," in *Science, Medicine and Dissent: Joseph Priestley (1733–1804)*, ed. R.G.W. Anderson and Christopher Lawrence (London: Wellcome Trust, 1987), pp. 39–53.

24. Henry Brougham, *Lives of Men of Letters and Science, Who Flourished in the time of George III*, 2 vols. (London, 1845–46), s.v. "Priestley," 1:402–28, on p. 402.

25. Ibid., p. 406

26. "Commemoration of the Centenary of the Birth of Dr. Priestley," *Philosophical Magazine* 2 (1833): 382–402.

27. Schaffer, "Self-Evidence."

28. Morus, "Sociology of Sparks."

29. Herschel, *Preliminary Discourse*, p. 151.

30. Ibid., pp. 70–71.

31. For Noad and his *Lectures*, see Thomas B. Greenslade Jr., "Henry Noad's 'Lectures on Electricity' and the State of Electrical Knowledge in 1844," *Rittenhouse* 2 (1987): 139–46.

32. Henry M. Noad, *A Course of Eight Lectures on Electricity, Galvanism, Magnetism, and Electro-magnetism* (London, 1839), pp. 380–81.

33. Collins, *Changing Order*.

34. "Electrical Society," *Literary Gazette* 21 (1837): 656.

35. *Transactions and Proceedings of the London Electrical Society* 1 (1841): 73–76, 140–41, 131, 132–33.

36. Clarke, "Description," p. 264.

37. Ibid. Clarke is discussed in the previous chapter.

38. See William Sturgeon, *Transactions and Proceedings of the London Electrical Society* 1 (1841): 151.

39. See, for example, Herschel, *Preliminary Discourse*.

40. See, for example, Charles V. Walker, "Description of a Constant Acid Battery, Constructed by the Electrotype Process; with General Observations on Electrotype Manipulation," *Proceedings of the London Electrical Society* 2 (1841): 26–33; George Mackrell, "Nitrate of Soda, Compared with Other Salts, Employed for Constant Batteries," *Proceedings of the London Electrical Society* 2 (1841): 232–37.

41. See Altick, *Shows of London*, chap. 27.

42. For Davenport, see Walter Rice Davenport, *Thomas Davenport: Pioneer Inventor* (Montpelier: Vermont Historical Society, 1929); C. Griglietta, *Electro-magnetism: A Brief Essay or Informal Lecture on Electro-magnetism* (Philadelphia, 1838). Also Thomas Davenport, *Autobiography of Thomas Davenport* (Unpublished MS, Vermont Historical Society, Montpelier).

43. William Ritchie, "Experimental Researches in Electro-magnetism and Magneto-electricity," *Philosophical Transactions* 123 (1833): 313–21; William Sturgeon, "Description of an Electro-magnetic Engine for Turning Machinery," *Annals of Electricity* 1 (1836–37): 75–78.

44. "Electrical Society," *Literary Gazette* 22 (1838): 458.

45. Previous chapter.

46. Noad, *Eight Lectures*, on pp. 381–382.

47. For a survey of the pies in which Whewell had a finger, see Fisch and Schaffer, *William Whewell*.

48. William Whewell, *History of the Inductive Sciences*, 3 vols. (London, 1837). The discussion of the mechanicochemical sciences appears in the first chapter of vol. 3.

49. For some of Whewell's worries on galvanism, see Schaffer, "History and Geography," pp. 201–31 n. 47.

50. Pollock is identified as a chemist with an establishment on Fenchurch Street, in Walker, "Inaugural Address," n. 18. Walker noted that Pollock was present at every meeting of the Electrical Society, serving as chairman of the final committee meeting.

51. "Electrical Society," *Literary Gazette* 21 (1837): 753–54.

52. "Electrical Society," *Literary Gazette* 21 (1837): 434.

53. "Electrical Society," *Literary Gazette* 21 (1837): 723.

54. For a full analysis of this historical episode, see Secord, "Extraordinary Experiment."

55. See Andrew Crosse, "Description of Some Experiments Made with the Voltaic Battery for the Purpose of Producing Crystals; in the Process of Which Experiments Certain Insects Constantly Appeared," *Transactions and Proceedings of the London Electrical Society* 1 (1841): 10–16.

56. Oliver Stallybrass, "How Faraday 'Produced Living Animalculae': Andrew Crosse and the Story of a Myth," *Proceedings of the Royal Institution* 41 (1967): 597–619.

57. MBLES, 7 April 1842.

58. Crosse, "Description"; Robert Were Fox, "On the Results of Mr. Fox's Experiments on the Production of Artificial Crystals by Voltaic Action," *Philosophical Magazine* 10 (1837): 171.

59. Alfred Smee, *Elements of Electro-biology, or the Voltaic Mechanism of Man; of Electro-pathology, Especially of the Nervous System, and of Electrotherapeutics* (London, 1849). For an account of similar "mimetic" experiments, see Galison and Assmus, "Artificial Clouds."

60. "Electrical Society," *Literary Gazette* 24 (1840): 184–85, on p. 185.

61. "Electrical Society," *Literary Gazette* 21 (1837): 530.

62. "Electrical Society," *Literary Gazette* 22 (1838): 392–93, on p. 393.

63. "Electrical Society," *Literary Gazette* 22 (1838): 54–56, on p. 54.

64. "Electrical Soirée," *Literary Gazette* 26 (1842): 295–96, on p. 296.

65. "Electrical Soirée," *Literary Gazette* 27 (1843): 352.

66. Robert Brain, *Going to the Fair* (Cambridge: Whipple Museum, 1993); Richards, *Commodity Culture*.

67. The *Literary Gazette* compared Gassiot with the astronomer Lord Rosse and (interestingly) Andrew Crosse in this regard: "No private individual in Great Britain stands higher in this respect than Mr. Gassiot; he ranks with Mr. Cross [*sic*] and Lord Rosse, the former an electrician, the latter a practical mechanic and chemist on a gigantic scale." "Electrical Soireé," p. 352.

68. Charles V. Walker, "An Account of Some Experiments with a Constant Voltaic Battery," *Transactions and Proceedings of the London Electrical Society* 1 (1841): 57–65. It should be noted that even at a cost of a shilling per cell, a battery of 160 cells would be well beyond the means of most electricians.

69. Ibid., p. 58.

70. George H. Bachhoffner, "On a Simple Voltaic Battery, in a Letter to William Sturgeon, Esq., &c., &c.," *Annals of Electricity* 1 (1836–37): 213–15.

71. John Shillibeer, "Description of a New Arrangement of the Voltaic Battery and Pole Director," *Annals of Electricity* 1 (1836–37): 224–25, on p. 225.

72. Walker, "Constant Voltaic Battery," p. 57.

73. Gassiot to Faraday, 3 January 1839, in Williams, *Selected Correspondence*, 1:328. See also *Correspondence*, 2:539–40.

74. J. F. Daniell, "On Voltaic Combinations," *Philosophical Transactions* 126 (1836): 107–24.

75. Papers from the *Philosophical Transactions* were frequently reprinted in other journals such as the *Philosophical Magazine*, which certainly facilitated their circulation. This did not happen in this case, however.

76. Walker, "Constant Voltaic Battery," p. 58.

77. Walker's description of the Voltameter specified the dimensions of the electrodes and the concentration of the electrolyte. His description of the galvanometer used specified the length and composition of the needle as well as the dimensions of the coil.

78. Morus, "Sociology of Sparks"; Schaffer, "Late Victorian Metrology"; Latour, *Science in Action*.

79. Walker, "Constant Voltaic Battery," p. 64.

80. Ibid., p. 65.

81. John P. Gassiot, "On a Remarkable Difference in the Heat Attained by the Electrodes of a Powerful Constant Battery," *Philosophical Magazine* 13 (1838): 436–37.

82. For Silliman, see George P. Fisher, *The Life of Benjamin Silliman, M.D., LL.D.*, 2 vols. (New York, 1866). For Silliman's *Journal*, see Simon Baatz, "Squinting at Silliman: Scientific Periodicals in the Early American Republic," *Isis* 82 (1991): 223–44.

83. William Sturgeon, "Miscellaneous Notices on Galvanic Results, in Letters Addressed to Prof. Silliman, October 4, 1838, and August, 6, 1839, from the Vicinity of London," *American Journal of Science* 39 (1840): 28–38.

84. Ibid., p. 31. My italics.

85. Ibid., p. 32. My italics.

86. Charles V. Walker, "Protest of Mr. Charles V. Walker," *American Journal of Science* 42 (1842): 383–86.

87. Ibid., p. 384.

88. Ibid.

89. Ibid., p. 385.

90. For witnesses, see Shapin and Schaffer, *Leviathan and the Air-Pump*.

91. Although he did have similar problems in the early 1820s. See chapter 1 and Morus, "Different Experimental Lives."

92. For Faraday's relationship with Anderson, see Ginn, "Philosophers and Artisans," chap. 6; Gladstone, *Faraday*, p. 47.

93. For salutary tales of the relinquishing of control in the laboratory, see Nye, "N-Rays"; Malcolm Ashmore, "The Theatre of the Blind: Starring a Promethean Prankster, a Phoney Phenomenon, a Prism, a Pocket, and a Piece of Wood," *Social Studies of Science* 23 (1993): 67–106.

94. For a fascinating account of another attempt to construct such a populist science culture during the first half of the nineteenth century, see Cooter, *Cultural Meaning of Popular Science*.

95. Kargon, *Science in Victorian Manchester*, on pp. 36–41.

96. In Sturgeon, *Scientific Researches*, on p. 48, he states that he withdrew his name from the list of members shortly after 4 December 1838, but gives no further explanation.

97. MBLES, 24 March 1841.

98. There is a report of the meeting in "London Electrical Society," *Literary Gazette* 25 (1841): 218.

99. MBLES, 20 July 1841.

100. Ibid., 7 April 1842.

101. Ibid., 19 July 1842.

102. Ibid., 8 April 1843.

103. William Robert Grove, "On a Voltaic Process for Etching Daguerreotype Plates," *Philosophical Magazine* 20 (1842): 18–24.

104. William Snow Harris to Michael Faraday, 26 October 1842, in Williams, *Selected Correspondence*, pp. 405–6 n. 74.

105. Charles V. Walker, "On the Action of Lightning Conductors," *Proceedings of the London Electrical Society* 2 (1843): 342–56.

106. Charles V. Walker, "Address to the Reader," *Electrical Magazine* 1 (1845). Walker's remarks are worth repeating: "The Publication of the *Proceedings of the London Electrical Society* having ceased, we were strongly urged by many, who were in the habit of reading them, against allowing the numerous channels of information, which had been opened out to us, to be dried up; men who were more acquainted with the under-working of that current, which was as yet but imperfectly represented in those pages, were aware that our resources had grown around us, and that our machinery was, as it were, in good working order for enabling us to be an organ by which the public might keep pace with the rapid advance of Electrical Science. The reiterated entreaties of our friends, and our own experience of the inconveniences attending the want of some such general magazine—a magazine wholly devoted to Electricity, and moving on with the swelling tide—have induced us to continue our intimacy with the lovers of Electricity, and again to exercise our humble abilities in furtherance of that science."

107. Gooding, "In Nature's School."

CHAPTER 5
THE RIGHT ARM OF GOD

1. William Sturgeon, *A Course of Twelve Elementary Lectures on Galvanism* (London, 1843).

2. For a recent discussion of the dispute between the two Italian philosophers, see Marcello Pera, *The Ambiguous Frog: The Galvani-Volta Controversy on Animal Electricity*, trans. Jonathan Mandelbaum (Princeton: Princeton University Press, 1992).

3. For the Royal Victoria Gallery, see Kargon, *Science in Victorian Manchester*.

4. William Sturgeon, *Lectures on Electricity, Delivered in the Royal Victoria Gallery* (London, 1842).

5. See, for example, [Edward Palmer], *Palmer's New Catalogue, with Three Hundred Engravings of Apparatus Illustrative of Chemistry, Pneumatics, Frictional and Voltaic Electricity, Elctro-magnetism, Optics &c.* (London, 1840).

6. Sturgeon, *Galvanism*, p. 4.

7. Ibid., p. 11.

8. Ibid., p. 18.

9. For Aldini, see Bern Dibner, "Aldini, Giovanni," in *Dictionary of Scientific Biography*, 1:107–8.

10. For an account of Volta's visit and of French responses to the voltaic pile, see Geoffrey Sutton, "The Politics of Science in Early Napoleonic France: The Case of the Voltaic Pile," *Historical Studies in the Physical Sciences* 11 (1981): 329–66.

11. "Abstract of the Late Experiments of Professor Aldini on Galvanism,"*Journal of Natural Philosophy* 3 (1802): 298–300.

12. Ibid., pp. 298–99.

13. Ibid., p. 299.

14. "Galvanism,"*Philosophical Magazine* 14 (1802): 364–68; "Galvanische Versuche, Angestellt am Körper eines Gehängten zu London am 17ten Jan. 1803 vom Professor J. Aldini," *Annalen der Physik* 18 (1804): 340–42.

15. "Galvanism," p. 367.

16. Ibid.

17. "Aldini on Galvanism," on p. 298.

18. Ruth Richardson, *Death, Dissection and the Destitute* (Harmondsworth: Penguin, 1988).

19. Andrew Ure, "An Account of Some Experiments Made on the Body of a Criminal Immediately after Execution, with Physiological and Practical Observations," *Quarterly Journal of Science* 6 (1819): 283–94.

20. Ibid., p. 289, p. 290, p. 290.

21. Ibid., p. 294.

22. For Davy and the battery, see Golinski, "Humphry Davy." For Davy more generally, see Golinski, *Science as Public Culture*.

23. Ure, "Experiments," p. 290.

24. Andrew Ure, *The Philosophy of Manufactures* (London, 1835).

25. Francis Maceroni, *Useful Knowledge for the People* (London, 1832).

26. Francis Maceroni, *Memoirs of the Life and Adventures of Colonel Maceroni*, 2 vols. (London, 1838). *DNB*, 12:514–12.

27. Francis Macerone [*sic*], "An Account of Some Remarkable Electrical Phenomena Seen in the Mediterranean, with Some Physiological Deductions," *Mechanics' Magazine* 15 (1831): 93–96, 98–100, on p. 94.

28. Ibid., p. 95.

29. Ibid.

30. W. H. W[eekes], "Economical and Convenient Method of Constructing Mercurial Troughs," *Mechanics' Magazine* 23 (1835): 393–94. For some background on Weekes, see Alan Twyman, *In Search of the Mysterious Dr. Weekes. A Fragment of Sandwich History* (Dover: Adams & Sons, 1988).

31. Ibid., p. 394.

32. Thomas Pine, "On the Probable Connexion between Electricity and Vegetation," *Mechanics' Magazine* 24 (1835–36): 99–104, on p. 99.

33. Ibid., p. 99.

34. Ibid., p. 100. Weekes had a long-standing interest in atmospheric electricity. A few years later as a member of the London Electrical Society he would submit reports on the electrical state of the atmosphere at every session.

35. Ibid., p. 101.

36. Ibid.

37. Ibid., p. 102.

38. Thomas Pine, "Electro-Vegetation," *Mechanics' Magazine* 24 (1835–36): 402–05. Sturgeon's letter is quoted on p. 403.

39. Richard Phillips, "A Brief Account of a Visit to Andrew Crosse, Esq.," *Annals of Electricity* 1 (1836–37): 135–45.

40. Ibid., p. 139.

41. For a brief account, see Morrell and Thackray, *Gentlemen of Science*, pp. 457–58.

For more details, see Secord, "Extraordinary Experiment."

42. Phillips, "Brief Account," p. 135. Phillips noted that Crosse's apparatus had cost almost three thousand pounds.

43. See previous chapter.

44. For biographical details, see *DNB*, 7:573. For information concerning the Fox family's extensive social circle, see Horace N. Pym, ed., *Memories of Old Friends: Being Extracts from the Journals and Letters of Caroline Fox*, 3d ed., 2 vols. (London, 1883).

45. Robert W. Fox, "On the Temperature of Mines," *Annals of Philosophy* 4 (1822): 440–48.

46. Robert W. Fox, "Theoretical Views of the Origin of Mineral Veins," *Annals of Electricity* 2 (1838): 166–94, on p. 175.

47. Ibid., p. 192.

48. Ibid., p. 193.

49. Ibid., p. 194.

50. William Sturgeon, "A General Outline of the Various Theories Which Have Been Advanced for the Explanation of Terrestrial Magnetism," *Annals of Electricity* 1 (1836–37): 117–123, on p. 123.

51. Sturgeon, "Electro-magnetical Experiments"; Sturgeon, "Electro- and Thermo-magnetical Experiments"; Sturgeon, "On Electromagnetism."

52. Sturgeon, "General Outline," on p. 123. See also Sturgeon, "On Electro-pulsations and Electro-momentum." The relationship between Sturgeon's experiments and his electrical cosmology is discussed at more length in chapter 2.

53. Maceroni, "Remarkable Electrical Phenomena," p. 95, referred to the earth as the "great PAN, or *whole*."

54. Fox, "Theoretical Views," p. 194.

55. Thomas S. Mackintosh, "Electrical Theory of the Universe," *Mechanic's Magazine* 24 (1835–36): 11–13, on p. 11.

56. Ibid., p. 11.

57. Thomas S. Mackintosh, "Electrical Theory of the Universe," *Mechanics' Magazine* 24 (1835–36): 227–34, on p. 228.

58. Ibid., p. 228.

59. Ibid.

60. Ibid., p. 234.

61. Thomas S. Mackintosh, "Electrical Theory of the Universe—Article III," *Mechanics' Magazine* 24 (1835–36): 419–30.

62. Ibid., p. 430.

63. Ibid.

64. Ibid.

65. Ursa Major, "Mackintosh's Electrical Theory of the Universe," *Mechanics' Magazine* 25 (1836): 92–94, on p. 93.

66. Kinclaven, "The Newtonian Philosophy," *Mechanics' Magazine* 25 (1836): 99–101, on p. 100.

67. Zeta, "Mackintosh's Electrical Theory of the Universe," *Mechanics' Magazine* 25 (1836): 148–49, on p. 149.

68. Ibid., p. 149.

69. Thomas S. Mackintosh, "On the Electrical Theory of the Universe—Mr. Mackintosh's Reply to Zeta," *Mechanics' Magazine* 24 (1836): 274–80, on p. 275.

70. Ibid., p. 275.

71. Thomas Pine, "On the First Excitement and Germination of Plants, with Remarks on the Franklinean Theory and That of Mr. Mackintosh," *Mechanics' Magazine* 26 (1837): 102–6.

72. Francis Maceroni, "On Mackintosh's Electrical Theory of the Universe, and Outlines of Another Theory," *Mechanics' Magazine* 26 (1837): 19–25, on p. 21.

73. "Mackintosh's Electrical Theory of the Universe," *New Moral World* 3 (1836–37): 239.

74. "Mr. Mackintosh's Lectures on His Electrical Theory of the Universe," *Mechanics' Magazine* 27 (1837): 9–10.

75. "Progress of Social Reform," *New Moral World* 4 (1837–38): 276.

76. "Progress of Social Reform," *New Moral World* 4 (1837–38): 353.

77. "Progress of Social Reform," *New Moral World* 4 (1837–38): 397–98.

78. Thomas S. Mackintosh, *Mackintosh's Electrical Theory of the Universe, of the Elements of Natural Philosophy, Combining Physics and Morals in One Science and Deducing the Actions and Reactions of Mind and Matter, from the Ultimate and Universal Forces of Attraction and Repulsion* (Manchester, n.d.).

79. Thomas S. Mackintosh, *The Electrical Theory of the Universe, or the Elements of Physical and Moral Philosophy*, 1st American ed., republished from British ed. of 1838 (Boston, 1846), p. v.

80. Ibid., p. 99.

81. Ibid., pp. 360–61.

82. Ibid., p. 371.

83. William Leithead, *Electricity: Its Nature, Operation, and Importance in the Phenomena of the Universe* (London, 1837). Leithead advo-

cated his views at meetings of the London Electrical Society as well. See previous chapter.

84. Such accounts were not universally welcomed. A review in the *Philosophical Magazine*, for example, dismissed the book as "neither more nor less than an electrical dream of Mr. Leithead's fancy." He was also accused of plagiarizing much of the text from Peter Mark Roget. "Reviews and Notices of New Books," *Philosophical Magazine* 12 (1838): 127–30.

85. James Murray, *Electricity as a Cause of Cholera, or Other Epidemics, and the Relation of Galvanism to the Action of Remedies* (Dublin, 1849).

86. Secord, "Extraordinary Experiment." Crosse's crystals are also discussed in the previous chapter.

87. Andrew Crosse, "On the Production of Insects by Voltaic Electricity," *Annals of Electricity* 1 (1836–37): 242–44, on pp. 242–43.

88. Secord, "Extraordinary Experiment."

89. Cornelia Crosse, *Memorials, Scientific and Literary, of Andrew Crosse, the Electrician* (London, 1857), p. 170.

90. Pym, *Memories of Old Friends*, p. 10.

91. Crosse, *Memorials*, p. 246.

92. Ibid., p. 247.

93. "The New Frankenstein," *Fraser's Magazine* 17 (1837): 21–30.

94. Ibid., p. 23.

95. Chris Baldick, *In Frankenstein's Shadow: Myth, Monstrosity and Nineteenth-Century Writing* (Oxford: Clarendon Press, 1987).

96. Francis Maceroni, "Production of Life by Galvanism—Animation of Horse Hairs!" *Mechanics' Magazine* 26 (1837): 441.

97. "Miscellaneous Communications," *Analyst* 8 (1838): 162–66.

98. "A Visit to Mr. Crosse, at His Residence on the Quantock Hills, Somerset," *Lancet* 1 (1836–37): 49–52. The article is mainly a long quotation from Sir Richard Phillips's letter, with a final editorial comment. See n. 39.

99. E. J. Hytch, "Mr. Crosse's Revivication of Insects Contained in Flint," *Lancet* 1 (1836–37): 710–11.

100. "Accidental Production of Animal Life—Mr. Crosse," *Magazine of Popular Science* 3 (1837): 145–48.

101. Henry M. Noad, *Lectures on Electricity, Comprising Galvanism, Magnetism, Electro-magnetism, Magneto- and Thermo-electric-*

ity (London, 1844); William H. Weekes, "Details of an Experiment in Which Certain Insects, Known as the Acarus Crossi, Appeared Incident to the Long-continued Operation of a Voltaic Current upon Silicate of Potassa, within a Close Atmosphere over Mercury," *Proceedings of the London Electrical Society* 1 (1842): 240–56.

102. MBLES, 7 April 1842. Institute of Electrical Engineers.

103. Quoted in Stallybrass, "How Faraday 'Produced Living Animalculae,'", from the *Times*, 4 March, 1837.

104. Secord, "Extraordinary Experiment."

105. *ERE*, 15th ser., 2:1–17. The paper was read to the Royal Society on 6 December 1838.

106. Ibid., p. 1.

107. Ibid., p. 12.

108. Ibid., p. 15.

109. Bence Jones, *Life and Letters*, 2:441n. The incident is also referred to in Gladstone, *Faraday*, pp. 150–51.

110. Iwan Rhys Morus, "Marketing the Machine: The Construction of Electrotherapeutics as Viable Medicine in Early Victorian England," *Medical History* 36 (1992): 34–52.

111. Willem Hackmann, "The Induction Coil in Medicine and Physics, 1835–1877," in *Studies in the History of Scientific Instruments*, ed. Christine Blondel, Françoise Parot, Anthony Turner, and Mari Williams (London: Roger Turner, 1989), pp. 235–50.

112. Clarke, "Description."

113. *Palmer's New Catalogue*, p. 46.

114. Sturgeon, *Galvanism*, p. v.

115. Crosse, *Memorials*, p. 60.

116. Joel Wiener, *Radicalism and Freethought in Nineteenth-Century Britain: The Life of Richard Carlile* (Westport, CT: Greenwood Press, 1983), pp. 251–52.

117. William H. Halse, "Wonderful Effects of Voltaic Electricity in Restoring Animal Life When the Sensorial Powers Have Entirely Ceased, or in Other Words, When Death in the Common Adaptation of the Term Has Actually Occurred," *Annals of Electricity* 4 (1840): 481–84.

118. William Hooper Halse, "The Power of Galvanism in Cases of Suspended Animation," *New Moral World* 9 (1841): 130.

119. William H. Halse, *The Extraordinary Remedial Efficacy of Medical Galvanism, When Scientifically Administered* (London, n.d.)

120. "Grand Display of Animal Magnetism!" broadsheet, Bakken Library and Museum of Electricity in Life, Minneapolis (dated 1843).

121. Sally Shuttleworth, "Female Circulation: Medical Discourse and Popular Advertising in the Mid-Victorian Era," in *Body/Politics: Women and the Discourses of Science*, ed. Mary Jacobus, Evelyn Fox Keller, and Sally Shuttleworth (London: Routledge, 1990), pp. 47–68 makes the point (correctly) in a specifically gendered context.

122. Halse, *Medical Galvanism*, p. 1.

123. Ibid., p. 6.

124. Ibid., p. 10.

125. For this aspect of "quack" culture, see Roy Porter, *Health for Sale: Quackery in England 1660–1850* (Manchester: Manchester University Press, 1989), pp. 228–30.

126. Halse, *Medical Galvanism*, p. 17.

127. Golding Bird, "On the Employment of Electro-magnetic Currents in the Treatment of Paralysis," *Lancet* 1 (1846): 649–51, on p. 649.

128. Golding Bird, *Lectures on Electricity and Galvanism, in Their Physiological and Therapeutical Relations* (London, 1849), p. 148.

129. Ibid., p. 163.

130. *DNB*, 18:398–99.

131. For the dispute, see Owsei Temkin, "Basic Science, Medicine, and the Romantic Era," in his *The Double Face of Janus and Other Essays in the History of Medicine* (Baltimore: Johns Hopkins University Press, 1977), pp. 345–72; Desmond, *Politics of Evolution*.

132. Shelley, *Frankenstein*, pp. xv–xx.

133. Smee, *Electro-biology*.

134. E.M. O[dling], *Memoir of the Late Alfred Smee, F.R.S.* (London, 1878).

135. Morus, "Different Experimental Lives."

136. Smee, *Electro-biology*, app.: "Electrobiological Map No. 1."

137. Alfred Smee, *Instinct and Reason: Deduced from Electro-biology* (London, 1850).

138. Ibid., pp. 245–46.

139. Ibid., p. 274.

140. Ibid.; Alfred Smee, *Principles of the Human Mind* (London, 1849), pp. 15–16.

141. *Literary Gazette* 33 (1849): 845.

142. "Elements of Electro-biology," *London Journal of Medicine* 1 (1849): 478–79, on p. 478.

143. [William B. Carpenter], "Smee and Wiglesworth on Physical Biology," *British and Foreign Medico-Chirurgical Review* 4 (1849): 371–82.

144. Ibid., p. 372.

145. Ibid., pp. 376–37.

146. "Instinct and Reason," *British and Foreign Medico-Chirurgical Review* 6 (1850): 522–24.

147. Ibid., p. 522.

148. Ibid., p. 524.

149. Alfred Smee, "Dr. Carpenter and the British and Foreign Medical Review," *Lancet* 2 (1850): 514.

150. Morus, "Marketing the Machine."

151. For "medical heretics," see Logie Barrow, "Why Were Most Medical Heretics at Their Most Confident around the 1840s? (the Other Side of Mid-Victorian Medicine)," in *British Medicine in an Age of Reform*, ed. Roger French and Andrew Wear (London: Routledge, 1991), pp. 165–85.

152. Alison Winter, "The Island of Mesmeria: The Politics of Mesmerism in Early Victorian Britain" (Ph.D. diss., University of Cambridge, 1992).

153. Location and the status of participants often determines the status of phenomena. See Schaffer, "Natural Philosophy and Public Spectacle"; Schaffer, "Self Evidence."

INTRODUCTION TO PART TWO

1. Gooding, "In Nature's School."

2. Patrick Joyce, *Democratic Subjects: The Self and the Social in Nineteenth Century England* (Cambridge: Cambridge University Press, 1994); Dror Wahrman, *Imagining the Middle Class: The Political Representation of Class in Britain, c. 1780–1840* (Cambridge: Cambridge University Press, 1995).

3. Morus, "Different Experimental Lives."

4. Peter Mathias, *The First Industrial Nation: The Economic History of Britain 1700–1914*, 2d ed. (London: Routledge, 1983).

5. The most obvious examples are the *Athenaeum* and the *Literary Gazette*.

6. William Whewell, "On the General Bearing of the Great Exhibition on the Progress of Art and Science," *Lectures on the Progress of Arts and Science, Resulting from the Great Exhibition in London* (London, 1856), pp. 3–25; Babbage, *Reflexions*; Charles Babbage, *The Exposition of 1851* (London, 1851).

7. Charles Babbage, *On the Economy of Machinery and Manufactures* (London, 1832); Ashworth, "Calculating Eye."

8. Schaffer, "Babbage's Intelligence."

9. Berg, *Machinery Question*.

10. Babbage, *Economy*.

11. Ibid., p. 174, p. 27, p. 54.

12. Schaffer, "Babbage's Intelligence."

13. Peter Barlow, *A Treatise on the Manufactures and Machinery of Great Britain* (London, 1836), p. 91.

14. Ibid., pp. 122–26.

15. Ure, *Philosophy of Manufactures*, pp. 13–14.

16. Peter Gaskell, *Artisans and Machinery* (London, 1836).

17. Ibid., pp. 357–58.

18. Babbage, *Economy*, p. 66.

19. Raymond Williams, *Culture and Society* (Harmondsworth: Penguin, 1961), pp. 85–98.

20. Baldick, *Frankenstein's Shadow*, esp. pp. 92–120.

21. Thomas Carlyle, "Chartism," in *Critical and Miscellaneous Essays, Collected Works*, 17 vols. (London, 1885–91), 13:24.

22. From "Sartor Resartus," quoted in Baldick, *Frankenstein's Shadow*, p. 105.

23. "The New Frankenstein."

24. Carlyle, *Latter-Day Pamphlets*, quoted in Baldick, *Frankenstein's Shadow*, p. 105.

25. Asa Briggs, *Victorian Things* (Harmondsworth: Penguin, 1990); Carolyn Marvin, *When Old Technologies Were New. Thinking about Electric Communication in the Late Nineteenth Century* (Oxford: Oxford University Press, 1988).

26. Percy Dunsheath, *A History of Electrical Engineering* (London: Faber & Faber, 1962); Robert M. Black, *The History of Electric Wires and Cables* (London: Peter Peregrinus, 1983).

27. Morrell and Thackray, *Gentlemen of Science*.

28. Cardwell, *Joule*.

29. For example, Jeffrey L. Kieve, *The Electric Telegraph: A Social and Economic History* (Newton Abbott: David & Charles, 1973).

30. For the technology of display, see Morus, "Currents."

31. Much of the chapter is based on Morus, "Marketing the Machine."

CHAPTER 6
THEY HAVE NO RIGHT TO LOOK FOR FAME

1. For the problematics of discovery, see Augustine Brannigan, *The Social Basis of Scientific Discoveries* (Cambridge: Cambridge University Press, 1981); Simon Schaffer, "Scientific Discovery and the End of Natural Philosophy," *Social Studies of Science* 16 (1986): 387–420.

2. For gentlemen, see Walter F. Cannon, "John Herschel and the Idea of Science," *Journal of the History of Ideas* 22 (1961): 215–39; Susan F. Cannon, *Science in Culture*; Morrell and Thackray, *Gentlemen of Science*; Marie Boas-Hall, *All Scientists Now* (Cambridge: Cambridge University Press, 1984). For a critique of the notion of gentlemanly homogeneity, see Pietro Corsi, *Science and Religion: Baden Powell and the Anglican Debate, 1800–1860* (Cambridge: Cambridge University Press, 1988); Boyd Hilton, *The Age of Atonement: The Influence of Evangelicalism on Social and Economic Thought, 1785–1865* (Oxford: Oxford University Press, 1988).

3. Henry Dutton, *The Patent System and Inventive Activity during the Industrial Revolution, 1750–1852* (Manchester: Manchester University Press, 1984), p. 2.

4. Christine MacLeod, *Inventing the Industrial Revolution: The English Patent System, 1660–1800* (Cambridge: Cambridge University Press, 1988), pp. 58–74.

5. Dutton, *Patent System*, pp. 69–85. See also Christine MacLeod, "Concepts of Invention and the Patent Controversy in Victorian Britain," in *Technological Change: Methods and Themes in the History of Technology*, ed. Robert Fox (London: Harwood Academic Publishers, 1995), pp. 137–53.

6. Dutton, *Patent System*, p. 110.

7. Ibid., p. 59.

8. "Preface," *Mechanics' Magazine* 11 (1829): iv.

9. Thomas Hodgskin, *Natural and Artificial Rights to Property Contrasted* (London, 1832), pp. 180–81.

10. [William Robert Grove], "Physical Science in England," *Blackwood's Magazine* 54 (1843): 514–25, on p. 521.

11. William Sturgeon, "Electrical Curiosities," *Annals of Electricity* 7 (1841): 377–81.

12. For an account of telegraphy's emergence from the laboratory, see Iwan Rhys Morus, "Telegraphy and the Technology of Display: The Electricians and Samuel Morse," *History of Technology* 13 (1991): 20–40.

13. W. James King, *The Development of Electrical Technology in the Nineteenth Century. The Arc-Light and Generator* (Washington, DC: Smithsonian Institution Press, 1964).

14. See Gooding, "He Who Proves Discov-

ers." Herschel suggested that the battery had been exhausted by previous experimenters before he could take his turn at using it. For an account of some efforts to overcome the problem, see Morus, "Sociology of Sparks."

15. Bachhoffner, "Simple Voltaic Battery."

16. Recall the collaborative experiment on battery assessment conducted under the auspices of the London Electrical Society, discussed in chap. 4.

17. Daniell, "Voltaic Combinations."

18. Thomas Spencer, *Working in Metal by Voltaic Electricity* (London, n.d.).

19. Thomas Spencer, "An Account of Some Experiments Made for the Purpose of Ascertaining How Far Voltaic Electricity May Be Usefully Applied to the Purpose of Working in Metal," *Annals of Electricity* 4 (1839–40): 258–79, on p. 258.

20. Morrell and Thackray, *Gentlemen of Science*.

21. Actually, William Carpmael, a London patent agent, the secretary to the Mechanical Sciences section that year. See ibid., p. 306.

22. Spencer, "An Account of Some Experiments," p. 259.

23. Morrell and Thackray, *Gentlemen of Science*, pp. 306–7.

24. *Mechanics' Magazine* 33 (1840): 20.

25. Note, *Annals of Electricity* 4 (1839–40): 277.

26. Spencer, *Working in Metal*; "An Account of Some Experiments."

27. Spencer, "An Account of Some Experiments."

28. "Novel Invention," *London Journal of Arts and Sciences* 15 (1840): 245–48; Thomas Spencer, "On the Mode of Producing Fac-Simile Copies of Medals, &c.—By the Agency of Voltaic Electricity," *London Journal of Arts and Sciences* 15 (1840): 306–11; "Electrotype," *London Journal of Arts and Sciences* 15 (1840): 101–3.

29. Samuel Cartwright, "On Electro-type from Engraved Copperplates," *Annals of Electricity* 5 (1840): 236–38; "On the Cultivation and Growth of Electrotypes," *Annals of Electricity* 5 (1840): 484–85.

30. Spencer, "An Account of Some Experiments," p. 261.

31. Ibid., p. 262.

32. Thomas Spencer, *Instructions for the Multiplication of Works of Art in Metal by Voltaic Electricity* (Glasgow, 1840), p. iv.

33. Ibid.

34. Alfred Smee, *Elements of Electrometallurgy*, 3d ed. (London, 1844), p. 351. For a full description of the Elkington patent, see *London Journal of Arts and Sciences* 19 (1842): 83–88.

35. William Ryland, "The Plated Wares and Electroplating Trades," in *The Resources, Products, and Industrial History of Birmingham and the Midland Hardware District*, ed. Samuel Timmins (London, 1866; reprint, Cass Library of Industrial Classics, 1967), pp. 477–98, on p. 490.

36. "Royal Institution," *Literary Gazette* 25 (1841): 73. This was a report on a Friday Evening Discourse on electrometallurgy, 19 March 1841.

37. Ryland, "Plated Wares," p. 490.

38. Smee, *Electrometallurgy*, pp. 351–52. Also, Sturgeon, "Electrical Curiosities," pp. 380–81.

39. *London Journal of Arts and Sciences* 20 (1842): 166–71.

40. Ibid., pp. 171–72.

41. Alfred Smee, *Elements of Electro-metallurgy, or the Art of Working in Metals by the Galvanic Fluid* (London, 1841), p. 147.

42. George Gore, "On the Relation of Science to Electro-plate Manufacture, Part I," *Popular Science Review* 1 (1862): 327–31, on p. 329.

43. Ryland, "Plated Wares," p. 491.

44. Ibid., p. 493.

45. Ibid., p. 491.

46. Ibid., p. 495. For a discussion of the large-scale survival of small-scale workshops well into the second half of the nineteenth century, see Raphael Samuel, "Workshop of the World: Steam Power and Hand Technology in Mid-Victorian Britain," *History Workshop* 3 (1977): 6–72.

47. There is a massive literature on the middle-class litany of science as self-improvement for the lower orders. See, for example, J. N. Hays, "Science and Brougham's Society," *Annals of Science* 20 (1964): 227–41; Steven Shapin and Barry Barnes, "Science, Nature and Control: Interpreting Mechanics' Institutes," *Social Studies of Science* 7 (1977): 31–74.

48. Ryland, "Plated Wares," p. 494.

49. Ure, *Philosophy of Manufactures*, p. 301.

50. George Head, *A Home Tour of the Manufacturing Districts of England in the Summer of 1835* (London, 1836); William Cooke Taylor, *A*

*Tour of the Manufacturing Districts of Lancashire* (London, 1842).

51. George Shaw, *A Manual of Electro-metallurgy* (London, 1842); Smee, *Electrometallurgy*; Spencer, *Instructions*; Charles Vincent Walker, *Electrotype Manipulation* (London, 1843).

52. Walker, *Electrotype Manipulation*, pp. 5–6.

53. Ibid., p. 6.

54. Ibid.

55. On the industrialization of photography and its relationship to the electroplating industry, see M. Susan Bargos and William B. White, *The Daguerreotype: Nineteenth Century Technology and Modern Science* (Washington, DC: Smithsonian Institution, 1991), pp. 49–51.

56. George Fisher, *Photogenic Manipulations* (London, 1844).

57. Smee, *Electrometallurgy*.

58. Faraday, quoted in a report of a lecture at the Royal Institution, in the *Literary Gazette*, 2 February 1839. The report is quoted in Gail Buckland, *Fox Talbot and the Invention of Photography* (London: Scolar Press, 1980), p. 39.

59. Grove, "Voltaic Process."

60. "Etching Daguerreotype Plates by a Voltaic Process," *Literary Gazette* 25 (1841): 548–49.

61. Grove, "Voltaic Process," p. 24.

62. Patricia Anderson, *The Printed Image and the Transformation of Popular Culture, 1790–1860* (Oxford: Oxford University Press, 1991).

63. Celina Fox, "The Engravers' Battle for Professional Recognition in Early Nineteenth Century London," *London Journal* 2 (1976): 3–32; *Graphic Journalism in England during the 1830s and 1840s* (New York: Garland Press, 1988); Gordon Fyfe, "Art and Reproduction: Aspects of the Relations between Painters and Engravers in London, 1760–1850," *Media, Culture and Society* 7 (1985): 399–425.

64. This could have an impact for philosophers as well. See Alex Soojung-Kim Pang, "Victorian Observing Practices, Printing Technology and Representation of the Solar Corona: The 1860s and 1870s," *Journal of the History of Astronomy* 25 (1994): 249–74.

65. Kellow Chesney, *The Victorian Underworld* (Harmondsworth: Penguin, 1970), pp. 291–300.

66. Peter Quenell, *London's Underworld* (Selections from Henry Mayhew, *London*

*Labour and the London Poor*, vol. 4) (London: Spring Books, 1950), p. 315.

67. Smee, *Electrometallurgy*, p. 345.

68. Both Charles Babbage and Andrew Ure in their canonical works saw machines as self-acting. Babbage, *Economy*; Ure, *Philosophy of Manufactures*. For a discussion of self-action in machines and laboratory apparatus, see Morus, "Correlation and Control."

69. Spencer, *Instructions*, p. iii.

70. [Joseph Robertson], "Books on Electrometallurgy," *Mechanics' Magazine* 36 (1842): 458–62.

71. Ibid., p. 458.

72. Alfred Smee, *Elements of Electro-metallurgy, or the Art of Working in Metals by the Galvanic Fluid*, 2d ed. (London, 1842).

73. [Robertson], "Books," p. 458.

74. Ibid., p. 459.

75. Henry Dircks, "Contributions towards a History of Electro-metallurgy," *Mechanics' Magazine* 40 (1844): 72–79. The entire exchange was published by Dircks in book form many years later. For convenience all future references will be to this source: Henry Dircks, *Contribution towards a History of Electro-metallurgy* (London, 1863). For brief biographical details on Dircks, see *DNB*, 5:1000–1001. Dircks was later to gain fame as the inventor of an optical illusion, commonly known as Pepper's Ghost, which was on display at the Polytechnic Institution for many years.

76. Dircks, *Contribution*, p. 2.

77. C. J. Jordan, "Engraving by Galvanism," *Mechanics' Magazine* 31 (1839): 163–64.

78. Dircks, *Contribution*, pp. 9–19.

79. Letter from Dancer to Dircks, 17 June, 1840, quoted in Dircks, *Contribution*, pp. 12–15, on p. 15.

80. Ibid., p. 15.

81. Ibid., pp. 19–20.

82. Ibid., pp. 20–21.

83. It is completely unclear why Dircks chose to attack Spencer in this manner. He did not, as far as can be ascertained, have any direct financial interest in the matter. Neither, or at least so he claimed, was he acquainted with Jordan. He was, however, acquainted with Spencer, and it seems clear from the tone of the debate that a great deal of very personal animosity was involved.

84. Dircks, *Contribution*, pp. 23–24.

85. Ibid., pp. 26–27. Nadgett was a character in Charles Dickens, *Martin Chuzzlewit*.

86. It may have been. It was nevertheless common practice during the period. In any case the abstract had, as Spencer pointed out, already been published in the *Liverpool Standard*.

87. Dircks, *Contribution*, p. 29.

88. Ibid., p. 31.

89. For mesmerism's problematics, see Winter, "Island of Mesmeria."

90. Dircks, *Contribution*, p. 31.

91. Ibid., p. 41.

92. Golding Bird, "Observations on the Electro-chemical Influence of Long-continued Electric Currents of Low Tension," *Philosophical Transactions* 128 (1837): 37–45.

93. Dircks, *Contribution*, p. 42.

94. Ibid., p. 50.

95. Ibid., p. 56, p. 71.

96. Smee, *Electrometallurgy*, 3rd ed., pp. xviii–xix.

97. [Robertson], "Books," p. 459.

98. "Royal Institution," *Literary Gazette* 26 (1842): 366.

99. Grove, *Progress*, p. 24.

100. Hays, "Science in the City."

101. Smee, *Electrometallurgy*, 3d ed., p. 348.

102. For Gassiot's soirees, see previous chapter. Also "Electrical Soirée," *Literary Gazette* 27 (1843): 352.

103. Noad, *Eight Lectures*, pp. 381–82.

104. See Secord, "Extraordinary Experiment"; Stallybrass, "How Faraday 'Produced Living Animalculae.'"

105. *London Journal of Arts* 31 (1847): 195–97, on pp. 195–96.

106. *London Journal of Arts* 32 (1848): 201–4.

107. "Staite's Improvements in Lighting, etc.," *Patent Journal and Inventor's Magazine* 4 (1848): 169–73.

108. William Robert Grove, "On the Application of Voltaic Ignition to Lighting Mines," *Philosophical Magazine* 27 (1845): 442–48.

109. "Electric Light," *Patent Journal* 6 (1849): 80.

110. For the importance of illumination in the nineteenth century, see Wolfgang Schivelbusch, *Disenchanted Night: The Industrialization of Light in the Nineteenth Century* (Berkeley and Los Angeles: University of California Press, 1988).

111. *Illustrated London News* 14 (1849): 293 (5 May 1849); King, *Development*, p. 337.

112. The source of such claims are later-nineteenth-century apologists for Faraday, typically arguing for a recognition of the role of "pure" science in the making of industrial progress. Donald Cardwell, "On Michael Faraday, Henry Wilde and the Dynamo," *Annals of Science* 49 (1992): 479–87.

113. William Sturgeon, "Improved Electromagnetic Apparatus"; Joule, "Short Account"; Thompson, *Electromagnet*, has a brief biography of Sturgeon as an appendix and also contains much historical information on electromagnets in the first chapter.

114. Joseph Henry, "On the Application of the Principle of the Galvanic Multiplier to Electro-magnetic Apparatus, and Also to the Development of Great Magnetic Power in Soft Iron, with a Small Galvanic Element," *American Journal of Science* 19 (1831): 400–8; Joseph Henry and Dr. Ten Eyck, "An Account of a Large Electro-magnet, Made for the Laboratory of Yale College," *American Journal of Science* 20 (1831): 201–3. For Henry, see Thomas Coulson, *Joseph Henry: His Life and Work* (Princeton: Princeton University Press, 1950).

115. W. James King, *The Development of Electrical Technology in the Nineteenth Century: The Electrochemical Cell and the Electromagnet* (Washington, DC: Smithsonian Institution Press, 1962), pp. 259–60; Richie, "Experimental Researches"; Sturgeon, "Electro-Magnetic Engine for Turning Machinery."

116. Davenport, *Thomas Davenport*; Thomas Davenport, "Autobiography of Thomas Davenport" (unpublished MS, Vermont Historical Society, Montpelier).

117. For example, "Davenport and Cook's Electro-magnetic Engine," *Mechanics' Magazine* 27 (1837): 159–60.

118. "Davenport's Electro-magnetic Engine," *Mechanics' Magazine* 27 (1837): 404–5.

119. Cardwell, *Joule*; Kargon, *Science in Victorian Manchester*. For Joule's brewing, see Heinz Otto Sibum, "Reworking the Mechanical Value of Heat: Instruments of Precision and Gestures of Accuracy in Early Victorian England," *Studies in History and Philosophy of Science* 26 (1995): 73–106.

120. Sturgeon, "Fifth Memoir."

121. Morus, "Sociology of Sparks."

122. Wise, "Work and Waste"; Cardwell, *Joule*; Sibum, "Mechanical Value of Heat"; Iwan Rhys Morus, "Industrious People: Biogra-

phy and Nineteenth Century Physics," *Studies in the History and Philosophy of Science* 21 (1990): 519–25.

123. Crosbie Smith and Norton Wise, "Measurement, Work and Industry in Lord Kelvin's Britain," *Historical Studies in the Physical Sciences* 17 (1986): 147–73; Norton Wise, "Mediating Machines," *Science in Context* 2 (1988): 77–113.

124. James P. Joule, "On the Use of Electromagnets Made of Iron Wire for the Electromagnetic Engine," *Annals of Electricity* 4 (1839–40): 58–62.

125. Ibid., p. 59.

126. James P. Joule, "Investigations in Magnetism and Electro-magnetism," *Annals of Electricity* 4 (1839–40): 131–37, on p. 134. Italics in original.

127. Ibid., p. 135.

128. James P. Joule, "Description of an Electro-magnetic Engine," *Annals of Electricity* 4 (1839–40): 203–5.

129. James P. Joule, "On Electro-magnetic Forces," *Annals of Electricity* 4 (1839–40): 474–81.

130. Ibid., p. 477.

131. Ibid., p. 481.

132. Sibum, "Mechanical Value of Heat."

133. James P. Joule and William Scoresby, "Experiments and Observations on the Mechanical Powers of Electro-magnetism, Steam, and Horses," *Philosophical Magazine* 28 (1846): 448–55.

134. [William Robert Grove], "On the Progress Made in the Application of Electricity as a Motive Power," *Literary Gazette* 28 (1844): 892.

135. On Grove in this context, see Morus, "Correlation and Control."

136. "The Applicability of Electro-magnetism as a Moving Power," *Inventor's Advocate and Journal of Industry* 2 (1840): 410–11, on p. 410.

137. Ibid., p. 411.

138. Uriah Clarke, "Electro-magnetic Locomotive Carriage," *Annals of Electricity* 5 (1840): 304–5.

139. Thomas Wright, "On a New Electro-Magnetic Engine," *Annals of Electricity* 5 (1840): 108–10.

140. Thomas Wright, "On Electro-magnetic Coil Machines," *Annals of Electricity* 5 (1840): 349–52.

141. *Edinburgh Witness*, quoted in "Electro-

magnetic Locomotive," *Railway Times* 5 (1842): 1012.

142. Ibid.

143. "Electro-magnetic Locomotive," *Railway Times* 5 (1842): 1259–60.

144. "Electro-magnetic Exhibition," *Railway Times* 5 (1842): 1273.

145. L., "Electro-magnetic Power," *Railway Times* 5 (1842): 1342.

146. PRO BONO PUBLICO, "Electro-magnetic Power," *Railway Times* 6 (1843): 136.

147. "Taylor's Electro-magnetic Engine," *Mechanics' Magazine* 32 (1840): 693–96, on p. 694.

148. Altick, *Shows of London*.

149. "Taylor's Electro-magnetic Engine," p. 695.

150. [Grove], "Physical Science."

151. "Applicability of Electro-magnetism," p. 410.

152. Smee, *Electrometallurgy*, 3d ed., pp. 111–16.

153. [Grove], "Progress."

154. William Robert Grove, "Experiments on Voltaic Reaction," *Philosophical Magazine* 23 (1843): 223–26.

155. Grove, *Correlation*. See Morus, "Correlation and Control."

156. Joule and Scoresby, "Experiments and Observations."

157. See chap. 5. Also Wise, "Work and Waste"; "Mediating Machines."

CHAPTER 7
To Annihilate Time and Space

1. Daniel Headrick, *The Tools of Empire: Technology and European Imperialism in the Nineteenth Century* (Oxford: Oxford University Press, 1981); *The Invisible Weapon: Telecommunications and National Politics, 1851–1945* (Oxford: Oxford University Press, 1991).

2. Crosbie Smith and M. Norton Wise, *Energy and Empire: A Biographical Study of Lord Kelvin* (Cambridge: Cambridge University Press, 1989); Bruce Hunt, *The Maxwellians* (Ithaca: Cornell University Press, 1991).

3. Bruce Hunt, "Michael Faraday, Cable Telegraphy and the Rise of Field Theory, *History of Technology* 13 (1991): 1–19; "The Ohm Is Where the Art Is: British Telegraph Engineers and the Development of Electrical Standards," in *Instruments*, ed. Albert Van Helden and Thomas Hankins, Special Issue,

*Osiris* 9 (1994): 48–63: Schaffer, "Late Victorian Metrology."

4. W. J. Copleston, *Memoir of Edward Copleston, D.D., Bishop of Llandaff* (London, 1851), p. 169.

5. "Preface," *Patent Journal and Inventor's Advocate* 10 (1850): iii–iv, on p. iv.

6. Ibid., p. iv.

7. Wolfgang Schivelbusch, *The Railway Journey: the Industrialization of Time and Space in the Nineteenth Century* (Berkeley and Los Angeles: University of California Press, 1986).

8. Latimer Clark, "Inaugural Address," *Journal of the Society of Telegraph Engineers* 4 (1875): 1–22.

9. Ibid., p. 2.

10. Ibid., p. 3.

11. Morus, "Telegraphy."

12. Noad, *Lectures on Electricity*.

13. Clark, "Inaugural Address"; Charles Bright, "Address of the President," *Journal of the Society of Telegraph Engineers* 16 (1887): 7–40.

14. Bright, "Address," p. 17.

15. Mathias, *First Industrial Nation*, pp. 252–65.

16. Schivelbusch, *Railway Journey*.

17. C. H. Greenhow, *An Exposition of the Dangers and Deficiencies of the Present Mode of Railway Construction* (London, 1846), pp. 5–6.

18. Dionysius Lardner, *Railway Economy* (London, 1851), p. 421.

19. Latour, *Science in Action*.

20. Nathan Rosenberg, *Exploring the Black Box: Technology, Economics and History* (Cambridge: Cambridge University Press, 1994), discusses the "path-dependence" of new technologies: their dependence on previously established systems.

21. For a detailed overview of the partnership and the controversy, see Geoffrey Hubbard, *Cooke, Wheatstone and the Invention of the Electric Telegraph* (London: Routledge & Kegan Paul, 1965).

22. J. J. Fahie, *A History of the Electric Telegraph to the Year 1837* (London, 1884).

23. William Fothergill Cooke to his Mother, 5 April 1836, printed in Cooke, *Extracts*, p. 5.

24. Ibid.

25. Cooke to his Mother, 26 April 836, in Cooke, *Extracts*, p. 6.

26. The pamphlet was not printed at the time but was published in William Fothergill Cooke, *The Electric Telegraph: Was It Invented by Professor Wheatstone?* 2 vols. (London, 1856–57), 2:239–64.

27. Ibid., p. 240.

28. Ibid., p. 250.

29. Ibid., p. 251.

30. Ibid., pp. 251–52.

31. Ibid., p. 253.

32. Ibid., p. 255.

33. Ibid., p. 259.

34. Ibid., p. 254.

35. Cooke to his Mother, 6 June 1836, in Cooke, *Extracts*, p. 8.

36. Cooke to his Mother, 7 October 1836, in Cooke, *Extracts*, pp. 10–11.

37. Cooke to his Mother, 17 November 1836, in Cooke, *Extracts*, pp. 11–12.

38. Cooke to his Mother, 24 November 1836, in Cooke, *Extracts*, pp. 13–14.

39. Cooke to his Mother, 28 November 1836, in Cooke, *Extracts*, p. 15.

40. Cooke to his Mother, 27 February 1837, in Cooke, *Extracts*, pp. 18–19. Clark was possibly the philosophical instrument-maker Edward M. Clarke.

41. Ibid. Wheatstone, of course, was professor of natural philosophy, not chemistry, and held his chair at King's College, London, not the London University.

42. Charles Wheatstone, "An Account of Some Experiments to Measure the Velocity of Electricity, and the Duration of Electric Light," *Philosophical Transactions* 124 (1834): 583–91.

43. These claims were later the subject of much dispute. See next section.

44. Cooke to his Mother, 14 March 1837, in Cooke, *Extracts*, pp. 20–22.

45. Kieve, *Electric Telegraph*, pp. 18–20.

46. Cooke to his Mother, 11 May 1837, in Cooke, *Extracts*, p. 29.

47. "Scientific and Miscellaneous Intelligence," *Railway Magazine* 2 (1837): 321–28, on p. 327.

48. Cooke to his Mother, 10 June 1837, in Cooke, *Extracts*, p. 35.

49. Cooke to his Mother, 2 July 1837, in Cooke, *Extracts*, pp. 36–37.

50. Ibid.

51. Ibid.

52. Cooke to his Mother, 4 July 1837, in Cooke, *Extracts*, p. 38.

53. Charles Wheatstone to William Fothergill Cooke, July 1837 [before 4 July 1837], Cooke Papers, Institution of Electrical Engineers (henceforward CPIEE), vol. 1.

54. Cooke to his Mother, 25 July 1837, in Cooke, *Extracts*, p. 41.

55. Charles Wheatstone to William Fothergill Cooke, 6 September 1837, CPIEE, vol. 1.

56. Robert Wilson to William Fothergill Cooke, 22 September 1837, 23 September 1837, CPIEE, vol. 1.

57. Extract from the Minutes of the London & Birmingham Railway, London Committee, 12 October 1837, CPIEE, vol. 2.

58. Proposals for Establishing an Electric Telegraph to unite London, Holyhead, Liverpool, Manchester, and Birmingham, n.d. prob. before September 1837, CPIEE, vol. 2.

59. Extract from the Minutes of the London & Birmingham Railway, London Committee, 29 November 1837, CPIEE, vol. 2.

60. Ibid., 14 December 1837, vol. 2.

61. Cooke to Betsy, 8 September 837, in Cooke, *Extracts*, pp. 43–44.

62. "Scientific and Miscellaneous Intelligence," *Railway Magazine* 3 (1837). 431–44, on p. 444.

63. Charles Wheatstone to William Fothergill Cooke, 15 December 1837, CPIEE, vol. 1.

64. "Scientific and Miscellaneous Intelligence," *Railway Magazine* 4 (1838): 119–26, on p. 120.

65. Fahie, *History*, pp. 402–3.

66. Wheatstone to Cooke, 20 January 1838, CPIEE, vol. 1.

67. Edward Davy to his Father, 23 January 1838, Davy Papers, Institution of Electrical Engineers (henceforward DPIEE).

68. Fahie, *History*, p. 418.

69. Ibid., p. 420.

70. Ibid.

71. Ibid., p. 421.

72. Charles Wheatstone to William Fothergill Cooke, 10 March 1838, 24 March 1838, CPIEE, vol. 1.

73. Edward Davy, "Draft Response," DPIEE.

74. Davy to his Father, 30 May 1838, 16 June 1838, DPIEE.

75. Davy to his Father, 30 May 1838, DPIEE.

76. Davy to his Father, 23 June 1838, DPIEE.

77. Ibid.

78. Ibid.

79. Davy to his Father, 4 July 1838; Bunn to Davy, 13 June 1838, DPIEE.

80. Davy to his Father, 4 July 1838, DPIEE.

81. Davy to his Father, 23 June 1838, DPIEE.

82. Davy to his Brother, ? August 1838, DPIEE.

83. Fahie, *History*.

84. Thomas Watson to Thomas Davy, 11 April 1840, DPIEE.

85. L.T.C. Rolt, *Isambard Kingdom Brunel* (Harmondsworth: Penguin, 1970).

86. Isambard Kingdom Brunel to William Fothergill Cooke, 22 September 1837, CPIEE, vol. 2.

87. Cooke to his Mother, 10 March 1838, 13 March 1838, 3 April 1838, in *Cooke, Extracts*, pp. 50–52.

88. Cooke to his Mother, 3 April 1838, in Cooke, *Extracts*, p. 52.

89. Cooke to his Mother, 18 April 1838, in Cooke, *Extracts*, p. 54.

90. Cooke to his Mother, 31 May 1838, in Cooke, *Extracts*, pp. 56–57.

91. Cooke to his Mother, 8 June 1838, in Cooke, *Extracts*, p. 57.

92. Kieve, *Electric Telegraph*, p. 30.

93. Cooke to his Sister, 28 November 1839, CPIEE, vol. 2.

94. "Galvanic Telegraph," *Inventor's Advocate* 1 (1839): 53.

95. "Electro-magnetic Telegraph," *Inventor's Advocate* 1 (1839): 281.

96. Cooke, *Electric Telegraph*, 2:255; "Electro-magnetic Telegraph," *Inventor's Advocate* 1 (1839): 281.

97. Cooke to his Sister, 28 November 1839, CPIEE, vol. 2.

98. *Railway Times* 5 (1842): 296.

99. "Electric Telegraph," *Inventor's Advocate* 3 (1840). 90–91; "Wheatstone and Cooke's Electric Telegraph," *Mechanic's Magazine* 33 (1840): 161–70.

100. Ibid., p. 161.

101. Ibid., pp. 166–70.

102. Ibid., p. 167.

103. Ibid., p. 168.

104. Ibid.

105. Ibid., p. 169.

106. Kieve, *Electric Telegraph*, p. 32.

107. William Fothergill Cooke to Charles Wheatstone, 26 November 1842, CPIEE, vol. 2.

108. Advertisement of "The Electric Telegraph—Cooke and Wheatstone Patentees," *Railway Times* 5 (1842): 145.

109. William Fothergill Cooke, *Telegraphic Railways; or, The Single Way, Recommended by Safety, Economy, and Efficiency, under the Safeguard and Control of the Electric Telegraph* (London, 1842).

110. Ibid., p. 1.

111. Ibid., p. 7.

112. Ibid., p. 25 By *express train* Cooke meant a train not scheduled on the timetable.

113. Ibid., p. 28. Similar remarks are scattered throughout the text; see, for example, p. 30, p. 32.

114. Ibid., p. 33.

115. *Railway Times* 5 (1843): 620.

116. *Railway Times* 5 (1843): 158–59.

117. *Railway Times* 5 (1843): 620; "Memoirs of Deceased Members," *Minutes of Proceedings of the Institution of Civil Engineers* 57 (1878): 294–309.

118. Extracted in "The Electric Telegraph," *Railway Magazine* 5 (1842): 988.

119. Ibid.

120. Cooke, *Electric Telegraph*, 2:250.

121. "Electric Telegraph," *Railway Magazine*.

122. Cooke, *Electric Telegraph*, 1:7.

123. Previous chapter.

124. Cooke, *Electric Telegraph*, vol. 1.

125. Ibid., pp. 16–17.

126. Cooke to Betsy, 8 September 1837, in Cooke, *Extracts*, p. 44.

127. Cooke to his Mother, 25 October 1837, in Cooke, *Extracts*, p. 45.

128. Cooke to his Mother, 11 May 1837, in Cooke, *Extracts*, p. 29.

129. Cooke, *Electric Telegraph*, 1:90.

130. Ibid., 2:64.

131. Ibid, 1:11–12.

132. Wheatstone to Cooke, 26 October 1840, printed in Cooke, *Electric Telegraph*, 2:113.

133. Ibid., p. 124, my italics. "My" would appear from context to be a mistake for "your."

134. Ibid., p. 121.

135. Charles Wheatstone, "An Account of Several New Instruments and Processes for Determining the Constants of a Voltaic Circuit," *Philosophical Transactions* 133 (1843): 303–27. Reprinted in *The Scientific Papers of Sir Charles Wheatstone* (London, 1879), pp. 97–133.

136. *Scientific Papers*, p. 97.

137. Ibid., p. 98.

138. Cooke, *Electric Telegraph*, 1:16–17.

139. Cooke, *Electric Telegraph*. It should be noted that this publication (by Cooke) reprinted the pamphlets written by both combatants.

140. "Mr. Cooke's Electric Telegraph," *Railway Times* 6 (1843): 594–97, on p. 594.

141. Ibid., p. 594.

142. For biographical details of Bain, see *DNB*, s.v. "Bain."

143. Alexander Bain, "Electro-magnetic Clock," *Inventor's Advocate* 4 (1841): 203.

144. John Lamb, "Claims to the Invention of the Electro-magnetic Clock," *Inventor's Advocate* 4 (1841): 234.

145. Alexander Bain, "Electro-magnetic Clock," *Inventor's Advocate* 4 (1841): 250–51, on p. 251.

146. John Lamb, "The Electrical Clock," *Inventor's Advocate* 4 (1841): 266.

147. Alexander Bain, "Electro-magnetic Clocks," *Inventor's Advocate* 4 (1841): 299–300, on p. 299.

148. Ibid., p. 300.

149. "Royal Institution," *Literary Gazette* 26 (1842): 366; "Original Correspondence," *Literary Gazette* 26 (1842): 404.

150. "Original Correspondence," *Literary Gazette* 26 (1842): 423.

151. John Finlaison, *An Account of Some Remarkable Applications of the Electric Fluid to the Useful Arts* (London, 1843). It is not clear that Finlaison's position was as nonpartisan as he wished to assert. According to Thomas Watson to Thomas Davy, 14 February 1843, DPIEE, he had been contacted by Bain who requested information concerning Edward Davy's dealings with Wheatstone since *Bain* was in the process of putting together a pamphlet attacking him.

152. Finlaison, *Account*.

153. Ibid., p. 41.

154. Ibid., p.109.

155. "Mr. Bain's Electro Magnetic Inventions," *Mechanic's Magazine* 39 (1843): 64–77.

156. William Fothergill Cooke, "The Electric Printing Press and Clock," *Mechanic's Magazine* 39 (1843): 108–10.

157. Ibid., p. 108, p. 109.

158. Kieve, *Electric Telegraph*, p. 32.

159. Broadsheet, 1845.

160. Kieve, *Electric Telegraph*, pp. 29–45.

161. Ada Lovelace to William Lovelace, 29 November 1844, in *Ada, the Enchantress of Numbers: A Selection from the Letters of Lord Byron's Daughter and Her Description of the*

*First Computer*, ed. Betty Alexandra Toole (Mill Valley, CA: Strawberry Press, 1992), pp. 302–3.

162. Harris, *Humbug*.

163. Wheatstone to Cooke, 2 August 1845, CPIEE, vol. 2; Cooke, *Electric Telegraph*, 1:232.

164. Articles of Agreement among Bidder, Cooke, and Ricardo, 28 November 1845; Indentures of Share Transfer, 23 December 1845, CPIEE, vol. 4.

165. Prospectus of the Electric Telegraph Company, September 1845, CPIEE, vol. 4.

166. *Patent Journal and Inventor's Magazine* 6 (1849): 216.

167. J. H. Hammerton, "The Electric Telegraph—Improved Modes of Insulation," *Mechanics' Magazine* 49 (1848): 272–74; Francis Wishaw, "Application of Gutta Percha to the Insulation of Electric Wires," *Mechanics' Magazine* 49 (1848): 309; William Baddeley, "On the Application of Gutta Percha as an Insulator of Electric Wires," *Mechanics' Magazine* 49 (1848): 309–10; Z. U., "The Use of Gutta Percha for Electro-telegraphic Purposes," *Mechanics' Magazine* 49 (1848): 310–11.

168. *Patent Journal and Inventor's Magazine* 7 (1849): 33–36, 49–50, 106–7. Bakewell was later to be one of the foremost of many nineteenth-century commentators and historians of the telegraph.

169. "Messrs. Brett and Little's Electric Telegraph," *Mechanics' Magazine* 47 (1847): 185–87.

170. "Rights of Inventors by Patent and by Registration—and Effect of Disclosure under an Obligation of Secrecy," *Mechanics' Magazine* 46 (1847): 522–25.

171. Ibid., p. 525.

172. "Court of Common Pleas: The Electric Telegraph Company vs. Brett & Little," *Times*, 23 February 1850, p. 7.

173. "Court of Common Pleas," *Times*, 26 February 1850, p. 7.

174. *Times*, 2 March 1850, p. 8. The attorney, Sidney Smith, was objecting to what he regarded as the *Times*'s misleading report of the court's deliberations.

175. Ibid.

176. "Alleged Misconduct of the Electric Telegraph Company," *Morning Herald*, 11 October 1849, p. 5.

177. Ibid.

178. Ibid.

179. Ibid.

180. "Criminal Information against the Liverpool Correspondents of the Morning Herald," *Morning Herald*, 26 November 1849; see also "Electric Telegraph Company vs. Wilmer and Smith," *Times*, 26 November 1849, p. 3.

181. Ibid.

182. Ibid.

183. Francis C. Mather, "The Railways, the Electric Telegraph and Public Order during the Chartist Period," *History* 38 (1953): 40–53.

184. "The Electric Telegraph," *Morning Herald*, 4 December 1849, p. 3.

185. "The Electric Telegraph Company vs. Willmer and Smith," *Times*, 20 November 1850, p. 3.

186. "Alleged Misconduct."

187. "Telegraphic Communication Commercially Considered," *London Journal of Arts, Sciences, and Manufactures* 37 (1850): 205–7, on p. 207.

188. *Times*, 4 January 1851, p. 5.

189. "The Corn Market and the Telegraph," *Mechanics' Magazine* 47 (1847): 184.

190. Extract from the Telegraph Book at Paddington, published in [Andrew Wynter], "The Electric Telegraph," *Quarterly Review* 95 (1854): 118–64, on p. 128. This was the article that prompted the recommencement of Cooke's priority dispute with Wheatstone.

191. "Time and the Electric Telegraph," *Mechanics' Magazine* 42 (1845): 416.

192. Derek Howse, *Greenwich Time* (Oxford: Oxford University Press, 1980).

193. "A Game of Chess by Telegraph," *Morning Herald*, 8 April 1845, p. 2.

194. *Morning Herald*, 15 April 1845.

195. For chess and intelligent machines, see Simon Schaffer, "Babbage's Dancer and the Impresarios of Mechanism," in *Cultural Babbage: Technology, Time and Invention*, ed. Francis Spufford and Jenny Uglow (London: Faber & Faber, 1996), pp. 53–80.

196. "The Electric Telegraph," *Patent Journal* 4 (1850): 229–31, on pp. 229–30.

197. "Tricks of the Electrics," *Punch* 27 (1854): 64.

198. For similar anxieties and concerns later in the century with the introduction of telephony, see Marvin, *When Old Technologies Were New*.

199. [Wynter], "Electric Telegraph," p. 133.

200. "The Electric Story-Teller," *Punch* 27 (1854): 143.

201. [Wynter], "Electric Telegraph," p. 132, p. 142.

202. On electricity and the organic character of its technologies, see Marshall McLuhan, *Understanding Media: The Extensions of Man* (New York: McGraw Hill, 1964).

203. Ibid., p. 151.

204. Mather, "Railways."

205. Iwan Rhys Morus, "The Electric Ariel: Telegraphy and Commercial Culture in Early Victorian England," *Victorian Studies* 39 (1996): 339–78.

CHAPTER 8
UNDER MEDICAL DIRECTION

1. SCRUTATOR, *Lancet* 1 (1859): 280.

2. Hector A. Colwell, *An Essay on the History of Electrotherapy and Diagnosis* (London: Heinemann, 1922).

3. Margaret Rowbottom and Charles Susskind, *Electricity and Medicine: History of Their Interaction* (San Francisco: San Francisco Press, 1984).

4. Ivan Waddington, *The Medical Profession in the Industrial Revolution* (Dublin: Gill & Macmillan, 1984).

5. For an outline of the chaotic state of the medical profession in the early nineteenth century, see M. Jeanne Peterson, *The Medical Profession in Mid-Victorian London* (Berkeley and Los Angeles: University of California Press, 1978), esp. chap. 1.

6. George Clark, *A History of the Royal College of Physicians*, 2 vols. (Oxford: Oxford University Press, 1966).

7. Zachary Cope, *The Royal College of Surgeons of England* (London: Anthony Blond, 1959).

8. W.S.C. Copeman, *The Worshipful Society of Apothecaries: A History, 1617–1904* (Oxford: Pergamon Press, 1967).

9. Ivan Waddington, "The Role of the Hospital in the Development of Modern Medicine: A Sociological Analysis," *Sociology* 7 (1973): 211–25.

10. John Woodward, *To Do the Sick No Harm: A Study of the British Voluntary Hospital System to 1875* (London: Routledge, 1974).

11. Peterson, *Medical Profession*, pp. 12–15.

12. Ivan Waddington, "General Practitioners and Consultants in early Nineteenth Century England," in *Health Care and Popular Medicine in Nineteenth Century England: Essays in the Social History of Medicine*, ed. John Woodward and David Richards (London: Croom Helm, 1977), pp. 164–88.

13. G. N. Stephen, *The British Medical Association and the Medical Profession* (London, 1914).

14. Iain McCalman, *Radical Underworld: Prophets, Revolutionaries and Pornographers in London, 1795–1840* (Cambridge: Cambridge University Press, 1988).

15. S. Squire Sprigge, *The Life and Times of Thomas Wakley, Founder and First Editor of the Lancet, Member of Parliament for Finsbury, and Coroner for West Middlesex* (London, 1899).

16. *Lancet* 1 (1831–32): 2. Quoted in Peterson, *Medical Profession*, p. 26.

17. Desmond, "Lamarckism and Democracy" and *Politics of Evolution.*

18. Alison Winter, "Ethereal Epidemic: Mesmerism and the Introduction of Inhalation Anaesthesia to Early Victorian London," *Social History of Medicine* 4 (1991): 1–27.

19. See chap. 5.

20. Samuel Wilks and G. T. Bettany, *A Biographical History of Guy's Hospital* (London, 1892).

21. For an account of Bird's chemical work, see Noel G. Coley, "The Collateral Sciences in the Work of Golding Bird," *Medical History* 13 (1969): 363–76.

22. For biographical details on Bird, see William Hale-White, "Golding Bird: Assistant Physician to Guy's Hospital 1843–1854," *Guy's Hospital Reports* 76 (1926): 1–20.

23. Minutes of Court of Committees, Guy's Hospital, H9/GY/A3/6/1, 29 August 1838, p. 236. Held at the Greater London Record Office.

24. Henry C. Cameron, *Mr. Guy's Hospital 1726–1948* (London: Longmans, 1954), pp. 105–87. For more details of his role at Guy's, see Amalie Kass and Edward Kass, *Perfecting the World: The Life and Times of Dr. Thomas Hodgkin 1798–1866* (New York: Harcourt Brace Jovanovich, 1988).

25. E. M. McInnes, *St. Thomas's Hospital* (London: George Allen & Unwin, 1963), p. 149.

26. See chap. 5.

27. See his biography in *DNB*, s.v. "Birch."

28. Golding Bird, "Report on the Value of Electricity as a Remedial Agent in the Treatment of Diseases," *Guy's Hospital Reports* 6 (1841): 81–120.

29. "The New Frankenstein." See chap. 5.

30. Morus, "Currents."

31. Heilbron, *Electricity*.

32. Bird, "Report," pp. 85–87.

33. Thomas Addison, "On the Influence of Electricity, as a Remedy in Certain Convulsive and Spasmodic Diseases," *Guy's Hospital Reports* 2 (1837): 493–507, on pp. 493–94. The choice of the prestigious Addison rather than the largely unknown Bird to make the first announcement of the electrifying room's activities was clearly strategic and an indication of the topic's potential sensitivity. Addison made his indebtedness to Bird quite clear.

34. Bird, "Report," pp. 88–89, p. 116; Addison, "Influence."

35. For a fascinating account of the early-nineteenth-century patient's perception of the hospital, see Richardson, *Death, Dissection and the Destitute*.

36. Winter, "Island of Mesmeria," chap. 2.

37. James Cholmely to Golding Bird, 23 April 1837, Golding Bird Correspondence, Royal College of Surgeons, 129.c.16. Cholmely was physician at Guy's from 1804 to 1837.

38. Bird, "Report," p. 93.

39. Benjamin Guy Babington, "On Chorea," *Guy's Hospital Reports* 6 (1841): 411–47.

40. Ibid., p. 447.

41. Henry Marshall Hughes, "Digest of One Hundred Cases of Chorea treated in the Hospital," *Guy's Hospital Reports* 4 (1846): 360–94. For Hughes, see *Munck's Roll*, 4: 36–37.

42. Ibid., p. 388.

43. William Withey Gull, "A Further Report on the Value of Electricity as a Remedial Agent," *Guy's Hospital Reports* 8 (1852): 81–143, on p. 81.

44. Coley, "Colateral Sciences," p. 365; Astley Cooper, *On the Anatomy of the Breast* (London, 1840).

45. Marshall Hall, "Memoirs on Some Principles of Pathology in the Nervous System," *Medico-Chirurgical Transactions* 22 (1839): 191–217, on pp. 200–201.

46. Wilks and Bettany, *Guy's Hospital*, pp. 428–29, for Williams. Williams had been a participant in the postmortem dissection of the radical publisher Richard Carlisle. See "Examination of the Body of Mr. Richard Carlisle," *Lancet* 1 (1842–43): 774.

47. Thomas Williams, "On the Laws of the Nervous Force, and the Function of the Roots of the Spinal Nerves," *Lancet* 2 (1847): 516–17.

48. Henry Letheby, "Nervous and Electrical Forces," *London Medical Gazette* 30 (1842): 809–10. This was an abstract of a paper delivered to the London Electrical Society on 16 August 1842.

49. Morus, "Currents."

50. For an account of enculturation and the importance of community membership for the acquisition and transmission of tacit skills and knowledge, see Collins, *Changing Order*.

51. Sturgeon, *Galvanism*, discussed in chap. 5.

52. Ornella Moscucci, *The Science of Woman: Gynaecology and Gender in England, 1800–1929* (Cambridge: Cambridge University Press, 1990).

53. Edward John Tilt, *On Diseases of Women and Ovarian Inflammation*, 2d ed. (London, 1853) pp. 84–85.

54. L. Stephen Jacyna, "Somatic Theories of Mind and the Interests of Medicine in Britain, 1850–1879," *Medical History* 26 (1982): 233–58, argues suggestively that the prevalence of such physicalist accounts of mind may be linked to the interests of the medical profession in maintaining and enhancing their positions as carers for the mentally ill. Establishing physical causes was a way of advocating physical (i.e., medical) rather than moral therapies.

55. "Reviews and Notices of New Books," *Lancet* 2 (1840–41): 197–98.

56. For Laycock's biography, see *DNB*, 11:744–45.

57. These were later published together in Thomas Laycock, *An Essay on Hysteria: Being an Analysis of Its Irregular and Aggravated Forms; Including Hysterical Hemorrhage, and Hysterical Ischuria* (Philadelphia, 1840).

58. Thomas Laycock, *A Treatise on the Nervous Disorders of Women; Comprising an Inquiry into the Nature, Causes, and Treatment of Spinal and Hysterical Disorders* (London, 1840).

59. Desmond, *Politics of Evolution*.

60. Laycock, *Hysteria*, p. 59.

61. Ibid., p. 69. Note the hierarchy: the brain is subservient to the reproductive organs.

62. Laycock, *Nervous Disorders of Women*, p. 76.

63. See, for example, some of the remarks quoted in Mary Poovey, "Scenes of an Indelicate Character: The Medical 'Treatment' of Victorian Women," *Representations* 14 (1986): 137–68; Jean L'Esperance, "Doctors and Women in Nineteenth Century Society: Sexuality and Role," in Woodward and Richards, *Health Care*, pp. 105–27; Michael Mason, *The Making of Victorian Sexuality* (Oxford: Oxford University Press, 1994).

64. Tilt, *Diseases of Women*, p. 56.

65. Thomas Laqueur, *Making Sex: Body and Gender from the Greeks to Freud* (Cambridge: Harvard University Press, 1990). A similar point is made by Londa Schiebinger, *The Mind Has No Sex: Women in the Origins of Modern Science* (Cambridge: Harvard University Press, 1989).

66. Laycock, *Hysteria*, p. 72.

67. Ibid., p. 76.

68. Laycock, *Nervous Disorders of Women*, p. 113.

69. [Thomas Laycock], "Woman in Her Psychological Relations," *Journal of Psychological Medicine and Mental Pathology* 4 (1851): 18–50, on p. 38. Laycock's authorship is acknowledged in Thomas Laycock, *Testimonials Submitted to the Honourable Patrons of the University of Edinburgh* (York, 1855).

70. Laycock, *Hysteria*, p. 113.

71. Marshall Hall, *Essays Chiefly on the Theory of Paroxysmal Diseases of the Nervous System* (London, 1849), p. 23.

72. Laycock, *Nervous Disorders of Women*, pp. 176–77.

73. Tilt, *Diseases of Women*, p. 54.

74. For an enlightening account of Victorian attitudes toward sex, see Mason, *Victorian Sexuality*.

75. Thomas William Nunn, *Inflammation of the Breast, and Milk Abscess* (London, 1853), pp. 3–4.

76. Marshall Hall, "On the Function of the Medulla Oblongata and Medulla Spinalis, and on the Excito-motory System of Nerves," *Proceedings of the Royal Society* 3 (1830–37): 463–64, on p. 464.

77. Laycock, *Nervous Disorders of Women*, p. 99. The remarks concerning the mode whereby electricity was transmitted strongly suggest that Laycock was familiar with Michael Faraday's most recent speculations.

78. Ibid., p. 100.

79. For passage points, see Latour, *Science in Action*.

80. Thomas Laycock, "On the Treatment of Cerebral Hysteria, and of Moral Insanity in Women, by Electro-galvanism," *Medical Times* (1850): 57–58, on p. 58.

81. Ibid.

82. Ibid.

83. Tilt, *Diseases of Women*, p. 9.

84. Ibid., p. 11.

85. Marshall Hall, "On a New and Lamentable Form of Hysteria," *Lancet* 1 (1850): 660–61.

86. Robert Brudenell Carter, *On the Pathology and Treatment of Hysteria* (London, 1853).

87. Ibid., p. 69.

88. Laycock, "Cerebral Hysteria," pp. 57–58.

89. Bird, "Treatment of Paralysis."

90. Ibid., p. 649.

91. Ibid.

92. For an account of the development of the induction coil, particularly as it related to medicine, see Hackmann, "Induction Coil."

93. Golding Bird, "Observations on Induced Electric Currents, with a Description of a Magnetic Contact-breaker," *Philosophical Magazine* 12 (1838): 18–22.

94. Henry Letheby, "A Description of a New Electro-magnetic Machine, Adapted So As to Give a Succession of Shocks in One Direction," *Medical Gazette* 32 (1844): 858–59. For Letheby. see *DNB*, 11:1010.

95. Ibid., p. 859.

96. John Crowch Christophers, "Anaesthesia Treated by Electro-magnetism," *Lancet* 2 (1846): 144–45.

97. Ibid., p. 145.

98. See, for example, the contents of trade catalogs such as Edward M. Clarke, *Laboratory of Science. List of Prices of Magnetical, Philosophical, Optical, and Chemical Instruments and Apparatus, Manufactured by Edward M. Clarke, Magnetical Instrument Maker &c.* (London, ca. 1840).

99. Smee, *Electro-biology*, p. 111.

100. *Directions for Using Kemp & Co.'s*

*Electro-magnetic Coil Machine, for the Medical Application of Galvanism* (Edinburgh, n.d., after 1846).

101. Ibid., p. 2.

102. Ibid., p. 3.

103. Ibid., p. 2.

104. J. H. Horne, *A Guide to the Correct Administration of Medical Galvanism, Being a Companion to the Improved Electro-galvanic Machines, Manufactured and Sold by Horne, Thornthwaite, and Wood* (London, n.d., prob. after 1845).

105. Ibid., p. i–iv.

106. Ibid., p. 5.

107. *Palmer's New Catalogue*, p. 36.

108. *Directions for Using Kemp & Co.'s Electro-magnetic Coil Machine*.

109. Handbill, quoted in [Charles V. Walker], "Facts(?) for the History of Electricity," *Electrical Magazine* 1 (1845): 551–52, on p. 552.

110. Ibid.

111. {Walker, Charles Vincent], "Galvanic Rings," *Electrical Magazine* 1 (1845): 607.

112. Christopher Lawrence, "Incommunicable Knowledge: Science, Technology and the Clinical Art in Britain 1850–1914," *Journal of Contemporary History* 20 (1985): 503–20.

113. Chap. 5.

114. For a description of such accounts, see Morus, "Marketing the Machine."

115. See chaps. 6 and 7.

116. Peterson, *Medical Profession*.

117. Mason, *Victorian Sexuality*, p. 163.

THE DISCIPLINING OF EXPERIMENTAL LIFE

1. Walker, "Inaugural Address."

2. Christopher Hobhouse, *1851 and the Crystal Palace*, 2d ed. (London: John Murray, 1950).

3. Sturgeon, "General Outline," p. 123.

4. Quoted in Cardwell, *Joule*, p. 45; Grove, *Correlation*.

5. Ryland, "Plated Wares," p. 494.

6. Cooke, *Electric Telegraph*, 2: 250.

7. Mather, "Railways."

8. "Dinner of the Institution of Electrical Engineers," *Electrician* 24 (1889): 12–15, on p. 13.

9. Ibid., p. 13.

10. Ibid.

11. McLuhan, *Understanding Media*, p. 247.

12. Carolyn Marvin has also drawn attention to the ways in which electricity could be used to culturally locate the body. She argues that electricity and its attendant technologies provided a whole range of new resources for manipulating, talking about, and understanding the human body in its relationships to both culture and nature. Marvin, *When Old Technologies Were New*.

# BIBLIOGRAPHY

Primary Sources

"Abstract of the Late Experiments of Professor Aldini on Galvanism." *Journal of Natural Philosophy* 3 (1802): 298–300.

"Accidental Production of Animal Life—Mr. Crosse." *Magazine of Popular Science* 3 (1837): 145–48.

Addison, Thomas. "On the Influence of Electricity, as a Remedy in Certain Convulsive and Spasmodic Diseases." *Guy's Hospital Reports* 2 (1837): 493–507.

"Alleged Misconduct of the Electric Telegraph Company." *Morning Herald,* 11 October 1849.

A. M. "Chemical Society." *Chemist* 1 (1824): 167.

"The Applicability of Electro-magnetism as a Moving Power." *Inventor's Advocate and Journal of Industry* 2 (1840): 410–11.

Babbage, Charles. *Reflexions on the Decline of Science in England.* London, 1830.

———. *On the Economy of Machinery and Manufactures.* London, 1832.

———. *The Exposition of 1851.* London, 1851.

Babington, Benjamin Guy. "On Chorea." *Guy's Hospital Reports* 6 (1841): 411–17.

Bachhoffner, George H. "On a Simple Voltaic Battery, in a Letter to William Sturgeon, Esq., &c., &c." *Annals of Electricity* 1 (1836–37): 213–15.

Baddeley, William. "On the Application of Gutta Percha as an Insulator of Electric Wires." *Mechanics' Magazine* 49 (1848): 309–10.

Bain, Alexander. "Electro-magnetic Clock." *Inventor's Advocate* 4 (1841): 203.

———. "Electro-magnetic Clock." *Inventor's Advocate* 4 (1841): 250–51.

———. "Electro-magnetic Clocks." *Inventor's Advocate* 4 (1841): 299–300.

Barlow, Peter. "A Curious Electro-magnetic Experiment." *Philosophical Magazine* 59 (1822): 241–42.

———. "An Account of Some Electro-magnetic Combinations, for Exhibiting Thermo-electric Phenomena, Invented by James Marsh of Woolwich." *Philosophical Magazine* 62 (1823): 321–27.

———. *A Treatise on the Manufactures and Machinery of Great Britain.* London, 1836.

Beamish, Richard. *Memoir of the Life of Sir Marc Isambard Brunel.* London, 1862.

Bence Jones, Henry. *The Life and Letters of Faraday.* 2 vols. London, 1870.

———. *The Royal Institution: Its Founder and First Professors.* London, 1871.

Bird, Golding. "Observations on the Electro-chemical Influence of Long-continued Electric Currents of Low Tension." *Philosophical Transactions* 128 (1837): 37–45.

———. "Observations on Induced Electric Currents, with a Description of a Magnetic Contact-breaker." *Philosophical Magazine* 12 (1838): 18–22.

———. "Report on the Value of Electricity as a Remedial Agent in the Treatment of Diseases." *Guy's Hospital Reports* 6 (1841): 81–120.

———. "On the Employment of Electro-magnetic Currents in the Treatment of Paralysis." *Lancet* 1 (1846): 649–51.

———. *Lectures on Electricity and Galvanism, in Their Physiological and Therapeutical Relations.* London, 1849.

Brande, William T. *A Manual of Chemistry.* London, 1819.

Bright, Charles. "Address of the President." *Journal of the Society of Telegraph Engineers* 16 (1887): 7–40.

"The British Association—Journal of a Week at Cambridge. Conclusion." *Literary Gazette* 17 (1833): 490–93.

Brougham, Henry. *Lives of Men of Letters and Science, Who Flourished in the Time of George III.* 2 vols. London, 1845–46.

Carlyle, Thomas. *Critical and Miscellaneous Essays.* In *Collected Works.* 17 vols. London, 1885–91.

[Carpenter, William B.]. "Smee and Wiglesworth on Physical Biology." *British and Foreign Medico-Chirurgical Review* 4 (1849): 371–82.

Carter, Robert Brudenell. *On the Pathology and Treatment of Hysteria.* London, 1853.

Cartwright, Samuel. "On Electro-type from Engraved Copperplates." *Annals of Electricity* 5 (1840): 236–38.

———. "On the Cultivation and Growth of Electro-types." *Annals of Electricity* 5 (1840): 484–85.

Christophers, John Crowch. "Anaesthesia Treated by Electro-magnetism." *Lancet* 2 (1846): 144–45.

Clark, Latimer. "Inaugural Address." *Journal of the Society of Telegraph Engineers* 4 (1875): 1–22.

Clarke, Edward M., "On a New Phenomenon in Magneto-electricity." *Philosophical Magazine* 6 (1835): 169–70.

———. "On Certain Optical Effects of the Magnetic-electrical Machine, and on an Apparatus for Decomposing Water by Its Means." *Philosophical Magazine* 6 (1835): 427–28.

———. "Remarks on a Peculiar State of Polarity Induced in Soft Iron by Voltaic Magnetism." *Philosophical Magazine* 7 (1835): 422–23.

———. "Description of E. M. Clarke's Magnetic Electric Machine." *Philosophical Magazine* 9 (1836): 262–66.

———. *List of Prices of Magnetical, Philosophical, Optical, and Chemical Instruments and Apparatus, Manufactured by Edward M. Clarke, Magnetician.* London, 1837?

———*Laboratory of Science. List of Prices of Magnetical, Philosophical, Optical, and Chemical Instruments and Apparatus, Manufactured by Edward M. Clarke, Magnetical Instrument Maker &c.* London, ca. 1840.

Clarke, Uriah. "Electro-magnetic Locomotive Carriage." *Annals of Electricity* 5 (1840): 304–5.

"Commemoration of the Centenary of the Birth of Dr. Priestley." *Philosophical Magazine,* 2 (1833): 382–402.

Cooke, William Fothergill. *Telegraphic Railways; or, The Single Way, Recommended by Safety, Economy, and Efficiency, under the Safeguard and Control of the Electric Telegraph.* London, 1842.

———. "The Electric Printing Press and Clock." *Mechanics' Magazine* 39 (1843): 108–10.

———. *The Electric Telegraph: Was It Invented by Professor Wheatstone?* 2 vols. London, 1856–57.

———. *Extracts from the Private Letters of the Late Sir William Fothergill Cooke, 1836–39, relating to the Invention and Development of the Electric Telegraph.* London, 1895.

Cooper, Astley. *On the Anatomy of the Breast.* London, 1840.

Copleston, W. J. *Memoir of Edward Copleston, D.D., Bishop of Llandaff.* London, 1851.

"The Corn Market and the Telegraph." *Mechanics' Magazine* 47 (1847): 184.

"Court of Common Pleas: The Electric Telegraph Company vs. Brett & Little." *Times*, 23 February 1850.

"Court of Common Pleas." *Times*, 26 February 1850.

"Criminal Information against the Liverpool Correspondents of the Morning Herald." *Morning Herald*, 26 November 1849.

Crosse, Andrew. "On the Production of Insects by Voltaic Electricity." *Annals of Electricity* 1 (1836–37): 242–44.

———. "Description of Some Experiments Made with the Voltaic Battery for the Purpose of Producing Crystals; in the Process of Which Experiments Certain Insects Constantly Appeared." *Transactions and Proceedings of the London Electrical Society* 1 (1841): 10–16.

Crosse, Cornelia. *Memorials, Scientific and Literary, of Andrew Crosse, the Electrician.* London, 1857.

———. "Science and Society in the Fifties." *Temple Bar* 93 (1891): 33–51.

Cunningham, Peter. *Handbook of London: Past and Present.* 2d ed. London, 1850.

Daniell, John F. "On Voltaic Combinations." *Philosophical Transactions* 126 (1836): 107–24.

"Davenport and Cook's Electro-magnetic Engine." *Mechanics' Magazine* 27 (1837): 159–60.

"Davenport's Electro-magnetic Engine." *Mechanics' Magazine* 27 (1837): 404–5.

Davy, Humphry. "New Experiments on Some of the Combinations of Phosphorus." *Philosophical Transactions* 108 (1818): 316–37.

*Diary.* See Martin, Thomas, ed. *Faraday's Diary.*

"Dinner of the Institution of Electrical Engineers." *Electrician* 24 (1889): 12–15.

Dircks, Henry. "Contributions towards a History of Electro-metallurgy." *Mechanics' Magazine* 40 (1844): 72–79.

———. *Contribution towards a History of Electro-metallurgy.* London, 1863.

*Directions for Using Kemp & Co.'s Electro-magnetic Coil Machine, for the Medical Application of Galvanism.* Edinburgh, n.d., after 1846.

"Electrical Society." *Literary Gazette* 21 (1837): 434.

"Electrical Society." *Literary Gazette* 21 (1837): 530.

"Electrical Society." *Literary Gazette* 21 (1837): 656.

"Electrical Society." *Literary Gazette* 21 (1837): 723.

"Electrical Society." *Literary Gazette* 21 (1837): 753–54.

"Electrical Society." *Literary Gazette* 22 (1838): 54–56.

"Electrical Society." *Literary Gazette* 22 (1838): 392–93.

"Electrical Society." *Literary Gazette* 22 (1838): 458.

"Electrical Society." *Literary Gazette* 24 (1840): 184–85.

"Electrical Society of London." *Annals of Electricity* 1 (1836–37): 415–17.

"Electrical Society of London." *Annals of Electricity* 2 (1838): 62–64.

"Electrical Soirée." *Literary Gazette* 26 (1842): 295–96.

"Electrical Soirée." *Literary Gazette* 27 (1843): 352.

"Electric Light." *Patent Journal* 6 (1849): 80.

"The Electric Story-Teller." *Punch* 27 (1854): 143.

"Electric Telegraph." *Inventor's Advocate* 3 (1840): 90–91.

"The Electric Telegraph." *Railway Magazine* 5 (1842): 988.

"The Electric Telegraph." *Morning Herald*, 4 December 1849.

"The Electric Telegraph." *Patent Journal* 4 (1850): 229–31.

"Electric Telegraph Company vs. Wilmer and Smith." *Times*, 26 November 1849.
"The Electric Telegraph Company vs. Willmer and Smith." *Times*, 20 November 1850.
"Electro-magnetic Exhibition." *Railway Times* 5 (1842): 1273.
"Electro-magnetic Locomotive." *Railway Times* 5 (1842): 1012.
"Electro-magnetic Locomotive." *Railway Times* 5 (1842): 1259–60.
"Electro-magnetic Telegraph." *Inventor's Advocate* 1 (1839): 281.
"Electrotype." *London Journal of Arts and Sciences* 15 (1840): 101–3.
"Elements of Electro-biology." *London Journal of Medicine* 1 (1849): 478–49.
Ellis, George. *Memoir of Sir Benjamin Thompson, Count Rumford*. Boston, 1871.
*ERE*. See Faraday, *Experimental Researches in Electricity*.
"Etching Daguerreotype Plates by a Voltaic Process." *Literary Gazette* 25 (1841): 548–49.
"Examination of the Body of Mr. Richard Carlile." *Lancet* 1 (1842–43): 774.
"Exhibition of Works of Popular Science." *Mechanics' Magazine* 17 (1832): 158–60.
Fahie, J. J. *A History of the Electric Telegraph to the Year 1837*. London, 1884.
Faraday, Michael. "Some Experiments and Observations on a New Acid Substance." *Quarterly Journal of Science* 3 (1817): 77–81.
———. "An Account of Some Experiments on the Escape of Gases through Capillary Tubes." *Quarterly Journal of Science* 3 (1817): 354–55.
———. "On Sirium." *Quarterly Journal of Science* 6 (1819): 112–15.
———. "On Sirium, or Vestium." *Quarterly Journal of Science* 7 (1819): 291–93.
———. "On Some New Electro-magnetical Motions, and on the Theory of Magnetism." *Quarterly Journal of Science* 12 (1821): 74–96.
[———]. "Historical Sketch of Electro-magnetism." *Annals of Philosophy* 2 (1821): 195–200, 274–90; 3 (1822): 1078–21.
———. "On Fluid Chlorine." *Philosophical Transactions* 113 (1823): 160–64.
———. "Reply to Dr. John Davy's Remarks on Certain Statements of Mr. Faraday Contained in His 'Researches in Electricity.'" *Philosophical Magazine* 7 (1835).
———. "On the History of the Condensation of Gases, in Reply to Dr. Davy, Introduced by Some Remarks on That of Electro-magnetic Rotation." *Philosophical Magazine* 8 (1836).
———. *Experimental Researches in Electricity*. 3 vols. London, 1839–55.
———. "A Speculation Touching Electric Conduction and the Nature of Matter." *Philosophical Magazine* 24 (1844): 136–44.
———. *Experimental Researches in Chemistry and Physics*. London, 1859.
Finlaison, John. *An Acount of Some Remarkable Applications of the Electric Fluid to the Useful Arts*. London, 1843.
Fisher, George. *Phototype Manipulation*. London, 1844.
Fisher, George P. *The Life of Benjamin Silliman, M.D., LL.D.* 2 vols. New York, 1866.
Fox, Robert W. "On the Temperature of Mines." *Annals of Philosophy* 4 (1822): 440–48.
———. "On the Results of Mr. Fox's Experiments on the Production of Artificial Crystals by Voltaic Action." *Philosophical Magazine* 10 (1837): 171.
———. "Theoretical Views of the Origin of Mineral Veins." *Annals of Electricity* 2 (1838): 166–94.
"Gallery of Practical Science." *Literary Gazette* 18 (1834): 578.
"Gallery of Practical Science." *Literary Gazette* 19 (1835): 715.
"Galvanic Telegraph." *Inventor's Advocate* 1 (1839): 53.
"Galvanische Versuche, Angestellt am Körper eines Gehängten zu London am 17ten Jan. 1803 vom Professor J. Aldini." *Annalen der Physik* 18 (1804): 340–42.

"Galvanism." *Philosophical Magazine* 14 (1802): 364–68.

"A Game of Chess by Telegraph." *Morning Herald*, 8 April 1845.

Gaskell, Peter. *Artisans and Machinery*. London, 1836.

Gassiot, John P. "On a Remarkable Difference in the Heat Attained by the Electrodes of a Powerful Constant Battery." *Philosophical Magazine* 13 (1838): 436–37.

Gladstone, John H. *Michael Faraday*. London, 1872.

Gore, George. "On the Relation of Science to Electro-plate Manufacture, Part I." *Popular Science Review* 1 (1862): 327–31.

Greenaway, Frank, ed. *Archives of the Royal Institution*. 15 vols. London: Royal Institution, 1971–76.

Greenhow, C. H. *An Exposition of the Dangers and Deficiencies of the Present Mode of Railway Construction*. London, 1846.

Griglietta, C. *Electro-magnetism: A Brief Essay or Informal Lecture on Electro-magnetism*. Philadelphia, 1838.

Grove, William Robert. *On the Progress of Physical Science since the Opening of the London Institution*. London, 1842.

———. "On a Voltaic Process for Etching Daguerreotype Plates." *Philosophical Magazine* 20 (1842): 18–24.

[———]. "Physical Science in England." *Blackwood's Magazine* 54 (1843): 514–25.

———. "Experiments on Voltaic Reaction." *Philosophical Magazine* 23 (1843): 223–26.

[———]. "On the Progress Made in the Application of Electricity as a Motive Power." *Literary Gazette* 28 (1844): 892.

———. "On the Application of Voltaic Ignition to Lighting Mines." *Philosophical Magazine* 27 (1845): 442–48.

———. *On the Correlation of Physical Forces*. London, 1846.

Gull, William Withey. "A Further Report on the Value of Electricity as a Remedial Agent." *Guy's Hospital Reports* 8 (1852): 81–143.

Haldane Gee, William W. "William Sturgeon." *Electrician* 35 (1895): 632–35.

Hall, Marshall. "On the Function of the Medulla Oblongata and Medulla Spinalis, and on the Excito-motory System of Nerves." *Proceedings of the Royal Society* 3 (1830–37): 463–64.

———. "Memoirs on Some Principles of Pathology in the Nervous System." *Medico-Chirurgical Transactions* 22 (1839): 191–217.

———. *Essays Chiefly on the Theory of Paroxysmal Diseases of the Nervous System*. London, 1849.

———. "On a New and Lamentable Form of Hysteria." *Lancet* 1 (1850): 660–61.

Halse, William H. "Wonderful Effects of Voltaic Electricity in Restoring Animal Life When the Sensorial Powers Have Entirely Ceased, or in Other Words, When Death in the Common Adaptation of the Term Has Actually Occurred." *Annals of Electricity* 4 (1840): 481–84.

———. "The Power of Galvanism in Cases of Suspended Animation," *New Moral World* 9 (1841); 130.

———. *The Extraordinary Remedial Efficacy of Medical Galvanism, When Scientifically Administered*. London, n.d.

Hammerton, J. H. "The Electric Telegraph—Improved Modes of Insulation." *Mechanic's Magazine* 49 (1848): 272–74.

Head, George. *A Home Tour of the Manufacturing Districts of England in the Summer of 1835*. London, 1836.

Henry, Joseph, "On the Application of the Principle of the Galvanic Multiplier to Elec-tro-magnetic Apparatus, and Also to the Development of Great Magnetic Power in Soft Iron, with a Small Galvanic Element." *American Journal of Science* 19 (1831): 400–408.

———. "Facts in Reference to the Sparks &c. from a Long Conductor Uniting the Poles of a Galvanic Battery." *Journal of the Franklin Institute* 15 (1835): 169–70.

———. "Memoir of Joseph Saxton, 1799–1873." *Biographical Memoirs of the National Academy of Science* 1 (1877): 219–316.

Henry, Joseph, and Dr. Ten Eyck. "An Account of a Large Electro-magnet, Made for the Laboratory of Yale College." *American Journal of Science* 20 (1831): 201–3.

Herschel, John F. W. *A Preliminary Discourse on the Study of Natural Philosophy*. London, 1830.

[Hodgskin, Thomas]. "Apology and Preface." *Chemist* 1 (1824): v–viii.

[———]. "Humphry Davy's Copper." *Chemist* 2 (1825): 78–79.

———. *Popular Political Economy*. London, 1827.

———. *Natural and Artificial Rights to Property Contrasted*. London, 1832.

Horne, J. H. *A Guide to the Correct Administration of Medical Galvanism, Being a Companion to the Improved Electro-galvanic Machines, Manufactured and Sold by Horne, Thornthwaite, and Wood*. London, n.d., prob. after 1845.

Hughes, Henry Marshall. "Digest of One Hundred Cases of Chorea Treated in the Hospital." *Guy's Hospital Reports* 4 (1846): 360–94.

Hytch, E. J. "Mr. Crosse's Revivication of Insects Contained in Flint." *Lancet* 1 (1836–37: 710–11.

"Instinct and Reason." *British and Foreign Medico-Chirurgical Review* 6 (1850): 522–24.

J. G. "Chemical Society." *Chemist* 1 (1824): 206.

Jordan, C. J. "Engraving by Galvanism." *Mechanics' Magazine* 31 (1839): 163–64.

Joule, James P. "On the Use of Electro-magnets Made of Iron Wire for the Electro-magnetic Engine." *Annals of Electricity* 4 (1939–40): 58–62.

———. "Investigations in Magnetism and Electro-magnetism." *Annals of Electricity* 4 (1839–40): 131–37.

———. "Description of an Electro-magnetic Engine." *Annals of Electricity* 4 (1839–40): 203–5.

———. "On Electro-magnetic Forces." *Annals of Electricity* 4 (1839–40): 474–81.

———. "A Short Account of the Life and Writings of the Late William Sturgeon." *Memoirs of the Manchester Literary and Philosophical Society* 14 (1857): 53–83.

Joule, James Prescott, and William Scoresby. "Experiments and Observations on the Mechanical Powers of Electro-magnetism, Steam, and Horses." *Philosophical Magazine* 28 (1846): 448–55.

Kinclaven. "The Newtonian Philosophy." *Mechanics' Magazine* 25 (1836): 99–101.

L. "Electro-magnetic Power." *Railway Times* 5 (1842): 1342.

Lamb, John. "Claims to the Invention of the Electro-magnetic Clock." *Inventor's Advocate* 4 (1841): 234.

———. "The Electrical Clock." *Inventor's Advocate* 4 (1841): 266.

Lardner, Dionysius. *Railway Economy*. London, 1851.

Laycock, Thomas. *An Essay on Hysteria: Being an Analysis of Its Irregular and Aggravated Forms; Including Hysterical Hemorrhage, and Hysterical Ischuria*. Philadelphia, 1840.

———. *A Treatise on the Nervous Disorders of Women; Comprising an Inquiry into the Nature, Causes, and Treatment of Spinal and Hysterical Disorders*. London, 1840.

————. "On the Treatment of Cerebral Hysteria, and of Moral Insanity in Women, by Electro-galvanism." *Medical Times*, n.s., 1 (1850): 57–58.

[————]. "Woman in Her Psychological Relations." *Journal of Psychological Medicine and Mental Pathology* 4 (1851): 18–50.

————. *Testimonials Submitted to the Honourable Patrons of the University of Edinburgh*. York, 1855.

*Lectures on the Progress of Arts and Science, Resulting from the Great Exhibition in London*. London, 1856.

Leithead, William. *Electricity: Its Nature, Operation, and Importance in the Phenomena of the Universe*. London, 1837.

Letheby, Henry. "Nervous and Electrical Forces." *London Medical Gazette* 30 (1842): 809–10.

————. "A Description of a New Electro-magnetic Machine, Adapted So As to Give a Succession of Shocks in One Direction." *Medical Gazette* 32 (1844): 858–59.

"London Chemical Society." *Chemist* 2 (1825): 162–68.

"London Electrical Society." *Literary Gazette* 25 (1841): 218.

Macerone [*sic*], Francis. "An Account of Some Remarkable Electrical Phenomena Seen in the Mediterranean, with Some Physiological Deductions." *Mechanics' Magazine* 15 (1831): 93–96, 98–100.

Maceroni, Francis. *Useful Knowledge for the People*. London, 1832.

————. "On Mackintosh's Electrical Theory of the Universe, and Outlines of Another Theory." *Mechanics' Magazine* 26 (1837): 19–25.

————. "Production of Life by Galvanism—Animation of Horse Hairs!" *Mechanics' Magazine* 26 (1837): 441.

————. *Memoirs of the Life and Adventures of Colonel Maceroni*. 2 vols. London, 1838.

Mackintosh, Thomas S. "Electrical Theory of the Universe." *Mechanics' Magazine* 24 (1835–36): 11–13.

————. "Electrical Theory of the Universe." *Mechanics' Magazine* 24 (1835–36): 227–34.

————. "Electrical Theory of the Universe—Article III." *Mechanics' Magazine* 24 (1835–36): 419–30.

————. "On the Electrical Theory of the Universe—Mr. Mackintosh's Reply to Zeta." *Mechanics' Magazine* 25 (1836): 274–80.

Mackintosh, Thomas S. *The Electrical Theory of the Universe, or the Elements of Physical and Moral Philosophy*. 1st American ed., republished from British ed. of 1838. Boston, 1846).

Mackintosh, Thomas S. *Mackintosh's Electrical Theory of the Universe, of the Elements of Natural Philosophy, Combining Physics and Morals in One Science and Deducing the Actions and Reactions of Mind and Matter, from the Ultimate and Universal Forces of Attraction and Repulsion*. Manchester, n.d.

"Mackintosh's Electrical Theory of the Universe." *New Moral World* 3 (1836–37): 239.

Martin, Thomas, ed. *Faraday's Diary*. 7 vols. London: G. Bell & Sons, 1932.

Mackrell, George. "Nitrate of Soda, Compared with Other Salts, Employed for Constant Batteries." *Proceedings of the London Electrical Society* 2 (1841): 232–37.

"Memoir of Deceased Members." *Minutes of Proceedings of the Institute of Civil Engineers* 57 (1878): 294–309.

Menebrea, L. F. "Sketch of the Analytical Engine Invented by Charles Babbage." Translated by Ada Lovelace. *Taylor's Scientific Memoirs* 3 (1843): 666–731.

"Messrs. Brett and Little's Electric Telegraph." *Mechanics' Magazine* 47 (1847): 185–87.

"Michael Faraday." *Fraser's Magazine* 13 (1836): 224.

"Miscellaneous Communications." *Analyst* 8 (1838): 162–66.

"Mr. Bain's Electro Magnetic Inventions." *Mechanics' Magazine* 39 (1843): 64–77.

"Mr. Cooke's Electric Telegraph." *Railway Times* 6 (1843): 594–97.

"Mr. Mackintosh's Lectures on His Electrical Theory of the Universe." *Mechanics' Magazine* 27 (1837): 9–10.

Murray, James. *Electricity as a Cause of Cholera, or Other Epidemics, and the Relation of Galvanism to the Action of Remedies.* Dublin, 1849.

Nasmyth, James. *Autobiography.* Edited by Samuel Smiles. London, 1883.

"National Gallery of Practical Science. Adelaide Street." *Literary Gazette* 17 (1833): 730.

"National Repository." *Mechanics' Magazine* 11 (1829): 58–60.

"New Exhibition Room, Adelaide Street, Strand." *Literary Gazette* 16 (1832): 378.

"The New Frankenstein." *Fraser's Magazine* 17 (1837): 21–30.

Noad, Henry M. *A Course of Eight Lectures on Electricity, Galvanism, Magnetism, and Electro-magnetism.* London, 1839.

———. *Lectures on Electricity, Comprising Galvanism, Magnetism, Electro-magnetism, Magneto- and Thermo-electricity.* London, 1844.

Nobili, Leopoldo, and Vincenzo Antinori. "On the Electro-motive Force of Magnetism: With Notes by Michael Faraday." *Philosophical Magazine* 11 (1832): 401–13.

"Novel Invention." *London Journal of Arts and Sciences* 15 (1840): 245–48.

Nowrojee, Jehanger, and Hirjeebhoy Merwanjee. *Journal of a Residence of Two Years and a Half in Great Britain.* London, 1841.

Nunn, Thomas William. *Inflammation of the Breast, and Milk Abscess.* London, 1853.

O[dling], E. M. *Memoir of the Late Alfred Smee, F.R.S..* London, 1878.

"Original Correspondence." *Literary Gazette* 26 (1842): 404.

"Original Correspondence." *Literary Gazette* 26 (1842): 423.

"Our Weekly Gossip." *Athenaeum* 11 (1838): 554–55.

Owen, Mrs. *The Life of Richard Owen.* 2 vols. London, 1894.

[Palmer, Edward], *Palmer's New Catalogue, with Three Hundred Engravings of Apparatus Illustrative of Chemistry, Pneumatics, Frictional and Voltaic Electricity, Electro-magnetism, Optics &c.* London, 1840.

"The Panopticon of Science and Arts." *Year-book of Facts in Science and Arts* 17 (1855): 9–11.

Peale, Franklin. "Notice of a Large Magnet." *Journal of the Franklin Institute* 6 (1830): 284.

Phillips, Richard. "A Brief Account of a Visit to Andrew Crosse, Esq." *Annals of Electricity* 1 (1836–37): 135–45.

Pine, Thomas. "On the Probable Connexion between Electricity and Vegetation." *Mechanic's Magazine* 24 (1835–36): 99–104.

———. "Electro-Vegetation." *Mechanics' Magazine* 24 (1835–36): 402–5.

———. "On the First Excitement and Germination of Plants, with Remarks on the Franklinean Theory and That of Mr. Mackintosh." *Mechanics' Magazine* 26 (1837): 102–6.

"Portable Electro-magnetic Apparatus." *Transactions of the Society of Arts* 41 (1823): 47–52.

"Preface." *Mechanics' Magazine* 11 (1829): i–iii.

"Preface." *Patent Journal and Inventor's Advocate* 10 (1850): iii–iv.

Priestley, Joseph. *The History and Present State of Electricity.* London, 1767.

PRO BONO PUBLICO. "Electro-magnetic Power." *Railway Magazine* 6 (1843): 136.

"Progress of Social Reform." *New Moral World* 4 (1836–37): 276.

"Progress of Social Reform." *New Moral World* 4 (1836–37): 353.

"Progress of Social Reform." *New Moral World* 4 (1836–37): 397–98.

Pym, Horace N., ed. *Memories of Old Friends: Being Extracts from the Journals and Letters of Caroline Fox.* 3d ed., 2 vols. London, 1883.

"Reviews and Notices of New Books." *Philosophical Magazine* 12 (1838): 127–30.

"Reviews and Notices of New Books." *Lancet* 2 (1840–41): 197–98.

"A Revival and Removal." *Mechanics' Magazine* 19 (1833): 336.

Ridolfi, Marquis of. "Method of Separating Platina from Other Metallic Substances Which Are Found with It in the State of Ore." *Quarterly Journal of Science* 1 (1816): 259–64.

"Rights of Inventors by Patent and by Registration—and Effect of Disclosure under an Obligation of Secrecy." *Mechanics' Magazine* 46 (1847): 522–25.

Ritchie, William. "Experimental Researches in Electro-magnetism and Magneto-electricity." *Philosophical Transactions* 123 (1833): 313–21.

[Robertson, Joseph?]. "Preface." *Mechanics' Magazine* 1 (1823): iii–iv.

[———]. "Preface." *Mechanics' Magazine* 17 (1832): iii–vi.

[———]. "Books on Electro-metallurgy." *Mechanics' Magazine* 36 (1842): 458–62.

"Royal Institution." *Literary Gazette* 25 (1841): 73.

"Royal Institution." *Literary Gazette* 26 (1842): 366.

Ryland, William. "The Plated Wares and Electroplating Trades." In Timmins, *The Resources, Products, and Industrial History of Birmingham and the Midland Hardware District*, pp. 477–98.

Saxton, Joseph. "Mr. J. Saxton on His Magneto-electric Machine; with Remarks on Mr. E. M. Clarke's Paper in the Preceding Number." *Philosophical Magazine* 9 (1836): 360–65.

"Scientific and Miscellaneous Intelligence." *Railway Magazine* 2 (1837): 321–28.

"Scientific and Miscellaneous Intelligence." *Railway Magazine* 3 (1837): 431–44.

"Scientific and Miscellaneous Intelligence." *Railway Magazine* 4 (1838): 119–26.

Shaffner, Taliaferro. *The Telegraph Manual: A Complete History and Description.* New York, 1859.

Shaw, George. *A Manual of Electro-metallurgy.* London, 1842.

Shelley, Mary. *Frankenstein, or the Modern Prometheus* (1818). Edited with an introduction by Marilyn Butler. London: Pickering & Chatto, 1993.

"Shewing How the Tories and Whigs Extend Their Patronage to Science and Literature." *Fraser's Magazine* 12 (1835): 703–9.

Shillibeer, John. "Description of a New Arrangement of the Voltaic Battery and Pole Director." *Annals of Electricity* 1 (1836–37): 224–25.

Smart, Benjamin H. *A Practical Grammar of English Pronunciation.* London, 1810.

———. *The Practice of Elocution.* London, 1819.

Smee, Alfred. *Elements of Electro-metallurgy, or the Art of Working in Metals by the Galvanic Fluid.* London, 1841.

———. *Elements of Electro-metallurgy, or the Art of Working in Metals by the Galvanic Fluid.* 2d ed. London, 1842.

———. *Elements of Electrometallurgy.* 3d ed. London, 1844.

———. *Elements of Electro-biology, or the Voltaic Mechanism of Man; of Electro-pathology, Especially of the Nervous System, and of Electrotherapeutics.* London, 1849.

———. *Principles of the Human Mind.* London, 1849.

Smee, Alfred. *Instinct and Reason: Deduced from Electro-biology.* London, 1850.

———. "Dr. Carpenter and the British and Foreign Medical Review." *Lancet* 2 (1850): 514.

*Society for the Illustration and Encouragement of Practical Science, Gallery for the Exhibition of Objects Blending Instruction with Amusement, Adelaide Street, and Lowther Arcade, West Strand. Catalogue for May, 1836.* 14th ed. London, 1836.

Spencer, Thomas. "An Account of Some Experiments Made for the Purpose of Ascertaining How Far Voltaic Electricity May Be Usefully Applied to the Purpose of Working in Metal." *Annals of Electricity* 4 (1839–40): 258–79.

———. "On the Mode of Producing Fac-Simile Copies of Medals, &c.—By the Agency of Voltaic Electricity." *London Journal of Arts and Sciences* 15 (1840): 306–11.

———. *Instructions for the Multiplication of Works of Art in Metal by Voltaic Electricity.* Glasgow, 1840.

———. *Working in Metal by Voltaic Electricity.* London, n.d.

"Staite's Improvements in Lighting, &c." *Patent Journal and Inventor's Magzine* 4 (1848): 169–73.

Sturgeon, William, "Electro-magnetical Experiments." *Philosophical Magazine* 63 (1824): 95–100.

———. "Electro- and Thermo-magnetical Experiments." *Philosophical Magazine* 63 (1824): 266–68.

———. "Description of a Rotative Thermo-magnetical Experiment." *Philosophical Magazine* 63 (1824): 269–71.

———. "On Electro-magnetism." *Philosophical Magazine* 64 (1824): 242–49.

———. "Account of an Improved Electro-magnetic Apparatus." *Annals of Philosophy* 12 (1826): 357–61.

———. "An Account of Some Experiments Made with the Large Magnet at the Exhibition Rooms, Adelaide-street." *Philosophical Magazine* 5 (1834): 376–77.

———. "Explanatory Facts." *Philosophical Magazine* 7 (1835): 231–34.

———. "On Electro-pulsations and Electro-momentum." *Philosophical Magazine*, n.s., 9 (1836): 132–36.

———. "On the Relation by Measure of Common and Voltaic Electricity." *Annals of Electricity* 1 (1836–37): 52–65.

———. "On the Electric Shock from a Single Pair of Voltaic Plates." *Annals of Electricity* 1 (1836–37): 67–75.

———. "Description of an Electro-magnetic Engine for Turning Machinery." *Annals of Electricity* 1 (1836–37): 75–78.

———. "A General Outline of the Various Theories Which Have Been Advanced for the Explanation of Terrestrial Magnetism." *Annals of Electricity* 1 (1836–37): 117–23.

———. "Remarks on Mr. Faraday's 'Ninth Series of Experimental Researches' with Experiments in Electricity." *Annals of Electricity* 1 (1836–37): 186–91.

———. "Remarks on the Preceeding Paper, with Experiments." *Annals of Electricity* 1 (1836–37): 367–76.

———. "An Experimental Investigation of the Laws Which Govern the Production of Electric Shocks &c. from a Single Voltaic Pair." *Annals of Electricity* 1 (1836–37): 192–98.

———. "Address, Delivered by W. Sturgeon, Esq., Lecturer on Experimental Philosophy at the Honourable East India Company's Military Seminary Addiscombe &c. &c. at a General Meeting of the London Electrical Society." *Annals of Electricity* 2 (1838): 64–72.

———. "Fifth Memoir on Experimental and Theoretical Researches in Electricity, Magnetism &c." *Annals of Electricity* 5 (1840): 121–35, 293–310.

———. "Miscellaneous Notices on Galvanic Results, in Letters Addressed to Prof. Silliman, October 4, 1838, and August 6, 1839, from the Vicinity of London." *American Journal of Science* 39 (1840): 28–38.

———. "Electrical Curiosities." *Annals of Electricity* 7 (1841): 377–81.

———. *Lectures on Electricity, Delivered in the Royal Victoria Gallery.* London, 1842.

———. *A Course of Twelve Elementary Lectures on Galvanism.* London, 1843.

———. *Scientific Researches, Experimental and Theoretical in Electricity, Magnetism, Galvanism, Electromagnetism and Electrochemistry.* Bury, 1850.

Taylor, William Cooke. *A Tour of the Manufacturing Districts of Lancashire.* London, 1842.

"Taylor's Electro-magnetic Engine." *Mechanics' Magazine* 32 (1840): 693–96.

"Telegraphic Communication Commercially Considered." *London Journal of Arts, Sciences, and Manufactures* 37 (1850): 205–7.

Thompson, Silvanus P. *The Electromagnet, and Electromagnetic Mechanism.* London, 1891.

———. *Michael Faraday: His Life and Work.* London, 1898.

Tilt, Edward John. *On Diseases of Women and Ovarian Inflammation.* 2d ed. London, 1853.

"Time and the Electric Telegraph." *Mechanics' Magazine* 42 (1845): 416.

Timmins, Samuel, ed. *The Resources, Products, and Industrial History of Birmingham and the Midland Hardware District.* London, 1866. Reprint, Cass Library of Industrial Classics, 1967.

"Tricks of the Electrics." *Punch* 27 (1854): 64.

Tyndall, John. *Faraday as a Discoverer.* London, 1868.

Ure, Andrew. "An Account of Some Experiments Made on the Body of a Criminal Immediately after Execution, with Physiological and Practical Observations." *Quarterly Journal of Science* 6 (1819): 283–94.

———. *The Philosophy of Manufactures.* London, 1835.

Ursa Major. "Mackintosh's Electrical Theory of the Universe." *Mechanics' Magazine* 25 (1836): 92–94.

"A Visit to Mr. Crosse, at His Residence on the Quantock Hills, Somerset." *Lancet* 1 (1836–37): 49–52.

Walker, Charles V. "An Account of Some Experiments with a Constant Voltaic Battery." *Transactions and Proceedings of the London Electrical Society* 1 (1841): 57–65.

———. "Description of a Constant Acid Battery, Constructed by the Electrotype Process, with General Observations on Electrotype Manipulation." *Proceedings of the London Electrical Society* 2 (1841): 26–33.

———. "Protest of Mr. Charles V. Walker." *American Journal of Science* 42 (1842): 383–86.

———. "On the Action of Lightning Conductors." *Proceedings of the London Electrical Society* 2 (1843): 342–56.

———. *Electrotype Manipulation.* London, 1843.

———. "Address to the Reader." *Electrical Magazine* 1 (1845).

[———]. "Facts(?) for the History of Electricity." *Electrical Magazine* 1 (1845): 551–52.

[———]. "Galvanic Rings." *Electrical Magazine* 1 (1845): 607.

———. "Inaugural Address." *Journal of the Society of Telegraph Engineers* 5 (1876): 1–22.

Watkins, Francis. "Observations on Mr. Sturgeon's Letter." *Philosophical Magazine* 6 (1835): 239.

Watkins, Francis. "On Magneto-electric Induction." *Philosophical Magazine* 7 (1835): 107–13.

Watkins, Francis. "Letter from Mr. Watkins." *Philosophical Magazine* 7 (1835): 335.

Watts, Isaac. *The Improvement of the Mind*. New ed. London, 1809.

W[eekes], W. H. "Economical and Convenient Method of Constructing Mercurial Troughs." *Mechanics' Magazine* 23 (1835): 393–94.

Weekes, William H.. "Details of an Experiment in Which Certain Insects, Known as Acarus Crossi, Appeared Incident to the Long-continued Operation of a Voltaic Current upon Silicate of Potassa, within a Close Atmosphere over Mercury." *Proceedings of the London Electrical Society* 2 (1842): 240–56.

Wheatstone, Charles. "An Account of Some Experiments to Measure the Velocity of Electricity, and the Duration of Electric Light." *Philosophical Transactions* 124 (1834): 583–91.

———. "An Account of Several New Instruments and Processes for Determining the Constants of a Voltaic Circuit." *Philosophical Transactions* 133 (1843): 303–27.

———. *The Scientific Papers of Sir Charles Wheatstone*. London, 1879.

"Wheatstone and Cooke's Electric Telegraph." *Mechanics' Magazine* 33 (1840): 161–70.

Whewell, William. *History of the Inductive Sciences*. 3 vols. London, 1837.

———. "On the General Bearing of the Great Exhibition on the Progress of Art and Science." In *Lectures on the Great Exhibition*, pp. 3–25.

Williams, Thomas. "On the Laws of the Nervous Force, and the Functions of the Roots of the Spinal Nerves." *Lancet* 2 (1847): 516–17.

Wishaw, Francis. "Application of Gutta Percha to the Insulation of Electric Wires." *Mechanics' Magazine* 49 (1848): 309.

Wright, Thomas. "On a New Electro-magnetic Engine." *Annals of Electricity* 5 (1840): 108–10.

———. "On Electro-magnetic Coil Machines." *Annals of Electricity* 5 (1840): 349–52.

[Wynter, Andrew]. "The Electric Telegraph." *Quarterly Review* 95 (1854): 118–64.

Yates, Edmund. *Recollections and Experiences*. 4th ed. London, 1885.

Zeta. "Mackintosh's Electrical Theory of the Universe." *Mechanics' Magazine* 25 (1836): 148–49.

———. "The Scientific Amusements of London." *London Polytechnic Magazine and Journal* 1 (1844): 225–34.

Z. U. "The Use of Gutta Percha for Electro-telegraphic Purposes." *Mechanics' Magazine* 49 (1848): 310–11.


SECONDARY SOURCES

Agassi, Joseph. *Faraday as a Natural Philosopher*. Chicago: University of Chicago Press, 1971.

Altick, Richard. *The English Common Reader: A Social History of the Mass Reading Public, 1800–1900*. Chicago: University of Chicago Press, 1957.

———. *The Shows of London*. Cambridge: Harvard University Press, 1978.

Anderson, Patricia. *The Printed Image and the Transformation of Popular Culture 1790–1860*. Oxford: Oxford University Press, 1991.

Anderson, R.G.W., and Christopher Lawrence, eds. *Science, Medicine and Dissent: Joseph Priestley (1733–1804)*. London: Wellcome Trust, 1987.

Anderson, R.G.W., J. A. Bennett, and W. F. Ryan. *Making Instruments Count: Essays on Historical Scientific Instruments Presented to Gerard L'Estrange Turner*. Aldershot: Variorum Press, 1993.

Ashmore, Malcolm. "The Theatre of the Blind: Starring a Promethean Prankster, a Phoney Phenomenon, a Prism, a Pocket, and a Piece of Wood." *Social Studies of Science* 82 (1993): 67–106.

Ashworth, William J.. "The Calculating Eye: Baily, Herschel, Babbage and the Business of Astronomy." *British Journal for the History of Science* 27 (1994): 409–41.

Baatz, Simon. "Squinting at Silliman: Scientific Periodicals in the Early American Republic." *Isis* 82 (1991): 223–44.

Baldick, Chris. *In Frankenstein's Shadow: Myth, Monstrosity and Nineteenth-Century Writing*. Oxford: Clarendon Press, 1987.

Bargos, M. Susan, and William B. White. *The Daguerreotype: Nineteenth Century Technology and Modern Science*. Washington, DC: Smithsonian Institution Press, 1991.

Barrow, Logie. "Why Were Most Medical Heretics at Their Most Confident around the 1840s? (the Other Side of Mid-Victorian Medicine)." In French and Wear, *British Medicine in an Age of Reform*, pp. 165–85.

Bathe, Dorothy, and Greville Bathe. *Jacob Perkins: His Inventions, His Times, and His Contemporaries*. Philadelphia, 1943.

Behagg, Clive. "Secrecy, Ritual and Folk Violence: The Opacity of the Workplace in the First Half of the Nineteenth Century." In Storch, *Popular Culture and Custom in Nineteenth Century England*, pp. 154–79.

———. "The Democracy of Work, 1820–1850." In Rule, *British Trade Unionism, 1780–1850*, pp. 162–77.

Bennet, James A. "Instrument Makers and the 'Decline of Science in England': The Effect of Institutional Change on the Elite Makers of the Early Nineteenth Century." In deClerq, *Nineteenth-Century Scientific Instruments and Their Makers*, pp. 13–27.

Berg, Maxine. *The Machinery Question and the Making of Political Economy, 1815–1848*. Cambridge: Cambridge University Press, 1980.

Berman, Morris. *Social Change and Scientific Organization: The Royal Institution, 1799–1844*. London: Heinemann, 1978.

Biagioli, Mario. "Galileo's System of Patronage." *History of Science* 28 (1990): 1–62.

———. *Galileo Courtier: The Practice of Science in the Culture of Absolutism*. Chicago: University of Chicago Press, 1993.

Black, Robert M. *The History of Electric Wires and Cables*. London: Peter Peregrinus, 1983.

Blondel, Christine, Françoise Parot, Anthony Turner, and Mari Williams, eds., *Studies in the History of Scientific Instruments*. London: Roger Turner, 1989.

Boase, Frederick. *Modern English Biography*. 7 vols. London, 1865.

Boas-Hall, Marie. *All Scientists Now*. Cambridge: Cambridge University Press, 1984.

Bowers, Brian, and Lenore Symons, eds. *Curiosity Perfectly Satisfyed: Faraday's Travels in Europe, 1813–1815*. London: Peregrinus Press, 1991.

Brain, Robert. *Going to the Fair*. Cambridge: Whipple Museum, 1993.

Brannigan, Augustine. *The Social Basis of Scientific Discoveries*. Cambridge: Cambridge University Press, 1981.

Briggs, Asa. *Victorian Things*. Harmondsworth: Penguin, 1990.

Brock, William H. "The London Chemical Society, 1824." *Ambix* 14 (1967): 133–39.

Brown, Ira V., ed. *Joseph Priestley, Selections from His Writings*. University Park: Pennsylvania State University Press, 1962.

Buckland, Gail. *Fox Talbot and the Invention of Photography*. London: Scolar Press, 1980.

Bud, Robert, and Susan Cozzens, eds. *Invisible Connections: Instruments, Institutions and Science*. Bellingham: SPIE Optical Engineering Press, 1992.

Burnett, J. E., and A. D. Morrison-Low. *"Vulgar and Mechanick": The Scientific Instrument Trade in Ireland, 1650–1921*. Dublin: Royal Dublin Society, 1989.

Cameron, Henry C. *Mr. Guy's Hospital, 1726–1948*. London: Longmans, 1954.

Cannon, Susan F. *Science in Culture: The Early Victorian Period*. New York: Science History Publications, 1978.

Cannon, Walter F. "John Herschel and the Idea of Science." *Journal of the History of Ideas* 22 (1961): 215–39.

Cantor, Geoffrey. "William Robert Grove, the Correlation of Forces, and the Conservation of Energy." *Centaurus* 19 (1976): 273–90.

———. *Michael Faraday: Scientist and Sandemanian*. London: Macmillan, 1991.

Cardwell, Donald. *The Organization of Science in England*. 2d ed. London: Heinemann, 1972.

———. *James Joule: A Biography*. Manchester: Manchester University Press, 1989.

———. "On Michael Faraday, Henry Wilde and the Dynamo." *Annals of Science* 49 (1992): 479–87.

Chesney, Kellow. *The Victorian Underworld*. Harmondsworth: Penguin, 1970.

Chilton, Donovan, and Noel G. Coley. "The Laboratories of the Royal Institution in the Nineteenth Century." *Ambix* 27 (1980): 173–203.

Clark, George. *A History of the Royal College of Physicians*. 2 vols. Oxford: Oxford University Press, 1966.

Coley, Noel G. "The Collateral Sciences in the Work of Golding Bird." *Medical History* 13 (1969): 363–76.

Collins, Harry. "The TEA Set: Tacit Knowledge and Scientific Networks." *Science Studies* 4 (1974): 165–86.

———. "The Seven Sexes: A Study in the Sociology of a Phenomenon or the Replication of Experiments in Physics." *Sociology* 9 (1975): 205–24.

———. *Changing Order: Replication and Induction in Scientific Practice*. London: Sage, 1985.

———, ed. *Knowledge and Controversy: Studies of Modern Natural Science*. Special Issue, *Social Studies of Science* 11, no 1 (1981).

Colwell, Hector A. *An Essay on the History of Electrotherapy and Diagnosis*. London: Heinemann, 1922.

Cooter, Roger. *The Cultural Meaning of Popular Science: Phrenology and the Organization of Consent in Nineteenth Century Britain*. Cambridge: Cambridge University Press, 1984.

Cope, Zachary. *The Royal College of Surgeons of England*. London: Anthony Blond, 1959.

Copeman, W.S.C., *The Worshipful Society of Apothecaries: A History, 1617–1904*. Oxford: Pergamon Press, 1967.

*Correspondence*. See James, *The Correspondence of Michael Faraday*.

Corsi, Pietro. *Science and Religion: Baden Powell and the Anglican Debate, 1800–1860*. Cambridge: Cambridge University Press, 1988.

Coulson, Thomas. *Joseph Henry: His Life and Work*. Princeton: Princeton University Press, 1950.

Davenport, Walter Rice. *Thomas Davenport: Pioneer Inventor*. Montpelier: Vermont Historical Society, 1929.

Dear, Peter. "Cultural History of Science: An Overview with Reflections." *Science, Technology and Human Values* 20 (1995): 150–70.

deClerq, P. R., ed. *Nineteenth-Century Scientific Instruments and Their Makers*. Amsterdam: Editions Rendopi B. V., 1985.

Desmond, Adrian. "Artisan Resistance and Evolution in Britain, 1819–1848." *Osiris* 3 (1987): 77–110.

———. "Lamarckism and Democracy: Corporations, Corruption and Comparative Anatomy in the 1830s." In Moore, *History, Humanity and Evolution*, pp. 99–130.

———. *The Politics of Evolution*. Chicago: University of Chicago Press, 1989.

Dodds, John W. *The Age of Paradox*. New York: Rinehart & Co., 1952.

Dunsheath, Percy. *A History of Electrical Engineering*. London: Faber & Faber, 1962.

Dutton, Henry. *The Patent System and Inventive Activity during the Industrial Revolution, 1750–1852*. Manchester: Manchester University Press, 1984.

Fara, Patricia. *Sympathetic Attractions: Magnetic Practices, Beliefs, and Symbolism in Eighteenth-Century England*. Princeton: Princeton University Press, 1996.

Ferguson, Eugene S., ed. *Early Engineering Reminiscences (1815–40) of George Escol Sellars*. Washington, DC: Smithsonian Institution Press, 1965.

Fisch, Menachem, and Simon Schaffer, eds. *William Whewell: A Composite Portrait*. Oxford: Clarendon Press, 1991.

Fleck, Ludwik. *Genesis and Development of a Scientific Fact*. Chicago: University of Chicago Press, 1979.

Foote, George. "Sir Humphry Davy and His Audience at the Royal Institution." *Isis* 43 (1952): 6–12.

Forgan, Sophie. "Faraday—From Servant to Savant: The Institutional Context." in Gooding and James, *Faraday Rediscovered*, pp. 51–67.

Fox, Celina. "The Engravers' Battle for Professional Recognition in Early Nineteenth Century London." *London Journal* 2 (1976): 3–32.

———. *Graphic Journalism in England during the 1830s and 1840s*. New York: Garland Press, 1988.

Fox, Robert, ed. *Technological Change: Methods and Themes in the History of Technology*. London: Harwood Academic Publishers, 1995.

Frazier, Arthur H. "Joseph Saxton's First Sojourn in Philadelphia, 1818–1831." *Smithsonian Journal of History* 3 (1968): 45–76.

French, Roger, and Andrew Wear, eds. *British Medicine in an Age of Reform*. London: Routledge, 1991.

Fullmer, June Z. "Humphry Davy's Adversaries." *Chymia* 8 (1962): 147–64.

———. "Humphry Davy and the Gunpowder Manufactory." *Annals of Science* 20 (1964): 165–94.

Fulton, John, and Elizabeth Thomson. *Benjamin Silliman, 1779–1864: Pathfinder in American Science*. New York: Schuman, 1947.

Fyfe, Gordon. "Art and Reproduction: Aspects of the Relations between Painters and Engravers in London, 1760–1850." *Media, Culture and Society* 7 (1985): 399–425.

Galison, Peter. *How Experiments End*. Chicago: University of Chicago Press, 1987.

Galison, Peter, and Alexi Assmus. "Artificial Clouds, Real Particles." In Gooding, Pinch, and Schaffer, *The Uses of Experiment*, pp. 225–74.

Gee, Brian. "Faraday's Plight and the Origins of the Magneto-electric Spark." *Nuncius* 5 (1990): 43–68.

Gee, Brian. "The Early Development of the Magneto-electric Machine." *Annals of Science* 50 (1993): 101–33.

Ginn, William. "Philosophers and Artisans: The Relationship between Men of Science and Instrument Makers in London, 1820–1860." Ph.D. diss., University of Kent at Canterbury, 1991.

Golinski, Jan. "Humphry Davy and the Lever of Experiment." In LeGrand, *Experimental Enquiries*, pp. 99–136.

———. "The Theory of Practice and the Practice of Theory: Sociological Approaches in the History of Science." *Isis* 81 (1990): 492–505.

———. *Science as Public Culture: Chemistry and Enlightenment in Britain, 1760–1820.* Cambridge: Cambridge University Press, 1992.

Gooding, David. "Final Steps to the Field Theory." *Historical Studies in the Physical Sciences* 11 (1981): 231–75.

———. "Experiment and Concept-Formation in Electromagnetic Science and Technology in England, 1820–1830." *History and Technology* 2 (1985): 151–76.

———. "He Who Proves Discovers: John Herschel, William Pepys and the Faraday Effect." *Notes and Records of the Royal Society* 39 (1985): 229–44.

———. "In Nature's School: Faraday as an Experimentalist." In Gooding and James, *Faraday Rediscovered*, pp. 105–35.

———. "Magnetic Curves and the Magnetic Field: Experimentation and Representation in the History of a Theory." In Gooding, Pinch, and Schaffer, *The Uses of Experiment*, pp. 183–223.

———. *Experiment and the Making of Meaning: Human Agency in Scientific Observation.* Dordrecht: Kluwer, 1990.

Gooding, David, and Frank A.J.L. James, eds. *Faraday Rediscovered: Essays on the Life and Work of Michael Faraday, 1791–1867.* London: Macmillan, 1985.

Gooding, David, Trevor Pinch, and Simon Schaffer, eds. *The Uses of Experiment.* Cambridge: Cambridge University Press, 1989.

Greenslade, Thomas B., Jr. "Henry Noad's 'Lectures on Electricity' and the State of Electrical Knowledge in 1844." *Rittenhouse* 2 (1987): 139–46.

Hackmann, Willem. "The Induction Coil in Medicine and Physics, 1835–1877." In Blondel, Parot, Turner, and Williams, *Studies in the History of Scientific Instruments*, pp. 235–50.

———. "Electricity from Steam: Armstrong's Hydroelectric Machine in the 1840s." In Anderson, Bennett, and Ryan, *Making Instruments Count*, pp. 146–73.

Hale-White, William. "Golding Bird: Assistant Physician to Guy's Hospital, 1843–1854." *Guy's Hospital Reports* 76 (1926): 1–20.

Harris, Neil. *Humbug: The Art of P. T. Barnum.* Boston: Little & Brown, 1973.

Hays, John. "Science and Brougham's Society." *Annals of Science* 20 (1964): 227–41.

———. "Science in the City: The London Institution, 1819–40." *British Journal for the History of Science* 7 (1974): 146–62.

———. "The London Lecturing Empire, 1800–50." In Inkster and Morrell, *Metropolis and Province*, pp. 91–119.

Headrick, Daniel. *The Tools of Empire: Technology and European Imperialism in the Nineteenth Century.* Oxford: Oxford University Press, 1981.

———. *The Invisible Weapon: Telecommunications and National Politics, 1851–1945.* Oxford: Oxford University Press, 1991.

Heilbron, John. *Electricity in the Seventeenth and Eighteenth Centuries: A Study of*

*Early Modern Physics*. Berkeley and Los Angeles: University of California Press, 1979.

*Henry Papers*. See Reingold, *The Papers of Jospeh Henry*.

Hilton, Boyd. *The Age of Atonement: The Influence of Evangelicalism on Social and Economic Thought, 1785–1865*. Oxford: Oxford University Press, 1988.

Hobhouse, Christopher. *1851 and the Crystal Palace*. 2d ed. London: John Murray, 1950.

Howse, Derek. *Greenwich Time*. Oxford: Oxford University Press, 1980.

Hubbard, Geoffrey. *Cooke, Wheatstone and the Invention of the Electric Telegraph*. London: Routledge & Kegan Paul, 1965.

Hughes, Jeff. "The Radioactivists: Community, Controversy and the Rise of Nuclear Physics." Ph.D. diss., University of Cambridge, 1993.

Hunt, Bruce. *The Maxwellians*. Ithaca: Cornell University Press, 1991.

―――. "Michael Faraday, Cable Telegraphy and the Rise of Field Theory." *History of Technology* 13 (1991): 1–19.

―――. "The Ohm Is Where the Art Is: British Telegraph Engineers and the Development of Electrical Standards." In van Helden and Hankins, *Instruments*, pp. 48–63.

Iliffe, Robert. "In the Warehouse: Privacy, Property and Priority in the Early Royal Society." *History of Science* 30 (1992): 30–68.

Inkster, Ian, and Jack Morrell, eds. *Metropolis and Province: Science and British Culture, 1780–1850*. London: Hutchinson, 1983.

Jacobus, Mary, Evelyn Fox Keller, and Sally Shuttleworth, eds. *Body/Politics: Women and the Discourses of Science*. London: Routledge, 1990.

Jacyna, L. Stephen. "Somatic Theories of Mind and the Interests of Medicine in Britain, 1850–1879." *Medical History* 26 (1982): 233–58.

James, Frank A.J.L. "The Optical Mode of Investigation: Light and Matter in Faraday's Natural Philosophy." In Gooding and James, *Faraday Rediscovered*, pp. 137–61.

―――. "Time, Tide and Michael Faraday." *History Today*, September 1991, 28–34.

―――. "Michael Faraday, the City Philosophical Society and the Society of Arts." *RSA Journal* 41 (1992): 192–99.

―――. "The Tales of Benjamin Abbott: A Source for the Early Life of Michael Faraday." *British Journal for the History of Science* 25 (1992): 229–40.

―――, ed. *The Correspondence of Michael Faraday*. Vol. 1. London: Peregrinus Press, 1991.

―――, ed. *The Correspondence of Michael Faraday*. Vol. 2. London: Institution of Electrical Engineers, 1993.

Joyce, Patrick. *Democratic Subjects: The Self and the Social in Nineteenth Century England*. Cambridge: Cambridge University Press, 1994.

Joyce, Patrick, ed. *The Historical Meanings of Work*. Cambridge: Cambridge University Press, 1987.

Kargon, Robert. *Science in Victorian Manchester: Enterprise and Expertise*. Manchester: Manchester University Press, 1977.

Kass, Amalie, and Edward Kass. *Perfecting the World: The Life and Times of Dr. Thomas Hodgkin, 1798–1866*. New York: Harcourt Brace Jovanovich, 1988.

Kieve, Jeffrey L. *The Electric Telegraph: A Social and Economic History*. Newton Abbott: David & Charles, 1973.

King, W. James. *The Development of Electrical Technology in the Nineteenth Century: The Arc-Light and Generator*. Washington, DC: Smithsonian Institution Press, 1964.

King, W. James. *The Development of Electrical Technology in the Nineteenth Century: The Electrochemical Cell and the Electromagnet*. Washington, DC: Smithsonian Institution Press, 1962.

Knight, David. "Davy and Faraday: Father and Son." In Gooding and James, *Faraday Rediscovered*, pp. 33–49.

————. *Humphry Davy: Science and Power*. Oxford: Blackwells, 1992.

Kusamitsu, Toshio. "Great Exhibitions before 1851." *History Workshop* 9 (1980): 70–89.

Lampe, David. *The Tunnel*. London: George Harrup, 1964.

Laqueur, Thomas. *Making Sex: Body and Gender from the Greeks to Freud*. Cambridge: Harvard University Press, 1990.

Latour, Bruno. *Science in Action: How to Follow Scientists and Engineers through Society*. Milton Keynes: Open University Press, 1987.

————. *We Have Never Been Modern*. Hemel Hempstead: Harvester Wheatsheaf, 1993.

Lawrence, Christopher. "Incommunicable Knowledge: Science, Technology and the Clinical Art in Britain, 1850–1914." *Journal of Contemporary History* 20 (1985): 503–20.

L'Esperance, Jean. "Doctors and Women in Nineteenth Century Society." In Woodward and Richards, *Health Care and Popular Medicine in Nineteenth Century England*, pp. 105–27.

LeGrand, Homer, ed. *Experimental Enquiries*. Dordrecht: Reidel, 1990.

MacKenzie, Donald, and Judy Wacjman, eds. *The Social Shaping of Technology: How the Refrigerator Got Its Hum*. Milton Keynes: Open University Press, 1985.

MacLeod, Christine. *Inventing the Industrial Revolution: The English Patent System, 1660–1800*. Cambridge: Cambridge University Press, 1988.

————. "Concepts of Invention and the Patent Controversy in Victorian Britain." In Fox, *Technological Change*, pp. 137–53.

Marvin, Carolyn. *When Old Technologies Were New: Thinking about Electric Communication in the Late Nineteen Century*. Oxford: Oxford University Press, 1988.

Marx, Karl. *Capital*. 3 vols. Harmondsworth: Penguin, 1976.

Mason, Michael. *The Making of Victorian Sexuality*. Oxford: Oxford University Press, 1994.

Mather, Francis C. "The Railways, the Electric Telegraph and Public Order during the Chartist Period." *History* 38 (1953): 40–53.

Mathias, Peter. *The First Industrial Nation: The Economic History of Britain, 1700–1914*. 2d ed. London: Routledge, 1983.

McCalman, Iain. *Radical Underworld: Prophets, Revolutionaries and Pornographers in London, 1795–1840*. Cambridge: Cambridge University Press, 1988.

McInnes, E. M. *St. Thomas's Hospital*. London: George Allen & Unwin, 1963.

McLuhan, Marshall. *Understanding Media: The Extensions of Man*. New York: McGraw-Hill, 1964.

Miller, David P. "The Royal Society of London, 1800–1835." Ph.D. diss., University of Pennsylvania, 1981.

————. "Between Hostile Camps: Sir Humphry Davy's Presidency of the Royal Society of London, 1820–1827." *British Journal for the History of Science* 16 (1983): 1–47.

————. "Method and the Micropolitics of Science: The Early Years of the Geological and Astronomical Societies." In Schuster and Yeo, *The Politics and Rhetoric of Scientific Method*.

———. "The Revival of the Physical Sciences in Britain, 1815–1840." *Osiris* 2 (1986): 107–34.

Moore, James, ed. *History, Humanity and Evolution*. Cambridge: Cambridge University Press, 1989.

Morrell, Jack, and Arnold Thackray. *Gentlemen of Science: The Early Years of the British Association for the Advancement of Science*. Oxford: Oxford University Press, 1981.

Morus, Iwan Rhys. "The Sociology of Sparks: An Episode in the History and Meaning of Electricity." *Social Studies of Science* 18 (1988): 387–417.

———. "The Politics of Power: Reform and Regulation in the Work of William Robert Grove." Ph.D. diss., University of Cambridge, 1989.

———. "Industrious People: Biography and Nineteenth Century Physics." *Studies in History and Philosophy of Science* 21 (1990): 519–25.

———. "Correlation and Control: William Robert Grove and the Construction of a New Philosophy of Scientific Reform." *Studies in History and Philosophy of Science* 22 (1991): 589–621.

———. "Telegraphy and the Technology of Display: The Electricians and Samuel Morse." *History of Technology* 13 (1991): 20–40.

———. "Different Experimental Lives: Michael Faraday and William Sturgeon." *History of Science* 30 (1992): 1–28.

———. "Marketing the Machine: The Construction of Electrotherapeutics as Viable Medicine in Early Victorian England." *Medical History* 36 (1992): 34–52.

———. "Currents from the Underworld: Electricity and the Technology of Display in Early Victorian England." *Isis* 84 (1993): 50–69.

———. "The Electric Ariel: Telegraphy and Commercial Culture in Early Victorian England." *Victorian Studies* 39 (1996): 339–78.

———. "Manufacturing Nature: Science, Technology and Victorian Consumer Culture." *British Journal for the History of Science* 29 (1996): 403–34.

Moscucci, Ornella. *The Science of Woman: Gynaecology and Gender in England, 1800–1929*. Cambridge: Cambridge University Press, 1990.

Nye Mary Jo. "N-Rays: An Episode in the History and Psychology of Science.": 125–56.

Pang, Alex Soojung-Kim. "Victorian Observing Practices, Printing Technology and Representation of the Solar Corona: The 1860s and 1870s." *Journal of the History of Astronomy*. 25 (1994): 249–74.

Pera, Marcello. *The Ambiguous Frog: The Galvani-Volta Controversy in Animal Electricity*. Translated by Jonathan Mandelbaum. Princeton: Princeton University Press, 1992.

Peterson, Jeanne. *The Medical Profession in Mid-Victorian London*. Berkeley and Los Angeles: University of California Press, 1978.

Phillipps, K. C. *Language and Class in Victorian England*. Oxford: Blackwells, 1984.

Pickering, Andrew. *The Mangle of Practice: Time, Agency and Science*. Chicago: University of Chicago Press, 1995.

Pinch, Trevor. *Confronting Nature: The Sociology of Solar-Neutrino Detection*. Dordrecht: Reidel, 1987.

Poovey, Mary. "Scenes of an Indelicate Character: The Medical 'Treatment' of Victorian Women." *Representations* 14 (1986): 137–68.

Porter, Roy. *Health for Sale: Quackery in England, 1660–1850*. Manchester: Manchester University Press, 1989.

Pritchard, J. L. *Sir George Cayley, the Inventor of the Aeroplane*. London: Max Parrish, 1961.

Prothero, Iorwerth. *Artisans and Politics in Early Nineteenth-Century London: John Gast and His Times*. London: Methuen, 1981.

Quenell, Peter. *London's Underworld*. (Selections from Henry Mayhew, *London Labour and the London Poor*, vol.4.) London: Spring Books, 1950.

Reingold, Nathan, ed. *The Papers of Joseph Henry*. 5 vols. Washington, DC: Smithsonian Institution Press, 1972–.

Richards, Thomas. *The Commodity Culture of Victorian England*. London: Verso, 1991.

Richardson, Ruth. *Death, Dissection and the Destitute*. Harmondsworth: Penguin, 1988.

Rolt, L.T.C. *Isambard Kingdom Brunel*. Harmondsworth: Penguin, 1970.

Rosenberg, Nathan. *Exploring the Black Box: Technology, Economics and History*. Cambridge: Cambridge University Press, 1994.

Ross, Sidney. "Scientist: The Story of a Word." *Annals of Science* 18 (1962): 65–85.

Rowbottom, Margaret, and Charles Susskind. *Electricity and Medicine: History of Their Interaction*. San Francisco: San Francisco Press, 1984.

Rudwick, Martin. "Charles Darwin in London: The Integration of Public and Private Science." *Isis* 73 (1982): 186–206.

———. *The Great Devonian Controversy: The Shaping of Gentlemanly Knowledge among Gentlemanly Specialists*. Chicago: University of Chicago Press, 1985.

Rule, John. "The Property of Skill in the Period of Manufacture." In Joyce, *The Historical Meanings of Work*, pp. 99–118.

———, ed. *British Trade Unionism, 1780–1850: The Formative Years*, London: Longmans, 1988.

Russell, Colin. *Science and Social Change, 1700–1900*. London: Macmillan, 1983.

Samuel, Raphael. "Workshop of the World: Steam Power and Hand Technology in Mid-Victorian Britain." *History Workshop* 3 (1977): 6–72.

Schaffer, Simon. "Natural Philosophy and Public Spectacle in the Eighteenth Century." *History of Science* 21 (1983): 1–43.

———. "Scientific Discovery and the End of Natural Philosophy." *Social Studies of Science* 16 (1986): 387–420.

———. "Priestley and the Politics of Spirit." In Anderson and Lawrence, *Science, Medicine and Dissent*, pp. 39–53.

———. "Glass Works: Newton's Prisms and the Uses of Experiment." In Gooding, Pinch, and Schaffer, *The Uses of Experiment*, pp. 67–104.

———. "The History and Geography of the Intellectual World: Whewell's Politics of Language." In Fisch and Schaffer, *William Whewell*, pp. 201–31.

———. "Late Victorian Metrology and Its Instrumentation: A Manufactory of Ohms." In Bud and Cozzens, *Invisible Connections*, pp. 23–56.

———. "Self-Evidence." *Critical Inquiry* 18 (1992): 327–62.

———. "Babbage's Intelligence: Calculating Engines and the Factory System." *Critical Inquiry* 21 (1994): 203–27.

———. "Babbage's Dancer and the Impresarios of Mechanism." In Spufford and Uglow, *Cultural Babbage*, pp. 53–80.

Schiebinger, Londa. *The Mind Has No Sex: Women in the Origins of Modern Science*. Cambridge: Harvard University Press, 1989.

Schivelbusch, Wolfgang. *The Railway Journey: The Industrialization of Time and Space in the Nineteenth Century*. Berkeley and Los Angeles: University of California Press, 1986.

———. *Disenchanted Night: The Industrialization of Light in the Nineteenth Century*. Berkeley and Los Angeles: University of California Press, 1988.

Schuster, John, and Richard Yeo, eds. *The Politics and Rhetoric of Scientific Method.* Dordrecht: Reidel, 1986.

Secord, Anne. "Science in the Pub: Artisan Botanists in Early Nineteenth Century Lancashire." *History of Science* 32 (1994): 269–315.

Secord, James A. *Controversy in Victorian Geology: The Cambrian-Silurian Dispute.* Princeton: Princeton University Press, 1986.

———. "Extraordinary Experiment: Electricity and the Creation of Life in Victorian England." In Gooding, Pinch, and Schaffer, *The Uses of Experiment,* pp. 337–83.

Shapin, Steven. "The History of Science and Its Sociological Reconstructions." *History of Science* 20 (1983): 157–211.

———. "Pump and Circumstance: Robert Boyle's Literary Technology." *Social Studies of Science* 14 (1984): 481–520.

———. "The House of Experiment in Seventeenth Century England." *Isis* 79 (1988): 373–404.

———. "The Mind Is Its Own Place: Science and Solitude in Seventeenth Century England." *Science in Context* 4 (1991): 191–218.

Shapin, Steven, and Barry Barnes. "Science, Nature and Control: Interpreting Mechanics' Institutes." *Social Studies of Science* 7 (1977): 31–74.

Shapin, Steven, and Simon Schaffer. *Leviathan and the Air-Pump: Hobbes, Boyle and the Experimental Life.* Princeton: Princeton University Press, 1985.

Shelley, Mary. *Frankenstein: Or the Modern Prometheus.* Edited with an introduction by Marilyn Butler. London: Pickering & Chatto, 1993.

Shuttleworth, Sally. "Female Circulation: Medical Discourse and Popular Advertising in the Mid-Victorian Era." In Jacobus, Keller, and Shuttleworth, *Body/Politics,* pp. 47–68.

Sibum, Heinz Otto. "Reworking the Mechanical Value of Heat: Instruments of Precision and Gestures of Accuracy in Early Victorian England." *Studies in History and Philosophy of Science* 26 (1995): 73–106.

Sinclair, Bruce. *Philadelphia's Philosopher Mechanics: A History of the Franklin Institute, 1824–1865.* Baltimore: Johns Hopkins University Press, 1974.

Smith, Crosbie, and M. Norton Wise. "Measurement, Work and Industry in Lord Kelvin's Britain." *Historical Studies in the Physical Sciences* 17 (1986): 147–73.

———. *Energy and Empire: A Biographical Study of Lord Kelvin.* Cambridge: Cambridge University Press, 1989.

Sprigge, S. Squire. *The Life and Times of Thomas Wakley, Founder and First Editor of the Lancet, Member of Parliament for Finsbury, and Coroner of West Middlesex.* London, 1899.

Spufford, Francis, and Jenny Uglow, eds. *Cultural Babbage: Technology, Time and Invention.* London: Faber & Faber, 1996.

Stallybrass, Oliver. "How Faraday 'Produced Living Animalculae': Andrew Crosse and the Story of a Myth." *Proceedings of the Royal Institution* 41 (1967): 597–619.

Stephen, G. N. *The British Medical Association and the Medical Profession.* London, 1914.

Stokes, M. Veronica. "The Lowther Arcade in the Strand." *London Topographical Record* 23 (1974): 119–28.

Storch, Robert, ed. *Popular Culture and Custom in Nineteenth-Century England.* London: Croom Helm, 1982.

Sutton, Geoffrey. "The Politics of Science in Early Napoleonic France: The Case of the Voltaic Pile." *Historical Studies in the Physical Sciences* 11 (1981): 329–66.

Taylor, E.G.R. *Mathematical Practitioners of Hanoverian England.* Cambridge: Cambridge University Press, 1966.

Temkin, Owsei. "Basic Science, Medicine, and the Romantic Era." In *The Double Face of Janus,* pp. 345–72.

———. *The Double Face of Janus and Other Essays in the History of Medicine.* Baltimore: Johns Hopkins University Press, 1977.

Thomas, John M. *Michael Faraday and the Royal Institution.* London: Adam Hilger, 1991.

Thompson, Edward P. *The Making of the English Working Class.* Harmondsworth: Penguin, 1968.

Toole, Betty Alexandra, ed. *Ada, the Enchantress of Numbers: A Selection from the Letters of Lord Byron's Daughter and Her Description of the First Computer.* Mill Valley, CA: Strawberry Press, 1992.

Twyman, Alan. *In Search of the Mysterious Mr. Weekes: A Fragment of Sandwich History.* Dover: Adams & Sons, 1988.

Van Helden, Albert, and Thomas Hankins, eds. *Instruments.* Special Issue, *Osiris* 9 (1994).

Waddington, Ivan. "The Role of the Hospital in the Development of Modern Medicine: A Sociological Analysis." *Sociology* 7 (1973): 211–25.

———. "General Practitioners and Consultants in Early Nineteenth Century England." In Woodward and Richards, *Health Care and Popular Medicine in Nineteenth Century England,* pp. 164–88.

———. *The Medical Profession in the Industrial Revolution.* Dublin: Gill & Macmillan, 1984.

Wahrman, Dror. *Imagining the Middle Class: The Political Representation of Class in Britain c. 1780–1840.* Cambridge: Cambridge University Press, 1995.

Wiener, Joel. *Radicalism and Freethought in Nineteenth-Century Britain: The Life of Richard Carlile.* (Westwood, CT: Greenwood Press, 1983.

Wilks, Samuel, and G. T. Bettany. *A Biographical History of Guy's Hospital.* London, 1892.

Williams, L. Pearce. *Michael Faraday.* London: Chapman & Hall, 1965.

———, ed. *The Selected Correspondence of Michael Faraday.* 2 vols. Cambridge: Cambridge University Press, 1971.

Williams, Raymond. *Culture and Society.* Harmondsworth: Penguin, 1961.

Winter, Alison. "Ethereal Epidemic: Mesmerism and the Introduction of Inhalation Anaesthesia to Early Victorian London." *Social History of Medicine* 4 (1991): 1–27.

———. "The Island of Mesmeria: The Politics of Mesmerism in Early Victorian Britain." Ph.D. diss., University of Cambridge, 1992.

Wise, M. Norton. "Mediating Machines." *Science in Context* 2 (1988): 77–113.

Wise, M. Norton (with the collaboration of Crosbie Smith). "Work and Waste: Political Economy and Natural Philosophy in Nineteenth-Century Britain." *History of Science* 27 (1989): 263–301, 391–449; 28 (1990): 221–61.

Wood, E. M. *A History of the Polytechnic.* London: MacDonald, 1965.

Woodward, John. *To Do the Sick No Harm: A Study of the British Voluntary Hospital System to 1875.* London: Routledge, 1974.

Woodward, John, and David Richards, eds. *Health Care and Popular Medicine in Nineteenth Century England: Essays in the Social History of Medicine.* London: Croom Helm, 1977.

ABOUT THE AUTHOR

Iwan Rhys Morus is currently Wellcome University Award Fellow and Lecturer in History and Philosophy of Science in the Department of Social Anthropology, Queen's University of Belfast. He has published numerous articles on early-nineteenth-century science and popular culture.

Ingram Content Group UK Ltd.
Milton Keynes UK
UKHW031809050523
421276UK00006B/103